T0210876

Lecture Notes in Bioinformatics 9289

Subseries of Lecture Notes in Computer Science

More information about this series at http://www.springer.com/series/5381

Mihai Pop · Hélène Touzet (Eds.)

Algorithms in Bioinformatics

15th International Workshop, WABI 2015
Atlanta, GA, USA, September 10–12, 2015
Proceedings

Springer

Editors
Mihai Pop
University of Maryland
College Park, MD
USA

Hélène Touzet
University of Lille
Lille
France

ISSN 0302-9743 ISSN 1611-3349 (electronic)
Lecture Notes in Bioinformatics
ISBN 978-3-662-48220-9 ISBN 978-3-662-48221-6 (eBook)
DOI 10.1007/978-3-662-48221-6

Library of Congress Control Number: 2015946588

LNCS Sublibrary: SL8 – Bioinformatics

Springer Heidelberg New York Dordrecht London

Printed on acid-free paper

Springer-Verlag GmbH Berlin Heidelberg is part of Springer Science+Business Media
(www.springer.com)

Preface

This proceedings volume contains papers presented at the Workshop on Algorithms in Bioinformatics 2015 (WABI 2015) that was held at Georgia Technological Institute, Atlanta, Georgia, USA, during September 10–12, 2015. WABI was held in conjunction with the ACM Conference on Bioinformatics, Computational Biology, and Health Informatics (ACM BCB). WABI is an annual conference series on all aspects of algorithms and data structures in molecular biology, genomics, and phylogeny data analysis that was first held in 2001. WABI 2015 was sponsored by the ACM Special Interest Group in Bioinformatics (ACM SIGBio), the European Association for Theoretical Computer Science (EATCS), and the International Society for Computational Biology (ISCB).

In 2015, a total of 57 manuscripts were submitted to WABI from which 26 were selected for presentation at the conference. Among them 23 were full papers not previously published in journals, and three were short abstracts for papers published simultaneously in journals. The papers were selected based on a thorough review from at least three independent reviewers as well as active discussions between members of the Program Committee. The selected papers cover a wide range of topics from networks, to phylogenetic studies, sequence and genome analysis, comparative genomics, and RNA structure. Extended versions of selected papers will be published in a thematic series in the journal *Algorithms for Molecular Biology* (AMB), published by BioMed Central.

We thank all the authors of submitted papers and the members of the Program Committee for their efforts that made this conference possible and the WABI Steering Committee for help and advice. In particular, we are indebted to the keynote speaker of the conference, Paola Bonizzoni, for her presentation. We are also grateful to Srinivas Aluru and the local Organizing Committee for their help in making WABI a success.

September 2015

Mihai Pop
Hélène Touzet

Organization

General Chair

Srinivas Aluru   Georgia Tech, USA

Program Chairs

Mihai Pop   University of Maryland, USA
Hélène Touzet   University of Lille, France

Program Committee

Tatsuya Akutsu	Kyoto University, Japan
Vikas Bansal	University of California San Diego, USA
Anne Bergeron	Université du Québec à Montreal, Canada
Daniel Brown	University of Waterloo, Canada
Rita Casadio	University of Bologna, Italy
Kun-Mao Chao	National Taiwan University, Taiwan
Ting Chen	University of Southern California, USA
Rayan Chikhi	University of Lille, France
Nadia El-Mabrouk	University of Montreal, Canada
Gang Fang	Mount Sinai School of Medicine, USA
Liliana Florea	Johns Hopkins University, USA
Anna Gambin	Warsaw University, Poland
Katharina Huber	University of East Anglia, UK
Rui Jiang	Tsinghua University, China
Carl Kingsford	Carnegie Mellon University, USA
Sergey Koren	BNBI/NBACC, USA
Vincent Lacroix	Université Lyon 1, France
Dominique Lavenier	CNRS/IRISA, France
Ming Li	University of Waterloo, Canada
Stefano Lonardi	UC Riverside, USA
Veli Mäkinen	University of Helsinki, Finland
Paul Medvedev	Pennsylvania State University, USA
Bernard Moret	EPFL, Lausanne, Switzerland
Burkhard Morgenstern	University of Göttingen, Germany
Vincent Moulton	University of East Anglia, UK
Niranjan Nagarajan	Genome Institute of Singapore, Singapore
William Stafford Noble	University of Washington, USA
Laurent Noé	University of Lille, France
Aïda Ouangraoua	University of Sherbrooke, Canada

Pierre Peterlongo	Inria Rennes-Bretagne-Atlantique, France
Adam Phillippy	BNBI/NBACC, USA
Nadia Pisanti	Università di Pisa, Italy
Yann Ponty	CNRS/LIX, École Polytechnique, France
Teresa Przytycka	NIH, USA
Ben Raphael	Brown University, USA
Marie-France Sagot	Inria Grenoble Rhône-Alpes and Université de Lyon 1, France
Cenk Sahinalp	Indiana University, USA
David Sankoff	University of Ottawa, Canada
Michael Schatz	Cold Spring Harbor Laboratory, USA
Russell Schwartz	Carnegie Mellon University, USA
Jared Simpson	Ontario Institute for Cancer Research, Canada
Mona Singh	Princeton University, USA
Andrew Smith	University of Southern California, USA
Jens Stoye	Bielefeld University, Germany
Krister Swenson	LIRMM, CNRS, Université de Montpellier, France
Haixu Tang	Indiana University, USA
Todd Treangen	BNBI/NBACC, USA
Jérôme Waldispühl	McGill University, Canada
Tandy Warnow	The University of Illinois at Urbana-Champaign, USA
Sebastian Will	University of Leipzig, Germany
Xiaohui Xie	University of California Irvine, USA
Dong Xu	University of Missouri-Columbia, USA
Yuzhen Ye	Indiana University, USA
Louxin Zhang	National University of Singapore, Singapore
Xuegong Zhang	Tsinghua University, China
Michal Ziv-Ukelson	Ben-Gurion University of the Negev, Israel

WABI Steering Committee

Bernard Moret	EPFL, Switzerland
Vincent Moulton	University of East Anglia, UK
Jens Stoye	Bielefeld University, Germany
Tandy Warnow	University of Illinois, USA

ACM BCB Organizing Committee

Srinivas Aluru	May D. Wang (Co-chair)
Mark Borodovski	Ying Xu
T.M. Murali	Christopher C. Yang
Yi Pan	Aidong Zhang (Co-chair)

Additional Reviewers

Hind Alhakami
Rumen Andonov
Djamal Belazzougui
Xin Chen
Dongyeon Cho
Matei David
Benjamin Decato
Piero Fariselli
Darya Filippova
Alexander Gawronski
Jean-Pierre Glouzon
John Halloran
Md Abid Hasan
Ermin Hodzic
Jan Hoinka
Guillaume Holley
Yoo-Ah Kim
Gunnar W. Klau
Mateusz Krzysztof Łącki
Manuel Lafond
Thierry Lecroq
Heewook Lee
Mark Leiserson
Claire Lemaitre
Yi Li
Jie Liu
Wenxiu Ma
Salem Malikic
Pier Luigi Martelli
Blazej Miasojedow
Paolo Milazzo
Ilia Minkin

Seyed Hamid Mirebrahim
Brian Ondov
Weihua Pan
Solon Pissis
Jianghan Qu
Sohini Ramachandran
Sebastien Roch
Kristoffer Sahlin
Leena Salmela
Alexander Schoenhuth
Jacob Schreiber
Raunak Shrestha
Blerina Sinaimeri
Ilan Smoly
Brad Solomon
Li Song
Michał Startek
Linda Sundermann
Ramesh Krishnan Pallavoor Suresh
Haixu Tang
Om Thakkar
Philip Uren
Gabriel Valiente
Juexin Wang
Mathias Weller
Bartek Wilczynski
Damian Wójtowicz
Yun Wu
Qiuming Yao
Zhaoxia Yu
Meng Zhou

Abstracts

Reference-free Compression of High Throughput Sequencing Data with a Probabilistic de Bruijn Graph

Gaëtan Benoit[1], Claire Lemaitre[1], Dominique Lavenier[1],
Erwan Drezen[1], Thibault Dayris[2], Raluca Uricaru[2],
and Guillaume Rizk[1]

[1] INRIA/IRISA/GenScale, Campus de Beaulieu, 35042 Rennes, France
[2] University of Bordeaux, CNRS/LaBRI, 33405, Talence

Abstract. Data volumes generated by next-generation sequencing (NGS) technologies is now a major concern for both data storage and transmission. This triggered the need for more efficient methods than general purpose compression tools, such as the widely used gzip method.

Most *de novo* methods either use a context-model to predict bases according to their context, followed by an arithmetic encoder, or re-order reads to maximize similarities between consecutive reads and therefore boost compression. However, by simply re-ordering reads read-pairing information is lost, thus many downstream analysis become impossible to run. Existing tools based on reads re-ordering, either lose the pairing information and achieve high compression, or provide an option to keep it at the cost of a much lower compression ratio.

We present a novel reference-free method meant to compress data issued from high throughput sequencing technologies, in both FASTA and FASTQ format. Our approach, implemented in the software LEON, employs techniques derived from existing assembly principles. The method is based on a reference probabilistic *de Bruijn Graph*, built *de novo* from the set of reads and stored in a Bloom filter. Each read is encoded losslessly as a path in this graph, by memorizing an anchoring kmer and a list of bifurcations. The same probabilistic *de Bruijn Graph* is used to perform a lossy transformation of the quality scores, which allows to obtain higher compression rates without losing pertinent information for downstream analyses. LEON was run on various real sequencing datasets (whole genome, exome, RNA-seq or metagenomics). In all cases, LEON showed higher overall compression ratios than state-of-the-art compression software.

On a *C. elegans* whole genome sequencing dataset, LEON divided the original file size by more than 20, corresponding to 0.67 bits/base for DNA and 0.24 bits/quality score. Leon can compress large datasets, for example a 733 GB FASTQ file (whole human genome sequenced at 102x depth) is compressed in 11h using 9.5 GB of ram and 8 CPU threads.

LEON is an open source software, distributed under GNU affero GPL License, available for download at http://gatb.inria.fr/software/leon/.

Keywords: Compression · *de Bruijn Graph* · NGS · Bloom filter

Genome Scaffolding with PE-Contaminated Mate-Pair Libraries

Kristoffer Sahlin[1], Rayan Chikhi[2], and Lars Arvestad[3]

[1] KTH Royal Institute of Technology, Science for Life Laboratory,
School of Computer Science and Communication, Solna, Sweden
[2] CNRS, CRIStAL, UMR 9189, 59650 Villeneuve d'Ascq, France
[3] Swedish e-Science Research Centre,
Science for Life Laboratory, and Department of Numerical Analysis
and Computer Science, Stockholm University, Stockholm, Sweden

Abstract. Scaffolding is often an essential step in a genome assembly process, in which contigs are ordered and oriented using read pairs from a combination of paired-ends libraries and longer-range mate-pair libraries. Although a simple idea, scaffolding is unfortunately hard to get right in practice. One source of problem is so-called PE-contamination in mate-pair libraries, in which a non-negligible fraction of the read pairs get the wrong orientation and a much smaller insert size than what is expected. This contamination has been discussed in previous work on integrated scaffolders in end-to-end assemblers such as Allpaths-LG and MaSuRCA but the methods relies on the fact that the orientation is observable, *e.g.*, by finding the junction adapter sequence in the reads. This is not always the case, making orientation and insert size of a read pair stochastic. Furthermore, work on modeling PE-contamination has so far been disregarded in stand-alone scaffolders and the effect that PE-contamination has on scaffolding quality has not been examined before.

We have addressed PE-contamination in an update of our scaffolder BESST. We formulate the problem as an Integer Linear Program (ILP) and use characteristics of the problem, such as contig lengths and insert size, to efficiently solve the ILP using a linear amount (with respect to the number of contigs) of Linear Programs. Our results show significant improvement over both integrated and standalone scaffolders. The impact of modeling PE-contamination is quantified by comparison with the previous BESST model. We also show how other scaffolders are vulnerable to PE-contaminated libraries, resulting in increased number of misassemblies, more conservative scaffolding, and inflated assembly sizes. BESST takes BAM files as input which makes it easily integrated in assembly pipelines. Source code and usage instructions are found at https://github.com/ksahlin/BESST.

Network Properties of the Ensemble of RNA Structures

Peter Clote

Boston College, Chestnut Hill, MA 02467, USA

Abstract. A neighbor of the RNA secondary structure s is obtained by removing, adding or shifting a base pair in s. Here, we describe the first efficient algorithm to compute the expected number of neighbors for the collection of all secondary structures of a given RNA sequence. This surprisingly complex algorithm permits a better understanding of kinetics of RNA folding when allowing defect diffusion, helix zippering, and related conformation transformations. Moreover, only when allowing shift moves does the network of secondary structures for certain RNAs satisfy the requirements of a small-world network.

RNA secondary structure kinetics is plays an essential role in certain biological processes, such as the *hok/sok* host-killing/suppression of killing (*hok/sok*) system that kills *E. coli* replicates if insufficient plasmids are transfered to the new daughter cell Nevertheless, RNA folding kinetics remains a difficult problem, since it is known that computation of optimal folding pathways is NP-complete [3].

Due to the biological importance of RNA folding kinetics, users generally run a secondary structure kinetics program, such as Kinfold, Kinefold, RNAKinetics. However, repeated simulations must be performed, each requiring lengthy computation times – for instance, the population occupancy curve for yeast phe-tRNA required 3 months of CPU time on a 2.4 GHz Intel Pentium 4 running linux [4]). Coarse-grained approaches using spectral methods also exist, such as Treekin, basin hopping with RNAlocmin, and Hermes.

Shift moves allow a transition from secondary structure s to structure t in which the base pair (i, j) of s is modified while fixing one base; i.e. base pair transitions of the form $(i,j) \rightarrow (i,k)$ or $(i,j) \rightarrow (k,j)$. Panels (a)-(d) of Fig. 1 depicts a particular type of shift move known as *defect diffusion*. Base pair addition, removal and shift moves constitute the default move set employed by the program Kinfold [2], with respect to which Wuchty [5] showed that the network of secondary structures of *E. coli* phe-tRNA (Sprinzl accession RF6280) is a *small-world network*.

The move set MS1 [resp. MS2] consists of base pair additions/removals [resp. additions/removals/shifts]. The network of the toy sequence ACGUACGU is illustrated in Fig. 1(e), and the distribution of the number of neighbors of each structure for the 32 nt selenocysteine insertion sequence fruA is depicted in Fig. 1(f). In this abstract, we describe an approach to efficiently compute the expected degree of an RNA network of secondary structures. Our work generalizes a recent paper [1], which describes a vastly simpler algorithm to compute the expected degree without consideration of shift

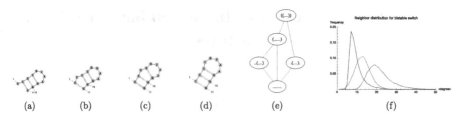

Fig. 1. *(a-d):* Example of successive shift moves, corresponding to *defect diffusion (e):* Network for the toy 8-mer ACGUACGU which has 5 nodes and 6 edges (hence 12 directed edges). The expected network degree is $\frac{12}{5} = 2.4$. Red edges indicate base pair addition or removal, while blue edges indicate shift moves. *(f):* Relative frequency for number of neighbors (degree) for the network of all secondary structures of the 25 nt bistable switch UGUACCGGAA GGUGCGAAUC UUCCG produced by exhaustive enumeration. The blue [resp. red resp. purple] curve corresponds to move set M2 [resp. (M2-M1) resp. M1]. Brute force analysis of the collection of all 83725 possible structures yields an expected network degree of 20.71 ± 6.91 [resp. 11.94 ± 3.93 resp. 8.77 ± 4.30] for move set MS2 [resp. MS2-MS1 resp. MS1] (Color figure online).

moves. Since our algorithm is surprisingly complex, we state the recursions for the RNA *homopolymer* model and leave the extensions to the general Turner nearest neighbor energy model to the journal version of this paper.

We now sketch the approach taken. Let $\mathbf{a} = a_1, \ldots, a_n$ be an arbitrary but fixed RNA sequence. For any $1 \leq i \leq j \leq n$, let $a[i, j]$ denote the subsequence $a_i, \ldots, a_j$. A secondary structure on $a[i, j]$ is a set of non-crossing base pairs (x, y), for $i \leq x < y \leq j$, where (a, b) and (c, d) are crossing if $a < c < b < d$. A base pair (x, y) is *external* if there is no base pair (u, v) for which $u < x < y < v$; a position x is *visible* if there is no base pair (u, v) such that $u \leq x \leq v$. The set of all secondary structures on $a[i, j]$ is denoted by $\mathbb{SS}[i,j]$. Define $Q_{i,j} = \sum_{s \in \mathbb{SS}[i,j]} \exp(-E(s)/RT) \cdot N(s)$, where $N(s)$ is the number of secondary structures t of $a[i, j]$ obtained from the structure s by the addition, deletion or shift of a base pair. The partition function for $a[i, j]$ is defined by $Z_{i,j} = \sum_{s \in \mathbb{SS}[i,j]} \exp(-E(s)/RT)$. It follows that the expected number of neighbors (network degree) is $\frac{Q_{1,n}}{Z_{1,n}}$.

For simplicity, we state the recursions for $Q_{1,n}$ and $Z_{1,n}$ for the *homopolymer model*, in which any two positions $1 \leq i < j \leq n$ can form a base pair, provided only that $i + 1 < j$. For the homopolymer model, there is no RNA sequence $\mathbf{a} = a_1, \ldots, a_n$, but rather only the interval $[1, n] = \{1, \ldots, n\}$. Thus we speak of a structure on $[i, j]$, rather than on $a[i, j]$. The energy of each structure in the homopolymer model is zero, so the probability of each structure s on $[i, j]$ equals one divided by the number of structures on $[i, j]$.

For $0 \leq n$, define Q_n to be the sum, taken over all structures s of $[1, n]$, of the number of base pair additions, removals or shifts of a base pair of s. Let Z_n denote the total number of homopolymer structures on $[1, n]$, where any two positions i, j can base-pair, as long as $j - i > 1$. Define $f(n, x)$ to be the number of secondary structures s for a length n homopolymer, such that s has x visible positions. Define $g(n, x)$ to be the number of secondary structures s for the length n homopolymer, such that s has x visible positions in the interval $[1, n - \theta - 1] = [1, n - 2]$, and position n is unpaired in s. Define the function E_n to be the number of *external base pairs* in all homopolymer structures on $[1, n]$. In the journal version of this paper, we give recursions for f, g, E and prove that

$$Q_n = Q_{n-1} + 2 \sum_{k=0}^{n-\theta-2} Z_{k-1} \cdot Z_{n-k-1} + 2 \left(E_{n-1} + E_{n-2} + \sum_{r=1}^{n-4} E_r \cdot Z_{n-r-3} \right) +$$

$$\sum_{x=2}^{n-\theta-1} x(x-1) \cdot g(n, x) + \sum_{k=1}^{n-\theta-1} (Z_{k-1} \cdot Q_{n-k-1}) + (Q_{k-1} \cdot Z_{n-k-1})$$

thus resulting in a cubic time algorithm to compute the expected network degree with respect to base pair additions, removals and shifts in the homopolymer case. The full paper also provides an even more complex extension to the Turner energy model.

Finally, we would like to thank the referees for valuable suggestions in a preliminary version of this abstract. This research was funded by the National Science Foundation grant DBI-1262439. Any opinions, findings, and conclusions or recommendations expressed in this material are those of the authors and do not necessarily reflect the views of the National Science Foundation.

References

[1] Clote, P.: Expected degree for RNA secondary structure networks. J. Comp. Chem. **36**(2), 103–117 (2015)
[2] Flamm, C., Fontana, W., Hofacker, I.L., Schuster, P.: RNA folding at elementary step resolution. RNA **6**, 325–338 (2000)
[3] Thachuk, C., Maňuch, J., Stacho, L., Condon, A.: NP-completeness of the direct energy barrier height problem. Nat. Comput. **10**(1), 391–405 (2011)
[4] Wolfinger, M., Svrcek-Seiler, W.A., Flamm, C., Stadler, P.F.: Efficient computation of RNA folding dynamics. J. Phys. A: Math. Gen. **37**, 4731–4741 (2004)
[5] Wuchty, S.: Small worlds in RNA structures. Nucleic. Acids. Res. **31**(3), 1108–1117 (2003)

Contents

BicNET: Efficient Biclustering of Biological Networks to Unravel Non-Trivial Modules

Rui Henriques$^{(\boxtimes)}$ and Sara C. Madeira

Inesc-ID, Instituto Superior Técnico, Universidade de Lisboa, Lisboa, Portugal
{rmch,sara.madeira}@tecnico.ulisboa.pt

Abstract. The discovery of dense biclusters in biological networks received an increasing attention in recent years. However, despite the importance of understanding the cell behavior, dense biclusters can only identify modules where genes, proteins or metabolites are strongly connected. These modules are thus often associated with trivial, already known interactions or background processes not necessarily related with the studied conditions. Furthermore, despite the availability of biclustering algorithms able to discover modules with more flexible coherency, their application over large-scale biological networks is hampered by efficiency bottlenecks. In this work, we propose BicNET (Biclustering NETworks), an algorithm to discover non-trivial yet coherent modules in weighted biological networks with heightened efficiency. First, we motivate the relevance of discovering network modules given by constant, symmetric and plaid biclustering models. Second, we propose a solution to discover these flexible modules without time and memory bottlenecks by seizing high efficiency gains from the inherent structural sparsity of networks. Results from the analysis of protein and gene interaction networks support the relevance and efficiency of BicNET.

1 Introduction

The increasing precision and completeness of biological networks from diverse organisms provide an unprecedented opportunity to understand the organization and dynamics of the cell [2]. In particular, the discovery of functional network modules has been largely used to characterize, discriminate and predict biological functions [2,25,28,29]. The task of discovering such modules can be mapped into the discovery of coherent regions in weighted graphs, where nodes represent the molecular units (typically genes, proteins or metabolites) and the edges' weights represent the strength of the interactions between the biological molecules. In this context, a large focus has been placed on the identification of dense regions [1,9,11,12], where each region is given by a statistically significant set of highly interconnected nodes. In recent years, a high number of biclustering algorithms has been proposed to discover dense regions from (bipartite) graphs by mapping them as adjacency matrices and searching for dense submatrices [1,3,9,22,25]. A bicluster is then given by two subsets of strongly connected nodes. Despite

© Springer-Verlag Berlin Heidelberg 2015
M. Pop and H. Touzet (Eds.): WABI 2015, LNBI 9289, pp. 1–15, 2015.
DOI: 10.1007/978-3-662-48221-6_1

the effectiveness of biclustering to model local interactions, the focus on dense regions comes with key drawbacks. First, such regions are usually associated with either trivial or already well-known putative modules. Second, the weights of the interactions associated with less studied genes, proteins and metabolites have lower confidence (with penalizations highly dependent on the organism under study) and may not reflect the true role of these molecular interactions in certain cellular processes [31]. In particular, the presence of (well-studied) regular/background cellular processes may mask the discovery of sporadic or less-trivial processes.

Although many biclustering algorithms are able to find flexible coherencies in (adjacency) matrices [23], two major challenges have been preventing their application to biological networks. First, the generalized lack of understanding on the relevance and biological meaning of network modules with flexible coherency (given by plaid models, for example). Second, the hard combinatorial nature of biclustering regions with flexible coherency, together with the high dimensionality of matrices derived from biological networks are often associated with memory and time bottlenecks, and/or undesirable restrictions on the structure and quality of biclusters. This work aims to answer these problems by: (1) pinpointing the biological relevance of modeling non-dense regions in a network, and (2) enabling the efficient learning of flexible biclustering models from large biological networks.

To address these challenges we propose the algorithm BicNET (Biclustering NETworks). BicNET integrates contributions from pattern-based biclustering algorithms [14, 15] for the exhaustive discovery of biclusters with parameterizable coherency and quality, and adapts their data structures and searches to explore efficiency gains from the inherent sparsity of biological networks. Furthermore, we motivate the relevance of finding non-dense yet coherent modules and provide a meaningful analysis of BicNET's outputs. Results gathered from synthetic and real data show: the relevance of the proposed efficiency principles for biclustering large (possibly dense) networks, and the effectiveness of BicNET to discover a complete set of non-trivial yet coherent and biologically significant modules.

The paper is organized as follows. Section 2 provides background on the target task of modeling functional modules given by regions with flexible coherency criteria and surveys major contributions from related work. Section 3 proposes the BicNET algorithm. Section 4 provides empirical evidence for the relevance of BicNET to unravel non-trivial yet relevant modules in synthetic and real networks. Finally, we draw conclusions and highlight directions for future work.

2 Background

Biclustering can be applied to different types of networks: homogeneous networks, given for instance by protein-protein interactions (PPI) and gene interactions (GI); and heteregeneous networks, capturing interactions between distinct molecular entities (proteins, protein complexes, metabolites, genes, etc.), between host and viral molecules, or between biological entities and certain terms/properties. These networks can be mapped into (bipartite) graphs for the subsequent discovery of highly interconnected regions associated with modules.

Definition 1. *Given a weighted bipartite graph with two sets of nodes $X=\{x_1,..,x_n\}$ and $Y=\{y_1,..,y_m\}$, and interactions $a_{ij}\in\mathbb{R}$ relating nodes x_i and y_j, **biclustering** aims to find a set of biclusters $\mathcal{B}=\{B_1,..,B_m\}$, where each bicluster $B_k=(I,J)$ is a subgraph (module) given by two subsets of nodes, $I\subseteq X \wedge J\subseteq Y$, satisfying specific criteria of coherency, quality, and significance.*

This task can be solved with traditional biclustering on real-valued matrices by mapping the bipartite graph into an adjacency matrix, where rows and columns are given by the nodes and the values by the weighted interactions. In this case, subsets of rows and columns define a bicluster associated with a network module with coherent interactions. The *structure* of a set of biclusters is defined by their number, size and positioning. Flexible structures are characterized by an arbitrary-high number of (possibly overlapping) biclusters. The *coherency* of a bicluster is defined by the observed correlation of values. Definition 2 introduces dense, constant, symmetric and plaid coherencies. The *quality* of a bicluster is defined by the type and amount of tolerated noise. The statistical *significance* of a bicluster determines the deviation of its probability of occurrence from expectations.

Definition 2. *Let the elements in a bicluster $a_{ij}\in(I,J)$ have specific coherency. A bicluster is **dense** when the average strength of its interactions, $\frac{1}{|I||J|}\Sigma_{i\in I}\Sigma_{j\in J}|a_{ij}|$, is significantly high. A **constant** coherency is observed when $a_{ij}=k_j$ where k_j is the expected strength of interactions between nodes in I and y_j node from J. In the presence of symmetries, $a_{ij}=k_jc_i$ where $c_i\in\{-1,1\}$. A **plaid** coherency considers cumulative contributions on the elements where biclusters/subgraphs overlap.*

Related Work on Biclustering Biological Networks. A large number of algorithms has been proposed to find modules in unweighted and/or weighted graphs mapped from homogeneous and/or heterogeneous biological networks [6,25,29]. In unweighted graphs, clique detection with Monte Carlo optimization [30], probabilistic motif discovery [5] and clustering on graphs [6] have been, respectively, applied to discover modules in PPIs (yeast), GIs (E. coli) and metabolic networks. In unweighted bipartite graphs, the densest regions correspond to bicliques. Bicliques can be efficiently mined using density-constrained biclustering [8], Motzkin-Straus optimization [11], formal concepts and pattern-based biclustering [3,22,25,34]. In weighted graphs, the density of a module is given by the average strength of interactions. Strength is either determined by a measure of confidence (when it is predicted from literature or diverse data sources) or by the functional correlation between nodes (when it is derived from experimental data). Densely weighted modules have been discovered with betweenness-based partitioning [6], graph flow-based clustering [27] and several biclustering approaches, including SAMBA [32], multi-objective searches [24] and pattern-based biclustering [1,9,10]. The application of these methods over homogeneous and viral-host PPIs show that protein complexes largely match the found modules [6, 24, 27].

The discovery of dense network modules has been largely accomplished with pattern-based biclustering algorithms [1, 3, 9, 10, 22, 25, 34] due to their intrinsic ability to exhaustively discover flexible structures of biclusters. Frequent patterns in discrete networks can be mapped[1] as biclusters with specific coherency strength determined by the number of symbols (ranges of weights) assigned to the interactions. In unweighted graphs, closed frequent itemset mining and association rule mining were applied to study interactions between proteins and protein complexes in yeast proteome network [34] and between HIV-1 and human proteins [22, 25]. More recently, association rules were also used to obtain a modular decomposition of positive and negative GIs ($a_{ij} \in \{-1, 0, 1\}$) [3]. In weighted graphs, Dao et. al [10] and Atluri et. al [1] relied on the loose antimonotone property of density to propose weight-sensitive pattern mining searches. DECOB [9], originally applied to PPIs and GIs from human and yeast, uses an additional filtering step to output of non-similar modules only.

Some of these works have been extended to discover discriminative modules, often referred as multigenic markers, for classification tasks such as function prediction [10, 22, 29]. Network-based (bi)clustering methods for function prediction have been comprehensively reviewed by Sharan et al. [29].

Related Work on Biclustering Modules with Flexible Coherency. Although the state-of-the-art is focused on the discovery of dense network modules, slight variants of this coherency have been proposed [1, 19, 32]. Despite the large availability of biclustering algorithms able to find biclusters with flexible coherency [23], empirical evidence shows that they are not prepared to deal with the sparsity and/or high-dimensionality of adjacency matrices mapped from networks. A first attempt towards this end was presented by Tomaino et al. [33] for small networks.

3   Solution

In what follows, we first show how biclustering can be applied to discover coherent modules following constant, symmetric and plaid models, possibly containing noisy and missing interactions. Second, we extend pattern-based searches to optimally handle the inherent structural sparsity of biological networks.

---

[1] Let $\mathcal{L}$ be a finite set of items, and P an itemset $P \subseteq \mathcal{L}$. A discrete matrix D is a set of transactions in $\mathcal{L}$, $\{P_1, .., P_n\}$. Let the *coverage* Φ_P of an itemset P be the set of transactions in D in which P occurs, $\{P_i \in D \mid P \subseteq P_i\}$, and its *support* sup_P be the coverage size, $|\Phi_P|$. Given D and a minimum support θ, the *frequent itemset mining* task aims to compute: $\{P \mid P \subseteq \mathcal{L}, sup_P \geq \theta\}$.

Given D, let a matrix A be the concatenation of D elements with their column indexes. Let Ψ_P of an itemset P in A be its indexes, and Υ_P be its original items in $\mathcal{L}$. A set of *biclusters* $\cup_k (I_k, J_k)$ can be derived from frequent itemsets $\cup_k P_k$ by mapping $(I_k, J_k) = (\Phi_{P_k}, \Psi_{P_k})$ to compose constant biclusters with coherency across rows $((I_k, J_k) = (\Psi_{P_k}, \Phi_{P_k})$ for column-coherency) with pattern Υ_P.

3.1   Network Modules with Flexible Coherency

Biclustering Weighted Graphs. For an effective application of state-of-the-art biclustering algorithms to (weighted) graphs derived from biological networks, two principles should be satisfied. First, the weighted graph should be mapped into a minimal bipartite graph. In heterogeneous networks, multiple bipartite graphs are created (each with two disjoint sets of nodes with heterogeneous interactions). The minimality requirement can be satisfied by identifying subsets of nodes with cross-set interactions but without intra-set interactions to avoid unnecessary duplicated nodes in the disjoint sets of nodes (see Fig. 1). This is essential to avoid the generation of large bipartite graphs and subsequent very large matrices.

Second, when targeting non-dense coherencies, two real-valued adjacency matrices need to be derived from the bipartite graph (a matrix with rows and columns mapped from the disjoint sets of nodes and its transpose) for an exhaustive space exploration. This is different from using all nodes as rows and columns in a single matrix and then filling the upper and lower triangular matrices, which can can lead to inconsistencies when a bicluster has elements from both the upper and lower triangular matrices. Also, the larger size and density of such matrix can significantly hamper the efficiency of the biclustering task. The few attempts to find non-dense biclusters in biological networks fail to satisfy this principle [33], thus delivering incomplete and often inconsistent solutions.

Pattern-based Biclustering. Under the satisfaction of the previous principles, a wide-range of biclustering algorithms can be applied to discover modules with flexible coherencies [23]. Yet, to our knowledge, only pattern-based biclustering [14–16] is able to guarantee an exhaustive yet efficient discovery of flexible structures of biclusters with parameterizable coherency and quality criteria. This provides the necessary context to measure the relevance and impact of discovering modules with non-dense coherency and noise-tolerance. In particular, we

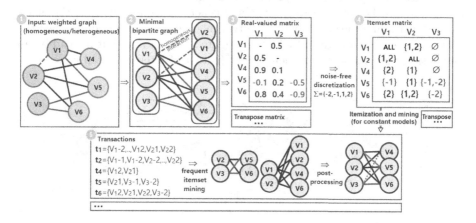

Fig. 1. Pattern-based biclustering of (heterogeneous) biological networks.

rely on BicPAM and BiP algorithms [13,15]. These algorithms, respectively, use frequent itemset mining and association rule mining to find biclusters with constant/symmetric and plaid coherencies. Furthermore, they integrate the dispersed contributions from previous pattern-based algorithms and address some of their limitations, providing key principles to surpass discretization problems (by introducing the possibility to assign multiple symbols to a single element) and robustly handle noise and missing values. Figure 1 provides a view on how transactions can be derived from (heterogeneous) biological networks for the discovery of constant modules (see [15] for details on the itemization, mining and postprocessing steps).

Constant Model. Given a bicluster defining a module with coherent interactions between two sets of nodes, the constant coherency (Definition 2) implies that the nodes in one set show a single type of interaction with the nodes in the remaining set. Illustrating, consider a set of interactions between genes and proteins, where their absolute weight defines the strength of the association and their sign determines whether the association corresponds to activation or repression mechanisms. The constant model guarantees that when a gene is associated with a group of proteins, it establishes the same type of interaction with all these proteins (such as heightened activation of the transcription of a complex of proteins). When analyzing the transposed matrix (by switching the disjoint sets of the bipartite graph), similar relations can be observed: a protein coherently affects a set of genes (softly repressing their expression, for example). The constant model can also disclose relevant interactions between homogeneous groups of genes, proteins and metabolites. Figure 2 provides an illustrative constant module.

The constant model can be further applied to networks with qualitative interactions capturing distinct types of regulatory relations, such as *binds*, *activates* or *enhances* associations, common in a wide-variety of PPIs [22,25].

The constant model is essential to guarantee that molecular units with non-necessarily high (yet coherent) influence on another set of molecular units are not excluded. The constant coherency is in general more flexible than the dense coherency, leading to the discovery of larger modules. The exception is when the dense coherency is not given by highly weighted interactions, but instead by all interactions independently of their weight (extent of interconnected nodes).

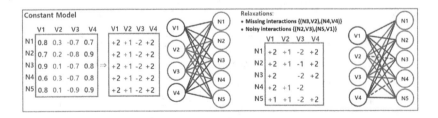

Fig. 2. Biclustering (noise-tolerant) modules with the constant model.

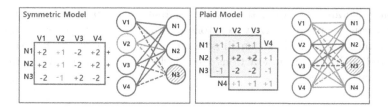

Fig. 3. Biclustering modules with the symmetric and plaid models.

Symmetric Model. The presence of symmetries is key to simultaneously capture activation and repression mechanisms associated with the interactions of a single node [15]. The symmetric model introduces a new degree of flexibility by enabling the discovery of more complex regulatory modules, where a specific gene/protein may show symmetric regulatory behavior according to the expected pattern, yet still respect the observed coherency. Figure 3 illustrates the symmetric model, where rows with symmetries are identified with dashed lines.

Plaid Model. The plaid assumption [13] is essential to describe overlapping regulatory behavior associated with cumulative effects in the strength of interactions between nodes that appear in multiple functional modules. Illustrating, consider that two genes interact in the context of multiple biological processes, a plaid model can consider their cumulative effect on their interaction's weights (based on the expected weight associated with each active process). This is also valid for the regulatory influence between proteins and for heterogeneous networks. The plaid assumption of GIs and PPIs also provides insights on the network topology and molecular functions, revealing hubs and core interactions (based on the amount of overlapping interactions), and between- and within-pathway interactions (based on the interactions inside and outside of the overlapping areas). Figure 3 illustrates a plaid model associated with two overlapping modules. These modules could not be discovered without a plaid assumption.

Handling Noisy and Missing Interactions. An undesirable restriction of exhaustive searches for dense modules is that they may exclude relevant nodes associated with a bicluster if those nodes do not interact with all of the nodes in one subset of nodes from the bicluster. Understandably, meaningful modules with missing interactions are common since the majority of existing biological networks are still largely incomplete. Pattern-based biclustering is able to recover missing interactions recurring to well-established and efficient postprocessing procedures (based on the merging and extension of the discovered modules) [15].

Furthermore, the scoring scheme of interactions might be prone to experimental noise, preprocessing biases and structural noise (particularly common for less studied and stable genes or proteins), not always reflecting the true interactions. Pattern-based biclustering also allows the assignment of multiple symbols to specific interactions [15], thus avoiding the exclusion of noisy interactions (see Fig. 1). Although default parameterizations are provided to guarantee an adequate tolerance to noise, the level of sparsity and noise of the discovered modules can be parametrically controlled using thresholds based on quality expectations. Figure 2 shows an illustrative coherent module with corrections associated with missing interactions (red dashed lines) and noisy interactions (red continuous lines).

3.2 BicNET: Efficient Biclustering of Biological Networks

Understandably, the task of discovering modules with the introduced coherencies is more complex than finding dense modules (complexity discussed in [15]). Empirical evidence shows that state-of-the-art biclustering algorithms are only scalable for biological networks up to a few hundreds of nodes (see Results). Nevertheless, a key property distinguishing biological networks from gene expression or clinical data is their underlying sparsity. Illustrating, some of the densest PPI and GI networks from well-studied organisms still have a density below 5 % (ratio of interconnected nodes after excluding nodes without interactions). While traditional biclustering depends on operations over matrices, pattern-based biclustering algorithms are prepared to mine transactions of varying length. This property makes pattern-based biclustering able to exclude missing interactions from searches and thus surpass memory and efficiency bottlenecks. Based on this observation, we propose BicNET (**BiC**lustering Biological **NET**works), a pattern-based biclustering algorithm for the discovery of network modules with non-trivial coherencies and robustness to noise. Additionally, BicNET relies on the following principles to explore further efficiency gains.

We propose a new data structure to efficiently preprocess data: an array, where each position (node from a disjoint set in the bipartite graph) has a list of pairs, each pair representing an interaction (corresponding node and the interaction weight). Discretization and itemization procedures are performed by linearly scanning this structure three times. Thus, their time and memory complexity is linear on the number of interactions.

Pattern-based searches commonly rely on bitset vectors due to the need to retrieve not only the frequent patterns but also their supporting transactions in order to compose biclusters. However, bitset vectors are costly in terms of memory, and the associated intersection operations are computationally expensive for large-scale networks. For this reason, we rely on the recently proposed F2G miner [17] and on revised implementations of Eclat and Charm miners where diffsets are used to address the bottlenecks of bitsets. These pattern-based searches guarantee an efficient discovery of constant, symmetric and plaid models.

Furthermore, the underlying pattern mining searches of BicNET are dynamically selected based on the properties of the network to optimize their efficiency. Horizontal versus vertical data formats [15] are selected based on the ratio of rows and columns from the mapped matrix. Apriori (candidate generation) versus pattern-growth (tree projection) searches [15] are selected based on network density (pattern-growth searches are preferable for dense networks). We also push the computation of similarities between all pairs of biclusters (the most expensive postprocessing procedure) into the mining step by checking similarities with distance operators on a compact data structure to store the frequent patterns.

4   Results and Discussion

Results are organized as follows. First, we compare the performance of BicNET against state-of-the-art biclustering algorithms using synthetic networks. Second, we use BicNET for the analysis of large-scale PPI and GI networks to show the relevance of discovering modules with flexible coherencies and parameterizable levels of noise and sparsity. BicNET is implemented in Java (JVM v1.6.0-24). Experiments were computed using an Intel Core i5 2.30 GHz with 6 GB of RAM.

Synthetic Data. Networks with planted biclusters were generated respecting the commonly observed topological statistics of biological networks. Variables:

- number of nodes, density and distributions of the weight (positive and negative ranges revealing the interaction strength);
- degree of noisy and missing interactions (from 0 % to 20 %).
- number, size (Uniform distribution on the number of nodes), shape (imbalance on the size of the disjoint sets of each subgraph), overlapping, and coherency (dense, constant, symmetric and plaid) of the planted biclusters:

	Network nodes (10 % density)					Network density (2000 nodes)			
	200	500	1000	2000	10000	1 %	5 %	10 %	25 %
♯ Hidden modules	5	10	15	20	30	3	5	10	20
♯ Nodes per module	[20,30]	[30,40]	[40,50]	[50,70]	[100,140]	[50,70]	[50,70]	[50,70]	[50,70]
% Interactions in modules	19,5 %	12,2 %	7,6 %	4,5 %	1,1 %	22,5 %	9,0 %	4,5 %	2,3 %

Real Data. We used four biological networks: GIs in yeast from DryGIN [21] and STRING v10 [31] databases, and two licensed PPIs in human and E. coli from STRING v10 [31] database. The scores in these networks show the expected strength of influence/physical interaction between genes/proteins (see Table 1 for statistics).

Performance Metrics. Given the set of planted modules $\mathcal{H}$ in a synthetic network, the accuracy of the retrieved modules $\mathcal{B}$ is given by two match scores

Table 1. Biological networks used to assess the relevance and efficiency of BicNET.

Type	Organism	♯Nodes	♯Interactions	Density	Notes on the weight of interactions
GI	Yeast	4455	191309	1.0%	Weights (65% negative) from double-mutant arrays [21].
GI	Yeast	6314	423335	1.1%	Known and predicted associations benchmarked
PPI	E. Coli	8428	3293416	4.6%	from multiple data sources and text mining,
PPI	Human	19247	8548002	2.3%	and combined through an integrative score [31].

(1): $MS(\mathcal{B}, \mathcal{H})$ defining the extent to what the found biclusters cover the hidden biclusters (*completeness*), and $MS(\mathcal{H}, \mathcal{B})$ reflecting how well the hidden biclusters are recovered (*precision*). We present the average of matches collected from 10 instantiations of synthetic networks. These accuracy criteria surpass the problems of Jaccard matches (only focused on one of the two subsets of nodes at a time [15]) and RNIA (loose matching criteria [15]). Efficiency and significance are used to complement this analysis.

$$\mathbf{MS}(\mathcal{B}, \mathcal{H}) = \frac{1}{|\mathcal{B}|} \Sigma_{(I_1, J_1) \in \mathcal{B}} max_{(I_2, J_2) \in \mathcal{H}} \sqrt{\frac{|I_1 \cap I_2|}{|I_1 \cup I_2|} \frac{|J_1 \cap J_2|}{|J_1 \cup J_2|}} \qquad (1)$$

4.1 Results on Synthetic Data

Figure 4 compares the efficiency of BicNET with state-of-the-art biclustering algorithms with flexible coherence criteria using networks with varying size and density and planted modules with constant coherency. We selected FABIA[2] [18], ISA [20], xMotifs [26], CC [7] and OPSM [4] to discover modules with flexible coherency. BicNET shows heightened efficiency levels. Understandably, as most of the remaining algorithms are only prepared to analyze (non-sparse) matrices, they show efficiency bottlenecks for even small networks. Furthermore, the majority is not able to accurately recover the planted modules as they cannot interpret missing interactions. Although SAMBA [32] and some pattern-based biclustering algorithms, such as BiMax and DECOB [9,25], are able to discover dense models efficiently, they are not prepared to discover modules with alternative coherence criteria.

Figure 5 zooms-in on the performance of BicNET by quantifying the efficiency gains in memory and time from using adequate data structures (replacing the need to use matrices) and searches (replacing the need to rely on bitset vectors). It also shows that the cost of assigning multiple symbols per interaction are moderate, despite resulting in an increased network density.

Figure 6 compares the performance of BicNET with peer algorithms for discovering dense network modules (hypercliques) in the presence of noisy and missing interactions. This analysis clearly shows that existing pattern-based searches for hypercliques have no tolerance to errors since their accuracy rapidly degrades

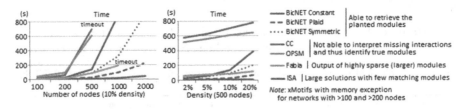

Fig. 4. Efficiency of flexible biclustering algorithms to discover constant modules in synthetic networks with varying size and density.

---

[2] Sparse prior equation with decreasing sparsity until able to retrieve a non-empty set of biclusters.

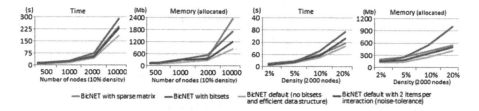

Fig. 5. Efficiency gains of BicNET when using sparse data structures, pattern mining searches providing robust alternatives to bitset vectors, and noise handlers.

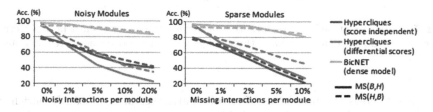

Fig. 6. Accuracy of BicNET against peer pattern-based searches to discover dense modules on networks (2000 nodes, 10 % density) with varying degree of noise and missings.

for an increased number of planted noisy/missing interactions. Thus, they are not able to deal with the natural incompleteness and scoring uncertainty associated with biological networks. On the other hand, the observed accuracy levels of BicNET demonstrate its robustness to noise (validating the importance of assigning multiple ranges of weights for some interactions) and to missing interactions (showing the effectiveness of BicNET's postprocessing procedures).

4.2 Results on Real Data

The biological significance of the modules discovered in real data was computed by assessing the over-representation of Gene Ontology (GO) terms with an hypergeometric test. A module is significant when its genes show enrichment for one or more terms by having a (Bonferroni corrected) p-value below 0.01. Figure 7 shows the properties of BicNET's solutions for the four biological networks in Table 1. 94 % of the modules discovered in DRYGIN's yeast GIs were significantly enriched. All the modules discovered in STRING's yeast GIs were significantly enriched. BicNET was able to discover the largest number of (non-similar and statistically significant) biclusters. The analysis of the enriched terms for these modules against the enriched terms found in other biclustering solutions supports the completeness, exclusivity and relevance of BicNET's solutions (Table 2). The significance of peer solutions from unweighted graphs is penalized by the inability to remove nodes with either low or non-coherent weights, while the significance of peer solutions focused on dense regions is additionally hampered by noise and discretization errors (Fig. 7).

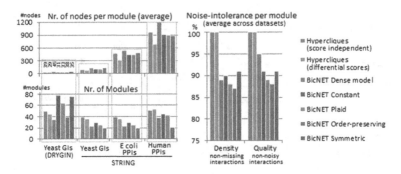

Fig. 7. Properties of BicNET's solutions with varying coherency against peer pattern-based searches for dense modules (hypercliques) in networks from DRYGIN and STRING.

Table 2 shows the properties of an illustrative set of significantly enriched modules. We can observe that such biclusters could hardly be discovered by peer methods due to their non-dense coherency. All of the illustrated modules show coherent patterns of interaction between nodes combining both differential and non-differential weights. The provided modules have an average of 5 to 10 % of missing interactions. BicNET is well positioned to find modules with varying size, coherency and quality. Illustrating, the constant modules D_6 and D_7 have, respectively, 23 and 47 nodes and distinct quality, being D_7 more tolerant to noisy interactions. Understandably, the number of nodes per module is naturally affected by the size and sparsity of the target network. Most of the discovered modules clearly show non-trivial yet meaningful correlations, whose relevance is pinpointed by the number of highly enriched terms after correction.

Table 3 lists some of the enriched terms for the modules in Table 2, showing their functional coherence and role to unravel putative biological processes. Interesting, some of the identified modules are part of an additive plaid model (with in-between condition [13]). Illustrating, modules D6 and S4 share, respectively, 21 % and 36 % of their interactions with modules D7 and S4 under a

Table 2. Exclusivity and relevance of BicNET solutions: properties of found modules.

	ID	Type	♯Nodes $|I| \times |J|$	Items	♯Terms $p<1E-15$	Notes
DRYGIN	D1	constant	18×9	{-4,..,4}	27	Module with coherent strong (-4) and soft (-1) negative interactions.
	D2	symmetric	4×9	{-3,..,3}	13	Varying levels of strong (mainly positive) interactions ({±3,±2}).
	D3	symmetric	5×6	{-2,-1,1,2}	12	Module with either all positive or negative interactions per "row"-node ({±1,±2}).
	D4	constant	7×5	{1,2}	12	Module with coherent strong (2) and soft (1) positive interactions.
	D5	symmetric	7×5	{-2,-1,1,2}	11	Module with either all positive or negative interactions per "row"-node ({±1,±2}).
	D6	constant	13×10	{-2,-1,1,2}	24	Module with mostly strong negative interactions per "row"-node.
	D7	constant	39×8	{-2,-1,1,2}	47	Noise-tolerant module with positive and negative interactions.
STRING	S1	constant	148×13	{1,2}	169	Noise-tolerant module with positive interactions of varying strength ({1,2}).
	S2	constant	80×18	{1,2,3}	98	Module with mostly of non-dense interactions ({1,2}).
	S3	constant	83×10	{1,2}	93	Module with non-dense positive interactions before postprocessing ({1}).
	S4	constant	50×20	{1,2,3}	70	Module with non-dense positive interactions ({1,2}) before postprocessing.
	S5	constant	45×31	{1,2,3}	76	Module with mostly dense interactions (weights in {2,3}).
	S6	constant	55×85	{1,2}	143	Module with mostly dense interactions ({2}).

Table 3. Illustrative set of biologically significant BicNET's modules: description of the highly enriched terms in the modules presented in Table 2.

	ID	Terms description (♯)	♯Terms $p<1E\text{-}15$	♯Nodes
DryGIN	D1	Histone modification; regulation of histones: H3-K79/H3-K4 methylation, H2B ubiquitination, etc. (5);	6	27
	D2	Gluconeogenesis; glutamate metabolic/catabolic processes (2); nicotinamide metabolism/biosynthesis (2);	6	13
	D3	Positive and negative regulation of transcription from RNA polymerase II; Invasive growth response to glucose limitation and hyperosmotic salinity response by regulating RNA polymerase II (5);	5	12
	D4	Meiotic anaphase I; activation of anaphase-promoting complex activity involved in meiotic cell cycle;	4	12
	D5	Negative reg. of phospholipid biosynthesis; lipid homeostasis; isopropylmalate and oxaloacetate transport;	4	11
	D6	Cotranslational protein targeting to membrane; protein insertion into mitochondrial membrane; protein import into peroxisome membrane; reg. sporulation; actin filament bundle assembly involved in cytokinesis;	5	25
	D7	Acetate fermentation, acetyl-CoA biosynthesis (from acetate), reg. transcription on exit from mitosis;	7	50
STRING	S1	Response to hypoxia; oxidation-dependent protein catabolic process; anaerobic respiration; age-dependent response to reactive oxygen species; cellular response to oxidative stress;	36	169
	S2	Positive & negative reg. of mitotic and nuclear cell cycle, DNA replication, budding cell apical bud growth;	16	98
	S3	Transport of aerobic e-, acetyl-CoA, vacuolar transm., amine (5); ribose phosphate & D-ribose processes (2);	22	93
	S4	Heterochromatin maintenance involved in chromatin silencing; sister chromatid segregation;	6	70
	S5	Cytoplasmic and mitochondrial translation (4); regulation of translational fidelity; ADP biosynthesis;	6	76
	S6	rRNA processing; separation, cleavage & maturation of SSU-rRNA (5); ribosomal (large subunit) biogenesis;	14	143

plaid assumption. Without this assumption, only smaller modules (excluding key nodes) could be obtained, resulting in a lower enrichment of their terms.

In a concluding note, when analyzing networks derived from knowledge-based repositories and literature (such as networks from STRING [31]), the flexibility of coherence and noise-robustness is critical to deal with uncertainty and regions where weights may be affected due to the unbalanced focus of research studies. When analyzing networks derived from data experiments (such as GIs from DRYGIN [21]), the discovery of modules with non-necessarily strong interactions (given by the constant model, for example) can be critical to identify less-predominant (yet key) biological processes, such as the ones associated with early stages of stimulation or disease.

5 Conclusions and Future Work

This work motivates and answers the task of biclustering large-scale biological networks to discover modules with flexible yet meaningful coherency and robustness to noise. In particular, we explored the relevance of mining non-trivial modules in both homogeneous and heterogeneous networks, and assessed the limits in efficiency of existing biclustering algorithms targeting non-dense models. Combining state-of-the-art contributions on pattern-based biclustering with efficient searches on networks, we propose BicNET algorithm for the exhaustive discovery of constant, symmetric and plaid models in biological networks. Additional strategies are further incorporated to retrieve modules with noisy and missing interactions, thus addressing the limitations of the existing exhaustive searches on networks. BicNET enables the analysis of dense networks with up to 50000 nodes. Results on synthetic and real networks confirm its efficiency and relevance to discover non-trivial (yet coherent and significant) modules.

Six possible directions are identified for future work: to consider further coherencies such as order-preserving and scale factors; enhance searches with scalability principles from pattern mining (data partitioning strategies and search for approximate patterns [14]); extend the proposed contributions for the integrative mining of network and expression data; explore the relevance of the plaid model to identify and characterize hubs; enlarge the experimental analyzes towards biological molecules with yet unclear roles; and embrace predictive tasks.

Acknowledgments. This work was supported by *FCT* under the project UID/CEC/ 50021/2013 and the PhD grant SFRH/BD/75924/2011 to RH.

References

1. Atluri, G., Bellay, J., Pandey, G., Myers, C., Kumar, V.: Discovering coherent value bicliques in genetic interaction data. In: IW on Data Mining in Bioinformatics (2010)
2. Barabasi, A.L., Oltvai, Z.N.: Network biology: understanding the cell's functional organization. Nat. Rev. Genet. **5**(2), 101–113 (2004)
3. Bellay, J., Atluri, G., Sing, T.L., Toufighi, K., Costanzo, M., et al.: Putting genetic interactions in context through a global modular decomposition. Genome Res. **21**(8), 1375–1387 (2011)
4. Ben-Dor, A., Chor, B., Karp, R., Yakhini, Z.: Discovering local structure in gene expression data: the order-preserving submatrix problem. In: RECOMB, pp. 49–57. ACM (2002)
5. Berg, J., Lässig, M.: Local graph alignment and motif search in biological networks. Nat. Acad. Sci. **101**(41), 14689–14694 (2004)
6. Chen, J., Yuan, B.: Detecting functional modules in the yeast protein protein interaction network. Bioinformatics **22**(18), 2283–2290 (2006)
7. Cheng, Y., Church, G.: Biclustering of expression data. In: ISMB, pp. 93–103. AAAI (2000)
8. Colak, R.: Towards finding the complete modulome: density constrained biclustering. Ph.D. thesis, Simon Fraser University (2008)
9. Colak, R., Moser, F., Chu, J.S.C., Schönhuth, A., Chen, N., Ester, M.: Module discovery by exhaustive search for densely connected, co-expressed regions in biomolecular interaction networks. PLoS One **5**(10), e13348 (2010)
10. Dao, P., Colak, R., Salari, R., Moser, F., Davicioni, E., Schnhuth, A., Ester, M.: Inferring cancer subnetwork markers using density-constrained biclustering. Bioinformatics **26**(18), i625–i631 (2010)
11. Ding, C., Zhang, Y., Li, T., Holbrook, S.: Biclustering protein complex interactions with a biclique finding algorithm. In: ICDM, pp. 178–187 (2006)
12. Georgii, E., Dietmann, S., Uno, T., Pagel, P., Tsuda, K.: Enumeration of condition-dependent dense modules in protein interaction networks. Bioinformatics **25**(7), 933–940 (2009)
13. Henriques, R., Madeira, S.: Biclustering with flexible plaid models to unravel interactions between biological processes. IEEE/ACM TCBB (2015). doi:10.1109/TCBB.2014.2388206
14. Henriques, R., Antunes, C., Madeira, S.C.: A structured view on pattern mining-based biclustering. Pattern Recognition (2015). http://www.sciencedirect.com/science/article/pii/S003132031500240X
15. Henriques, R., Madeira, S.: Bicpam: pattern-based biclustering for biomedical data analysis. Algorithms Mol. Biol. **9**(1), 27 (2014)

16. Henriques, R., Madeira, S.C.: Pattern-based biclustering with constraints for gene expression data analysis. In: 17th Portuguese Conference on Artificial Intelligence (EPIA-2015), Computational Methods in Bioinformatics and Systems Biology (CMBSB), Coimbra, Portugal. LNAI. Springer, Heidelberg (2015)
17. Henriques, R., Madeira, S.C., Antunes, C.: F2g: efficient discovery of full-patterns. In: ECML/PKDD IW on New Frontiers to Mine Complex Patterns. Springer-Verlag (2013)
18. Hochreiter, S., et al.: FABIA: factor analysis for bicluster acquisition. Bioinformatics 26(12), 1520–1527 (2010)
19. Ideker, T., Ozier, O., Schwikowski, B., Siegel, A.F.: Discovering regulatory and signalling circuits in molecular interaction networks. Bioinformatics 18(suppl 1), S233–S240 (2002)
20. Ihmels, J., Bergmann, S., Barkai, N.: Defining transcription modules using large-scale gene expression data. Bioinformatics 20(13), 1993–2003 (2004)
21. Koh, J.L.Y., Ding, H., Costanzo, M., Baryshnikova, A., Toufighi, K., Bader, G.D., Myers, C.L., Andrews, B.J., Boone, C.: Drygin: a database of quantitative genetic interaction networks in yeast. Nucleic Acids Res. 38(suppl 1), D502–D507 (2010)
22. MacPherson, J.I., Dickerson, J., Pinney, J., Robertson, D.: Patterns of HIV-1 protein interaction identify perturbed host-cellular subsystems. PLoS Comput. Biol. 6(7), e1000863 (2010)
23. Madeira, S.C., Oliveira, A.L.: Biclustering algorithms for biological data analysis: a survey. IEEE/ACM TCBB 1(1), 24–45 (2004)
24. Maulik, U., Mukhopadhyay, A., Bhattacharyya, M., Kaderali, L., Brors, B., Bandyopadhyay, S., Eils, R.: Mining quasi-bicliques from HIV-1-human protein interaction network: a multiobjective biclustering approach. IEEE/ACM TCBB 10(2), 423–435 (2013)
25. Mukhopadhyay, A., Maulik, U., Bandyopadhyay, S.: A novel biclustering approach to association rule mining for predicting HIV-1 human protein interactions. PLoS ONE 7(4), e32289 (2012)
26. Murali, T.M., Kasif, S.: Extracting conserved gene expression motifs from gene expression data. Pac. Symp. Biocomput. 8, 77–88 (2003)
27. Pereira-Leal, J.B., Enright, A.J., Ouzounis, C.A.: Detection of functional modules from protein interaction networks. Proteins Struct. Funct. Bioinf. 54(1), 49–57 (2004)
28. Segal, E., Wang, H., Koller, D.: Discovering molecular pathways from protein interaction and gene expression data. Bioinformatics 19(suppl 1), i264–i272 (2003)
29. Sharan, R., Ulitsky, I., Shamir, R.: Network-based prediction of protein function. Mol. Syst. Biol. 3(1), 88 (2007)
30. Spirin, V., Mirny, L.A.: Protein complexes and functional modules in molecular networks. Natl. Acad. Sci. 100(21), 12123–12128 (2003)
31. Szklarczyk, D., Franceschini, A., Wyder, S., Forslund, K., Heller, D., Huerta-Cepas, J., Simonovic, M., Roth, A., Santos, A., Tsafou, K.P., et al.: String v10: protein-protein interaction networks, integrated over the tree of life. Nucleic Acids Res. 43, D447–D452 (2014). p.gku1003
32. Tanay, A., Sharan, R., Shamir, R.: Discovering statistically significant biclusters in gene expression data. Bioinformatics 18, 136–144 (2002)
33. Tomaino, V., Guzzi, P.H., Cannataro, M., Veltri, P.: Experimental comparison of biclustering algorithms for PPI networks. In: BCB, pp. 671–676. ACM (2010)
34. Xiong, H., Heb, X.F., Ding, C., Zhang, Y., Kumar, V., Holbrook, S.R.: Identification of functional modules in protein complexes via hyperclique pattern discovery. Pac. Symp. Biocomput. 10, 221–232 (2005)

Simultaneous Optimization of both Node and Edge Conservation in Network Alignment via WAVE

Yihan Sun[1,2,3], Joseph Crawford[1(✉)], Jie Tang[2], and Tijana Milenković[1]

[1] Department of Computer Science and Engineering, Interdisciplinary Center for
Network Science and Applications, and ECK Institute for Global Health,
University of Notre Dame, Notre Dame, IN, USA
{jcrawfo7,tmilenko}@nd.edu, yihans@cs.cmu.edu
[2] Department of Computer Science and Technology,
Tsinghua University, Beijing, China
jietang@tsinghua.edu.cn
[3] Computer Science Department,
Carnegie Mellon University, Pittsburgh, Pennsylvania

Abstract. Network alignment can be used to transfer functional knowledge between conserved regions of different networks. Existing methods use a node cost function (NCF) to compare nodes across networks and an alignment strategy (AS) to find high-scoring alignments with respect to total NCF over all aligned nodes (or node conservation). Then, they evaluate alignments via a measure that is different than node conservation used to guide alignment construction. Typically, one measures edge conservation, but only after alignments are produced. Hence, we recently directly maximized edge conservation while constructing alignments, which improved their quality. Here, we aim to maximize both node and edge conservation during alignment construction to further improve quality. We design a novel measure of edge conservation that (unlike existing measures that treat each conserved edge the same) weighs conserved edges to favor edges with highly NCF-similar end-nodes. As a result, we introduce a novel AS, **W**eighted **A**lignment **V**ot**E**r (WAVE), which can optimize any measures of node and edge conservation. Using WAVE on top of well-established NCFs improves alignments compared to existing methods that optimize only node or edge conservation or treat each conserved edge the same. We evaluate WAVE on biological data, but it is applicable in any domain.

1 Introduction

1.1 Motivation

Network alignment aims to find topologically or functionally similar regions between different networks. It has applications in different areas, including computational biology [1–7], ontology matching [8–11], pattern recognition [12,13],

© Springer-Verlag Berlin Heidelberg 2015
M. Pop and H. Touzet (Eds.): WABI 2015, LNBI 9289, pp. 16–39, 2015.
DOI: 10.1007/978-3-662-48221-6_2

social networks [14,15], language processing [16], and others [17–20]. Our study focuses mainly on the computational biology domain.

Protein-protein interaction (PPI) networks have been the main focus of network alignment research among all biological networks. PPI network alignment can be used to transfer biological knowledge from the network of a poorly studied species to the network of a well studied species. This is of importance because not all cellular processes can easily be studied via biological experiments. For example, studying aging in human has to rely on across-species transfer of aging-related knowledge from model species [21]. And network alignment can be (and has been) used for this [6,7]. However, the problem is computationally intractable, as the underlying subgraph isomorphism problem is NP-complete [22]. Thus, network alignment methods are heuristics.

Network alignment can be local or global. Local network alignment aims to align well local network regions [23–31]. As such, it often fails to find large conserved regions between networks. Hence, majority of recent research has focused on global network alignment [1–7,32–42], which can find large conserved regions between networks. Typically, global network alignment aims to generate one-to-one node mapping between two networks [41] (although exceptions exist that produce many-to-many node mappings or that align more than two networks [3], but such methods are out of the scope of our study).

Of one-to-one global network alignment methods, many consist of two algorithmic components, namely, a node cost function (NCF) and an alignment strategy (AS) [7,42]. NCF captures pairwise similarities between nodes in different networks, and AS then searches for good alignments based on the NCF information. It has already been recognized that when two methods of this two-component NCF-AS type are compared, to fairly evaluate the methods, one should mix and match the different methods' NCFs and ASs, because NCF of one method and AS of another method could lead to a new method that is actually superior to the original methods [7,42].

We base our work on established state-of-the-art NCFs of existing methods. Then, we propose a novel AS, **W**eighted **A**lignment **V**otEr (WAVE), which when used on top of the established NCFs leads to a new superior method for global network alignment. And while we evaluate our new method in the computational biology domain, the method is easily applicable in any domain.

1.2 Related Work

We focus on NCFs of two popular existing methods, MI-GRAAL [4] and GHOST [5], and we aim to improve with our new WAVE AS upon these methods' ASs.

MI-GRAAL improves upon its predecessors, GRAAL [1] and H-GRAAL [2], by using the same NCF but by combining their ASs (see below). MI-GRAAL's NCF computes topological similarity between extended network neighborhoods of two nodes [43–46]. It does so by relying on the concept of small induced subgraphs called graphlets (e.g., a triangle or a square) [47,48], which are used to summarize the topology of up to 4-deep network neighborhood of a node into its graphlet degree vector (GDV) [43,49,50]. Then, GDV-similarity is used as

MI-GRAAL's NCF, which compares nodes' GDVs to compute their topological similarity. MI-GRAAL also allows for integration of other node similarity measures into its NCF, such as protein sequence similarity. We showed [7] that MI-GRAAL's NCF is superior to another, Google PageRank algorithm-based NCF, which is used by IsoRank [32] and IsoRankN [3]. Regarding AS [42], MI-GRAAL combines GRAAL's greedy seed-and-extend AS with H-GRAAL's optimal AS that uses the Hungarian algorithm to solve linear assignment problem of maximizing total NCF over all aligned nodes.

GHOST's NCF is conceptually similar to MI-GRAAL's, as it also assumes two nodes from different networks to be similar if their neighborhoods are similar. However, the mathematical and implementation details of the two NCFs are different. Namely, GHOST's NCF takes into account a node's k-hop neighborhood, (in this study, $k = 4$). Then, its NCF computes topological distance (or equivalently, similarity) between two nodes by comparing the nodes' "spectral signatures". We recently fairly compared MI-GRAAL's GDV-similarity-based NCF with GHOST's "spectral signature"-based NCF within our above mix-and-match framework, concluding that MI-GRAAL's NCF is superior or comparable to GHOST's NCF, depending on data [42]. Hence, since none of the two NCFs was dominant in all cases, we consider both NCFs in our study. Just as MI-GRAAL, GHOST also allows for integration of protein sequence information into its NCF. Regarding AS, GHOST is also a seed-and-extend algorithm, like MI-GRAAL. However, GHOST's AS considers the quadratic (instead of linear) assignment problem. When we evaluated the two ASs, their performance was data-dependent [42]. Hence, we consider both ASs in our study.

There exist additional more recent network alignment methods [41], both those that also belong to the category of NCF-AS methods, such as NETAL [36], and those that do not, such as MAGNA [6]. These methods became available close to completion of our study, and as such, we were not able to include them into the design of our new method. (Hence, NETAL implements a different NCF compared to NCFs of MI-GRAAL and GHOST, along with a different AS compared to ASs of MI-GRAAL, GHOST, and WAVE.) However, we still consider these methods in our evaluation. Importantly, our goal is to show that when we use under an existing NCF (such as MI-GRAAL's or GHOST's) our new WAVE AS, we get alignments of higher quality compared to when using an existing AS (such as MI-GRAAL's or GHOST's) on the same NCF. This would be sufficient to illustrate the superiority of WAVE. If in the process we also improve upon the more recent methods, such as those that use a different NCF and especially those that do not belong to the NCF-AS category, that would further demonstrate WAVE's superiority.

1.3   Our Contributions and Significance

We introduce WAVE, a novel, general, and as we will show well-performing AS, which can be combined with any NCF. WAVE is applicable to any domain. In this study, we evaluate it on biological networks.

Its novelty and significance is as follows. The existing ASs use NCF scores to rapidly identify from possible alignments the high-scoring alignments with respect to the overall NCF (henceforth also referred to as node conservation). But, their alignment accuracy is then evaluated with some other measure that is different than NCF used to construct the alignments [6]. Typically, one measures the amount of conserved (i.e., aligned) edges. Hence, a recent attempt aimed to directly maximize edge conservation during alignment construction [6]. Here, we aim to optimize both node and edge conservation while constructing an alignment, as also recognized by a recent effort [36]. In the process, unlike the existing methods that treat each conserved edge the same, we aim to favor conserved edges with NCF-similar end nodes over those with NCF-dissimilar end nodes. And we design WAVE with these goals in mind.

We combine WAVE with NCF of MI-GRAAL as well as with NCF of GHOST. We denote the resulting network aligners as M-W and G-W, respectively. We compare M-W and G-W against the original MI-GRAAL (henceforth also denoted by M-M) and GHOST (henceforth also denoted by G-G), which use MI-GRAAL's NCF and AS and GHOST's NCF and AS, respectively. Further, we compare M-W and G-W with a new method introduced recently [42], which is the combination of GHOST's NCF and MI-GRAAL's AS (henceforth also denoted by G-M). This allows us to test the performance of WAVE against the performance of MI-GRAAL's and GHOST's ASs, under each of MI-GRAAL's and GHOST's NCF. We note that we cannot compare M-W and G-W against the combination of MI-GRAAL's NCF and GHOST's AS (i.e., M-G), as the current implementation of GHOST does not allow for plugging MI-GRAAL's NCF into GHOST's AS [42]. Finally, we compare M-W and G-W against the very recent NETAL and MAGNA methods.

We evaluate all methods on synthetic and real-world PPI networks, relying on established data and performance measures [1,2,4–7]. We find that WAVE AS is overall superior to the existing ASs, especially in terms of topological alignment quality. Also, WAVE overall performs comparably to or better than NETAL and MAGNA, especially on synthetic data. This further validates WAVE, because NETAL implements a newer and thus possibly more efficient NCF compared to NCFs of M-W or G-W, which might give NETAL unfair advantage over WAVE.

2 Methods

2.1 Data

We evaluate WAVE on two popular network sets [1,2,4–7]: (1) "synthetic" networks with known node mapping, and (2) real-world networks with unknown node mapping.

The "synthetic" data consists of a high-confidence yeast PPI network [51] with 1,004 nodes and 8,323 PPIs, and of five noisy networks constructed by adding to the high-confidence network a percentage of low-confidence PPIs from the same data set [51]; we vary the percentage from 5 % to 25 % in increments of 5 %. We align the original high-confidence network to each of the five noisy networks, resulting in five network pairs to be aligned. Since we know the correct node correspondence, we can measure to what extent an aligner correctly reconstructs the correspondence.

The real-world set contains binary (yeast two-hybrid, Y2H) PPI networks of four species: *S. cerevisiae* (yeast/Y), with 3,321 nodes and 8,021 edges, *D. melanogaster* (fly/F), with 7,111 nodes and 23,376 edges, *C. elegans* (worm/W), with 2,582 nodes and 4,322 edges, and *H. sapiens* (human/H), with 6,167 nodes and 15,940 edges. We align each pair of the networks, resulting in six pairs. If we aimed to predict new biological knowledge, we would have evaluated our method on additional PPIs, such as those obtained via affinity purification followed by mass spectrometry (AP/MS). However, since our main focus is method evaluation, of all PPIs, we focus on binary Y2H PPIs because: (1) they have been argued to be of higher quality than literature-curated PPIs supported by a single publication [49,52], and (2) the same Y2H networks have already been used in many existing studies [1,2,4–7]. Ultimately, what is important for a fair evaluation is that all methods are tested on the same data, be it Y2H, AP/MS, or other PPIs [7].

When we combine within NCF nodes' topological similarity scores with their sequence similarity scores (see below), for the latter, we rely on BLAST bit-values from the NCBI database [53]. When we evaluate biological alignment quality with respect to functional enrichment of the aligned nodes (see below), we rely on Gene Ontology (GO) data [54] to evaluate the biological alignment quality. We use same data versions as in our recent work [7,42].

2.2   Combining Topological and Sequence Information Within NCF

We compute the linear combination of topological node similarity scores s_t and sequence node similarity scores s_s of nodes u and v as: $s(u, v) = \alpha s_t(u, v) + (1 - \alpha)s_s(u, v)$. We vary α from 0.0 to 1.0 in increments of 0.1. We do this for all combinations of MI-GRAAL's, GHOST's, and WAVE's NCFs and ASs. When we compare WAVE to recent NETAL and MAGNA, since current implementations of NETAL and MAGNA do not support inclusion of sequence information, for these methods, we only study topology-based alignments (corresponding to α of 1).

2.3   Evaluation of Alignment Quality

If we align graph $G(V_G, E_G)$ to graph $H(V_H, E_H)$ (where $|V_G| \leq |V_H|$) via an injective function $f : V_G \rightarrow V_H$, let us denote with E'_G this edge set: $E'_G = \{(f(u), f(v))|u \in V_G, v \in V_G, (u, v) \in E_G\}$. Also, let us denote with E'_H the edge set of the subgraph of H that is induced on nodes from V_H that are images of nodes from V_G. $E'_H = \{(f(u), f(v))|u \in V_G, v \in V_G, (f(u), f(v)) \in E_H\}$. With these notations in mind, we next define alignment quality measures that we use.

Topological Alignment Quality Measures
Node correctness (NC). Given a known true node mapping (which is typically not available in real-world applications), NC is the percentage of node pairs that

are correctly mapped by an alignment. If $f^* : V_G \rightarrow V_H$ is the correct node mapping of G to H and $f : V_G \rightarrow V_H$ is an alignment produced by the aligner, $NC = \frac{|\{u \in V_G : f^*(u) = f(u)\}|}{|V_G|} \times 100\%$ [1].

Edge Correctness (EC). EC represents the percentage of edges from G, the smaller network (in terms of the number of nodes), which are aligned to edges from H, the larger network [1]. Formally, $EC = \frac{|E'_G \cap E'_H|}{|E_G|} \times 100\%$, where the numerator is the number of conserved edges.

Induced Conserved Structure (ICS). ICS is defined as $ICS = \frac{|E'_G \cap E'_H|}{|E'_H|} \times 100\%$. It was introduced because EC fails to penalize for misaligning edges in the larger network, i.e., E'_H, as EC is defined with respect to edges in E_G only [5]. Hence, ICS accounts for this. However, ICS now fails to penalize for misaligning edges in the smaller network, i.e., E_G, as it is defined with respect to edges in E'_H only. Hence, the following measure, S^3, was introduced recently to penalize for misaligning edges in both the smaller and the larger network [6].

Symmetric Substructure Score (S^3). S^3 is defined as $S^3 = \frac{|E'_G \cap E'_H|}{|E_G| + |E'_H| - |E'_G \cap E'_H|} \times 100\%$ [6]. Thus, S^3 keeps advantages of both EC and ICS while addressing their drawbacks. S^3 was already shown to be the superior of the three measures [6]. Thus, we discard EC and ICS measures from further consideration, and instead, we report results for S^3.

The size of the *largest connected common subgraph (LCCS)* [1]. In addition to counting aligned edges via S^3 measure, it is important that the aligned edges cluster together to form large, dense, and connected subgraphs, rather than being isolated. In this context, a connected common subgraph (CCS) is defined as a connected subgraph (not necessarily induced) that appears in both networks [2]. We measure the size of the largest CCS (LCCS) in terms of the number of nodes as well as edges, as defined in the MAGNA paper [6].

In summary, we focus on NC, S^3, and LCCS. The larger their values, the better the topological alignment quality.

Biological Alignment Quality Measures. To transfer function from well annotated network regions to poorly unannotated ones, which is the main motivation behind network alignment in computational biology, alignment should be of good biological quality, mapping nodes that perform similar function.

Gene Ontology Enrichment (GO). One could measure GO, the percentage of aligned protein pairs in which the two proteins *share* at least one GO term, out of all aligned protein pairs in which both proteins are annotated with at least one GO term [6,42]. In this case, complete GO annotation data is used, independent of GO evidence code.

Experimental GO (Exp-GO). However, since many GO annotations have been obtained via sequence comparison, and since the aligners use sequence information within their NCF, it is important to test the aligners when considering only GO annotation data with experimental evidence codes. This avoids the circular argument of evaluating alignment quality with respect to the same data that was used to construct the alignments [1,2,4,6,7,42]. Thus, we discard GO measure from further consideration, and instead, we report results for Exp-GO.

In summary, we focus Exp-GO. The larger its value, the better the biological alignment quality.

2.4   Our Methodology

Problem Definition. Existing network alignment methods aim to maximize either node conservation or edge conservation. Further, they treat each conserved edge the same. Here, we aim to simultaneously maximize both node and edge conservation, while favoring conserved edges whose end nodes are highly similar. Given a measure of node conservation (denoted as Node Alignment Quality, NAQ) and a measure of edge conservation (denoted as Edge Alignment Quality, EAQ), our goal is to optimize the following expression (denoted as Alignment Quality, AQ):

$$AQ(G, H, f) = \beta_n NAQ(G, H, f) + \beta_e EAQ(G, H, f), \qquad (1)$$

where β_n and β_e are parameters used to balance between NAQ and NEQ. We note that a previous study [35] proposed a similar objective function; however, in our study, we define a new way to measure EAQ (see below).

As a proof of concept, we use the following measures as NAQ and EAQ (although any other measure can be used instead). We use the sum of NCF scores over all aligned pairs as our NAQ, which we denote as weighted node conservation (WNC). We design a novel measure of edge conservation as our EAQ, as follows. Similar to EC, ICS, and S^3, this new measure counts the number of conserved edges, but unlike EC, ICS, or S^3 that treat each conserved edge the same, our new measure weighs each conserved edge by the NCF-based similarity of its end nodes, so that aligning an edge with highly similar end nodes is preferred over aligning an edge with dissimilar end nodes. We denote our new EAQ measure as weighted edge conservation (WEC).

Formally, we define WNC and WEC as follows. Given a pairwise node similarity matrix s with respect to the given NCF, we denote similarity between $u \in V_G$ and $v \in V_H$ in this matrix as s_{uv}. Also, we represent the injection $f : V_G \rightarrow V_H$ as a matrix $y_{|V_G| \times |V_H|}$, where $y_{ij} = 1$ if and only if $f(i) = j$ and $y_{ij} = 0$ otherwise. Thus, the matrix satisfies the following three constraints:

$$y_{ij} \in \{0,1\}, \quad \forall i \in V_G, \forall j \in V_H; \qquad \sum_{l=1}^{|V_H|} y_{il} \leq 1, \quad \forall i \in V_G; \qquad \sum_{l=1}^{|V_G|} y_{lj} \leq 1, \quad \forall j \in V_H \quad (2)$$

Then:

$$WNC = \sum_{i \in V_G} \sum_{j \in V_H} y_{ij} s_{ij} \qquad (3)$$

To formally define WEC, recall the definitions of EC, ICS, and S^3 (Sect. 2.3). All three measures have the same numerator, which we can now rewrite as:

$$|E'_G \cap E'_H| = \frac{1}{2} \sum_{i \in V_G} \sum_{j \in V_H} \sum_{k \in \mathcal{N}_i} \sum_{l \in \mathcal{N}_j} y_{ij} y_{kl} \qquad (4)$$

Here, $\mathcal{N}_i$ denotes the neighborhood of node i, i.e., the set of nodes connected to i. Since each conserved edge will be counted twice, the $\frac{1}{2}$ constant corrects for this.

Now, to leverage the weight of conserved edges by the NCF-based similarity of its end nodes (see above), we define WEC as follows:

$$WEC = \sum_{i \in V_G} \sum_{j \in V_H} \sum_{k \in \mathcal{N}_i} \sum_{l \in \mathcal{N}_j} y_{ij} y_{kl} s_{kl} \qquad (5)$$

With WNC as our NAQ and WEC as our EAQ, formally, our problem is to find a matrix y that satisfies Eq. 2 and maximizes the following objective function:

$$
\begin{aligned}
AQ(G, H, y) &= \beta_n NAQ + \beta_e EAQ = \beta_n WNC + \beta_e WEC \\
&= \beta_n \sum_{i \in V_G} \sum_{j \in V_H} y_{ij} s_{ij} + \beta_e \sum_{i \in V_G} \sum_{j \in V_H} \sum_{k \in \mathcal{N}_i} \sum_{l \in \mathcal{N}_j} y_{ij} y_{kl} s_{kl} \qquad (6)
\end{aligned}
$$

Optimizing the WNC part in Eq. 6 is solvable in polynomial time (e.g., by using Hungarian algorithm for maximum bipartite weighted matching). However, optimizing the whole function on general graphs is NP-hard. We propose WAVE to solve this problem, while allowing for trade off between node conservation and edge conservation (as the two might not always agree).

Weighted Alignment VotEr (WAVE). Initially, we evaluate different values of β_n and β_e and thus the effect of these parameters on WAVE's results. Since (as we will show in Sect. 3.2) equally favoring WNC and WEC (i.e., setting the two parameters to the same value) in general yields best results, in all subsequent analyses, we set $\beta_n = \beta_e = 1$. Given this parameter value choice, we can rewrite Eq. 6 as:

$$AQ(G, H, y) = \sum_{(i,j) \in V_G \times V_H} y_{ij} \left(s_{ij} + \sum_{(k,l) \in \mathcal{N}_i \times \mathcal{N}_j} y_{kl} s_{kl} \right) \qquad (7)$$

Next, we use set $A = \{(u, v) \,|\, u \in V_G, v \in V_H, y_{uv} = 1\}$ to denote our alignment, so our objective function has set A as a variable. Then, we use a greedy approach to maximize the objective function, as follows. We start with an empty alignment set A_0. In each step t, given the current alignment A_{t-1}, we calculate the marginal gain of adding an available node pair (u, v) (in the sense that so far v and u are both unaligned) into A. (For a function $f(S)$ with variable

S as a set, the marginal gain of adding an element e into S is defined as $f(S \cup \{e\}) - f(S)$.) That is, we calculate: $AQ(A_{t-1} \cup \{(u,v)\}) - AQ(A_{t-1})$. Then, we align the pair (u^*, v^*) with the highest marginal gain, i.e., $A_t = A_{t-1} \cup \{(u^*, v^*)\}$. To calculate the marginal gain efficiently, we keep the current marginal gain of each node pair and update it in each step. The marginal gain of the node pair (u, v) to AQ is s_{uv} at the beginning (when A is empty, if we align this pair, we can only get s_{uv} in WNC part). In each step, note that if we align two nodes $u \in V_G$ and $v \in V_H$, the side effect is that, in the following steps, when we align another pair of nodes $u' \in \mathcal{N}_u, v' \in \mathcal{N}_v$, both the similarity of (u, v) and (u', v') will be counted once more by the correctly linked edge, namely, the edge $(u, u') \in E_G$ and $(v, v') \in E_H$. Thus, the marginal gain of (u', v') will be $s_{uv} + s_{u'v'}$ more after (u, v) is aligned.

Intuitively, this process is like voting. When a pair of nodes is aligned, this node pair has a chance to vote for their neighbors: when u and v are aligned, all other node pairs in $\mathcal{N}_u \times \mathcal{N}_v$ receive a weighted vote (with weight $s_{uv} + s_{u'v'}$) from (u, v), and the weight consists of two parts: (1) the "authority" of the voter, i.e., s_{uv}, (2) the "certainty" of the votee, i.e., $s_{u'v'}$.

The weight for the initial votes of each node pair is the original s_{uv} (which forms the WNC part in the objective function). In every round of WAVE, node pair (u^*, v^*) with the highest vote is aligned, and (u^*, v^*) then vote for all the pairs in $\mathcal{N}_{u^*} \times \mathcal{N}_{v^*}$. The current vote that a node pair gets from its aligned neighbors is the marginal gain to objective function of aligning them.

For WAVE's pseudocode, see Algorithm 1. For its implementation, visit: http://nd.edu/~cone/WAVE/WAVE.zip

3   Results and Discussion

We denote the five aligners resulting from mixing and matching NCFs of MI-GRAAL and GHOST with ASs of MI-GRAAL, GHOST, and WAVE as M-M, M-W, G-M, G-G, and G-W (Sect. 1.3).

Recall that a key novelty of WAVE is that while optimizing edge conservation (in addition to node conservation), WAVE weighs each conserved edge to favor aligning edges with highly NCF-similar end nodes. Thus, to evaluate whether weighing conserved edges leads to better alignments, we first compare the performance of the edge-weighted versions of WAVE (i.e., M-W(W) and G-W(W)) and its edge-unweighted versions (i.e., M-W(U) and G-W(U)). As we will show, the edge-weighted versions are superior.

Further, we evaluate the effect of the different parameters (i.e., β_n and β_e) on WAVE's results. As we will show, assigning the same value to the two parameters, i.e., equally favoring node and edge conservation, is superior to other parameter variations.

Next, using the edge-weighted versions of WAVE with $\beta_n = 1$ and $\beta_e = 1$, we evaluate the five aligners (M-M, M-W, G-M, G-G, and G-W) against each other. Also, we evaluate WAVE (the best of M-W and G-W) against NETAL and MAGNA. By comparing M-M and M-W, we can directly and fairly evaluate ASs

Algorithm 1. Weighted Alignment VotEr (WAVE) pseudocode

Input: $G = (V_G, E_G)$, $H = (V_H, E_H)$, s_{uv} $((u,v) \in V_G \times V_H)$
Output: Alignment $f : V_G \to V_H$

1: **for** $(u,v) \in V_G \times V_H$ **do**
2:     $vote_{u,v} \leftarrow \beta_n \times s_{u,v}$;
3: **end for**
4: **for** $u \in V_G$ **do**
5:     $visitedSrc_u \leftarrow false$;
6: **end for**
7: **for** $v \in V_H$ **do**
8:     $visitedTar_v \leftarrow false$;
9: **end for**
10: **for** $round = 1$ to $|V_G|$ **do**
11:     $(u^*, v^*) \leftarrow \arg\max_{unaliged(u,v)} vote_{u,v}$;
        where $unaligned(u,v)$ means both $vistedSrc_u$ and $visitedTar_v$ are
        false
12:     $visitedSrc_{u^*} \leftarrow true$;
13:     $visitedTar_{v^*} \leftarrow true$;
14:     $f(u^*) \leftarrow v^*$;
15:     **for** * $(u,v) \in \mathcal{N}_{u^*} \times \mathcal{N}_{v^*}$ **do**
16:         $vote_{u,v} \leftarrow vote_{u,v} + \beta_e \times (s_{u,v} + s_{u^*,v^*})$
17:     **end for**
18: **end for**
     Return f

---

of MI-GRAAL and WAVE under MI-GRAAL's NCF. By comparing G-M, G-G, and G-W, we can directly and fairly evaluate ASs of MI-GRAAL, GHOST, and WAVE under GHOST's NCF. If WAVE AS produces better alignments compared to the existing methods' ASs under both of the existing NCFs, this would indicate WAVE's superiority. If WAVE also produces better alignments compared to NETAL and MAGNA, this would even further demonstrate WAVE's superiority. However, this is not a strict requirement, as the two new methods either implement both different (newer, and thus possibly superior) NCF than any of M-W and G-W as well as different AS (in case of NETAL), which might give them an unfair advantage, or they work on different principles (in case of MAGNA) and could be thus viewed as complementary to WAVE.

For each combination of network pair, value of α (denoting topological versus sequence information within NCF), and alignment quality measure (Sect. 2), we do the following. First, to extract the most out of each source of biological information, it would be beneficial to know how much of new biological knowledge can be uncovered solely from topology before integrating it with other sources of biological information, such as protein sequence information [1,2,4]. Thus, we first compare the different edge-weighted and edge-unweighted versions of WAVE, the different combinations of β_n and β_e parameter values, and the different NCF-AS methods, on topology-only alignments (corresponding to α of 1 within NCF). Also, since NETAL and MAGNA also produce topology-only alignments, here, we can compare WAVE to these methods. Second, we examine different contributions of topology versus sequence information in NCF (by varying α), and for each method, we choose the best value of α, i.e., the method's *best alignment*. We do this when comparing the different edge-weighted and edge-unweighted versions of WAVE, as well as the five NCF-AS methods to each other. On the other hand, we do not do this when comparing the different combinations of β_n and β_e parameter values, due to the large number of evaluation tests required in this analysis. Also, since current implementations of NETAL and MAGNA do not allow for inclusion of sequence information, we cannot directly compare WAVE to these methods when adding sequence information into NCF. However, since in real-life applications one should give the best-case advantage to each method, we do compare best alignments of WAVE with topology-only alignments of NETAL and MAGNA, and we do consider this as comparison of the methods' best alignments.

For "synthetic" (noisy yeast) networks with known node mapping, we report alignment quality with respect to NC, S^3, LCCS, and Exp-GO. For real-world PPI networks of different species with unknown node mapping, we report alignment quality with respect to S^3, LCCS, and Exp-GO.

3.1   Comparison of Edge-Weighted and Edge-Unweighted Versions of WAVE

Here, we compare the edge-weighted and edge-unweighted versions of WAVE. We find that weighing conserved edges in general improves alignment quality (Figs. 1 and 2, as well as Figs. A.1 and A.2 in the Appendix), as follows.

Networks with Known Node Mapping

Topological Alignments. Weighing conserved edges improves alignment quality of topology-only alignments under both MI-GRAAL's and GHOST's NCFs, since the edge-weighted version of WAVE is comparable or superior to the edge-unweighted version in the majority of cases across all alignments and all alignment quality measures (Fig. 1).

Best Alignments. Here, under MI-GRAAL's NCF, the edge-weighted version of WAVE is comparable or superior to the edge-unweighted version across all alignments and all alignment quality measures (Fig. 2). Under GHOST's NCF, the edge-weighted version is comparable or superior to the edge-unweighted version in the majority of cases. Thus, the edge-weighted version is even more preferred by best alignments compared to topology-only alignments.

Networks with Unknown Node Mapping

Topological Alignments. Here, the edge-weighted version of WAVE is comparable or superior to the edge-unweighted version under MI-GRAAL's NCF for two out of three alignment quality measures (Fig. A.1 in the Appendix). Under GHOST's NCF, the edge-weighted version of WAVE is rarely favored in this

Table 1. Improvements of an edge-weighted version of WAVE over its edge-unweighted counterpart, over all evaluation tests in which the edge-weighted version is the superior one. The results are shown in terms of the minimum ("Min"), maximum ("Max") and average ("Avg") improvement over all such tests, along with the corresponding standard deviation ("Stdev").

Aligner / Network set	Topology-only alignments				Best alignments			
	Min	Max	Avg	Stdev	Min	Max	Avg	Stdev
M-W / "Synthetic" (noisy yeast) networks	0.03 %	11.79 %	4.71 %	4.24 %	0.17 %	4.11 %	1.41 %	1.21 %
M-W / Real-world PPI networks of different species	1.69 %	6.51 %	3.51 %	1.73 %	1.22 %	6.51 %	2.60 %	1.85 %
G-W / "Synthetic" (noisy yeast) networks	0.03 %	11.84 %	3.22 %	4.7 %	0.04 %	3.11 %	0.61 %	1.11 %
G-W / Real-world PPI networks of different species	5.16 %	8.30 %	6.73 %	2.22 %	5.16 %	8.30 %	6.73 %	2.22 %
Average across all aligners and network sets	1.73 %	9.56 %	4.54 %	3.24 %	1.65 %	5.51 %	2.84 %	1.60 %

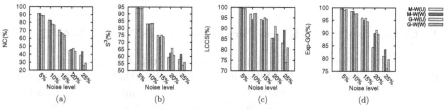

(a)   (b)   (c)   (d)

Fig. 1. Comparison of the edge-weighted and edge-unweighted versions of WAVE on topology-only alignments of "synthetic"(noisy yeast) networks with respect to (a) NC, (b) S^3, (c) LCCS, and (d) Exp-GO. For analogous results for real-world PPI networks of different species, see Fig. A.1 in the Appendix.

evaluation test. Nonetheless, the edge-weighted version is still favored over all evaluation tests.

Best Alignments. The edge-weighted version is preferred under MI-GRAAL's NCF for all three alignment quality measures and under GHOST's NCF for one of the measures, since in these cases the edge-weighted version is comparable or superior to the edge-unweighted version in the majority of cases (Fig. A.2 in the Appendix).

In summary, over both network sets (with known and unknown node mapping), both topology-only and best alignments, and all alignment quality measures, the edge-weighted version of WAVE is overall (though not always) superior to the edge-unweighted version. Over all cases in which we do observe superiority of the edge-weighted version over the edge-unweighted version, the level of superiority ranges from 1.73 % to 9.56 % (with the average of 4.54 %) for topology-only alignments and from 1.65 % to 5.51 % (with the average of 2.84 %) for best alignments (Table 1). Interestingly, superiority of the edge-weighted version of WAVE becomes more pronounced with increase of noise in the data, especially for topology-only alignments (we base this conclusion only on "synthetic" (noisy yeast) networks for which we know the level of noise in the data). Because the edge-weighted version of WAVE is overall the superior one, in the following sections, we use the edge-weighted version.

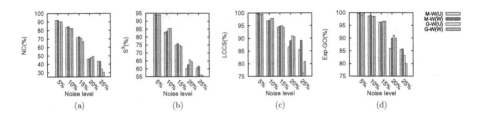

Fig. 2. Comparison of the edge-weighted and edge-unweighted versions of WAVE on best alignments of "synthetic" (noisy yeast) networks with respect to **(a)** NC, **(b)** S³, **(c)** LCCS, and **(d)** Exp-GO. For analogous results for real-world PPI networks of different species, see Fig. A.2 in the Appendix.

3.2   Comparison of Different Parameter Values Within WAVE

When we vary values of β_n and β_e parameters, corresponding to levels of node and edge conservation considered within WAVE, we find that assigning the same value to the two parameters overall leads to the best results (Fig. 3). This holds for M-W and G-W combined (Fig. 3(a)) as well as for M-W only (Fig. 3(b)) and G-W only (Fig. 3(c)).

Our above conclusion comes from the following observations. The combination in which only β_n is used (i.e., $\beta_n = 1$ and $\beta_e = 0$) is always inferior to any other combination of the parameter values. That is, considering only node conservation within WAVE (which is what most of the existing methods do) is inferior, and accounting for edge conservation improves results. The remaining combinations of parameter values that use some level of edge conservation (i.e., $\beta_n = 0.75$ and $\beta_e = 0.25$, $\beta_n = 0.5$ and $\beta_e = 0.5$, $\beta_n = 0.25$ and $\beta_e = 0.75$, and $\beta_n = 0$ and $\beta_e = 1$) are overall comparable to each other, with slight superiority of $\beta_n = 0.5$ and $\beta_e = 0.5$, especially for G-W (Fig. 3(c)). Even for all versions of WAVE (Fig. 3(a)), we argue that the combination of $\beta_n = 0.5$ and $\beta_e = 0.5$ is overall superior. Namely, even though the combination of $\beta_n = 0$ and $\beta_e = 1$ is ranked as the first best combination in most of the cases, the combination of $\beta_n = 0.5$ and $\beta_e = 0.5$ is following very closely (Fig. 3(a)). Further, the combination of $\beta_n = 0.5$ and $\beta_e = 0.5$ is ranked as the second or third best in more cases than the combination of $\beta_n = 0$ and $\beta_e = 1$; in other words, the combination of $\beta_n = 0$ and $\beta_e = 1$ is ranked as *the worst* (i.e., fourth) in more cases than the combination of $\beta_n = 0.5$ and $\beta_e = 0.5$ (or any other combination that considers some level of edge conservation). For these reasons, in our study, we have adopted this overall superior combination of $\beta_n = 0.5$ and $\beta_e = 0.5$ (or equivalently $\beta_n = 1$ and $\beta_e = 1$), which equally favors node and edge conservation.

Due to a large number of tests involved into evaluating different combinations of β_n and β_e values, all experiments in this section have been performed only on topology-only alignments of "synthetic" (noisy yeast) networks.

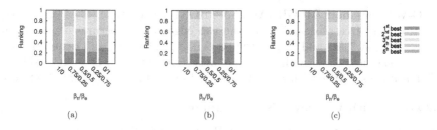

Fig. 3. Overall ranking of each tested β_n/β_e combination over all network pairs in the "synthetic" (noisy yeast) network data set and over all alignment quality measures, for topology-only alignments of: **(a)** both M-W and G-W combined, **(b)** M-W only, and **(c)** G-W only. The ranking is expressed as a percentage of all cases (i.e., all evaluation tests) in which the combination ranks as the kth best method. That is, the more cases in which a given combination achieves a higher ranking, the better the combination. For example, in panel (c), the combination of $\beta_n = 0.5$ and $\beta_e = 0.5$ (or equivalently, $\beta_n = 1$ and $\beta_e = 1$) is superior to all other combinations, since is ranked the highest (i.e., as the 1st best method) in most of the cases.

3.3 Comparison of Five NCF-AS Methods

Here, we compare M-M, M-W, G-M, G-G, and G-W, to test whether WAVE AS improves upon ASs of MI-GRAAL and GHOST under the same (MI-GRAAL's or GHOST's) NCF.

Networks With Known Node Mapping

Topological Alignments. WAVE is always superior to the existing methods (M-W is superior to M-M, and G-W is superior to G-M and G-G), for all noise levels and alignment quality measures, under both MI-GRAAL's and GHOST's NCFs (Figs. 4 (a) and 5).

WAVE in general works better under MI-GRAAL's NCF than under GHOST's NCF, as M-W is overall superior to G-W. WAVE (at least one of M-W and G-W) beats both MI-GRAAL and GHOST (all of M-M, G-M, and G-G) in 20/20=100 % of all cases (Figs. 4 (a) and 5). These results hold across all noise levels.

Best Alignments. Here, we give the best-case advantage to each method by selecting its optimal α parameter value. Under MI-GRAAL's NCF, WAVE is always superior (M-W is better than M-M), for all noise levels and alignment quality measures (Figs. A.3 (a) and A.4 in the Appendix).

Under GHOST's NCF, WAVE is always superior to MI-GRAAL's AS (G-W is better than G-M), and WAVE is superior to GHOST's AS (G-W is better than G-G) with respect to two of the four measures (edge-based S^3 and LCCS), while GHOST's AS is superior (G-G is better than G-W) with respect to the other two measures (node-based NC and Exp-GO) (Figs. A.3 (a) and A.4 in the Appendix). Hence, WAVE and GHOST's AS are comparable overall.

Again, WAVE in general works better under MI-GRAAL's NCF than under GHOST's, as M-W is overall superior to G-W. WAVE (at least one of M-W and G-W) beats both MI-GRAAL and GHOST (all of M-M, G-M, and G-G) in $6/10=60\%$ of cases dealing with the two edge-based measures of alignment quality (Figs. A.3 (a) and A.4 in the Appendix). The ranking of the different methods does not change with increase of noise level with respect to NC and Exp-GO, but it does change with respect to S^3 and LCCS for the highest noise levels.

Networks With Unknown Node Mapping

Topological Alignments. Under MI-GRAAL's NCF, WAVE is always superior (M-W is better than M-M) with respect to S^3, it is almost always superior with respect to LCCS, and it is sometimes superior with respect to Exp-GO (Figs. 4 (b) and 6). Hence, here WAVE seems to be favored by topological alignment quality measures.

Under GHOST's NCF, WAVE is superior to MI-GRAAL's AS (G-W is better than G-M) in almost all cases, for each of S^3, LCCS, and Exp-GO (Figs. 4 (b) and 6). Also, here WAVE is overall superior to GHOST's AS (G-W is better than G-G) with respect to Exp-GO but not with respect to S^3 or LCCS (Figs. 4 (b) and 6).

WAVE in general works better under MI-GRAAL's NCF than under GHOST's NCF, as M-W is overall superior to G-W. WAVE (at least one of M-W and G-W) beats both MI-GRAAL and GHOST (all of M-M, G-M, and G-G) in $14/18=78\%$ of all cases (Figs. 4 (b) and 6).

Best Alignments. Under MI-GRAAL's NCF, WAVE is always superior (M-W is better than M-M) with respect to S^3, and it is almost always superior with respect to LCCS as well as Exp-GO (Figs. A.3 (b) and A.5 in the Appendix). Hence, here WAVE is even more superior than for topological alignments only.

Under GHOST's NCF, WAVE is superior to MI-GRAAL's AS (as G-W is better than G-M) in most cases for each of S^3 and Exp-GO, and in some cases for LCCS. Also, here WAVE is overall superior to GHOST's AS (G-W is better than G-G) with respect to Exp-GO but not with respect to S^3 or LCCS (Figs. A.3 (b) and A.5 in the Appendix).

Again, WAVE works better under MI-GRAAL's NCF than under GHOST's AS, as M-W is superior to G-W. WAVE (at least one of M-W and G-W) beats both MI-GRAAL and GHOST (all of M-M, G-M, and G-G) in $13/18=72\%$ of all cases (Figs. A.3 (b) and A.5 in the Appendix).

The fact that WAVE in general works better under MI-GRAAL's NCF than under GHOST's NCF further adds to our recent finding that MI-GRAAL's NCF is superior to other NCFs [7,42].

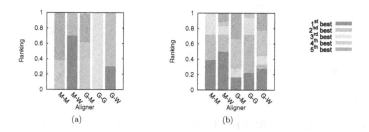

(a)   (b)

Fig. 4. Representative results for overall ranking of each NCF-AS method over all network pairs in a given data set and over all alignment quality measures. The ranking is expressed as a percentage of all cases (i.e., all evaluation tests) in which the given method ranks as the kth best method. That is, the more cases in which a given method achieves a higher ranking, the better the method. For example, in panel (a), M-W is the highest scoring of all methods shown on x-axis, since it is ranked the highest (i.e., as the 1st best method) in most of the cases. **(a)** Results for the five NCF-AS methods on topology-only alignments of "synthetic" (noisy yeast) networks. For equivalent results for best alignments, see Fig. A.3 (a) in the Appendix. **(b)** Results for the five NCF-AS methods on topology-only alignments of real-world PPI networks of different species. For equivalent results for best alignments, see Fig. A.3 (b) in the Appendix. Details (per network pair and alignment quality measure) for panels (a)-(b) are shown in Figs. 5 and 6, respectively. Recall that M-M and G-G are MI-GRAAL and GHOST.

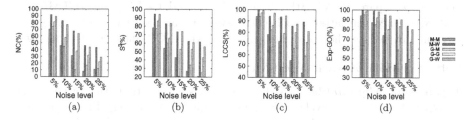

(a)   (b)   (c)   (d)

Fig. 5. Comparison of the five NCF-AS methods on topology-only alignments of "synthetic" (noisy yeast) networks with respect to: **(a)** NC, **(b)** S^3, **(c)** LCCS, and **(d)** Exp-GO. For analogous results for best alignments, see Fig. A.4 in the Appendix.

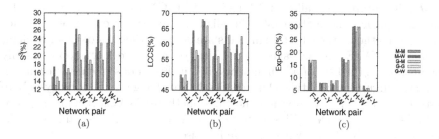

(a)   (b)   (c)

Fig. 6. Comparison of the five NCF-AS methods on topology-only alignments of real-world PPI networks of different species with respect to: **(a)** S^3, **(b)** LCCS, and **(c)** Exp-GO. For analogous results for best alignments, see Fig. A.5 in the Appendix.

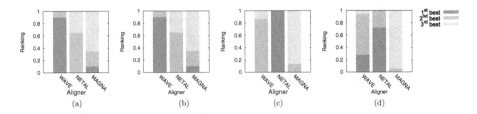

Fig. 7. Overall ranking of WAVE (the best of M-W and G-W) against the recent methods (NETAL and MAGNA) over all network pairs in a given data set and over all alignment quality measures. The ranking is expressed as a percentage of all cases (i.e., all evaluation tests) in which the given method ranks as the kth best method. That is, the more cases in which a given method achieves a higher ranking, the better the method. Results are shown for: **(a)** topology-only alignments of "synthetic" (noisy yeast) networks, **(b)** best alignments of "synthetic" (noisy yeast) networks, **(c)** topology-only alignments of real-world PPI networks of different species, and **(d)** best alignments of real-world PPI networks of different species. Details (per network pair and alignment quality measure) for panels (a)-(d) are shown in Fig. A.6 in the Appendix, Fig. 8, Fig. A.7 in the Appendix, and Fig. 9, respectively. Recall that M-M and G-G are MI-GRAAL and GHOST.

3.4   Comparison of WAVE with Very Recent Methods

Here, we compare WAVE (the best of M-W and G-W) with NETAL and MAGNA. Since the latter two became available at completion of our study, we could not include their novelties (e.g., NETAL's NCF) into our methodology.

Networks With Known Node Mapping

Topological Alignments. WAVE is always superior to both NETAL and MAGNA, for all noise levels and alignment quality measures (Figs. 7 (a) and A.6 in the Appendix). Only in $2/20 = 10\%$ of all cases, MAGNA is superior: with respect to S^3 for two largest noise levels. But this is not surprising, as MAGNA optimizes S^3. Overall, the ranking of the different methods does not change with increase in noise level.

Best Alignments. Recall that NETAL and MAGNA do now allow for inclusion of sequence information into the alignment construction process. So, for these two methods, their best alignments are actually their topology-only alignments. For WAVE, on this "synthetic" (noisy yeast) network set, topology-only alignments are the best of all alignments (i.e., inclusion of sequence information decreases alignment quality). Thus, results do not change from topology-only to best alignments when comparing the three methods: WAVE remains superior to NETAL and MAGNA (Figs. 7 (b) and 8). Again, overall, the ranking of the methods does not change with increase in noise level.

Networks With Unknown Node Mapping

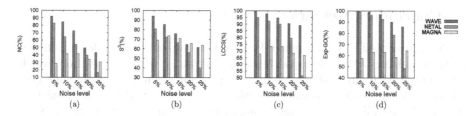

Fig. 8. Comparison of WAVE (the best of M-W and G-W) with very recent network alignment methods on best alignments of "synthetic" (noisy yeast) networks with respect to: **(a)** NC, **(b)** S^3, **(c)** LCCS, and **(d)** Exp-GO. For analogous results for topology-only alignments of "synthetic" (noisy yeast) networks, see Fig. A.6 in the Appendix.

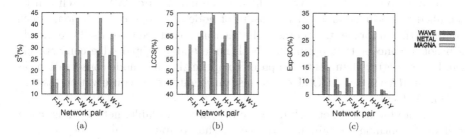

Fig. 9. Comparison of WAVE (the best of M-W and G-W) with very recent network alignment methods on best alignments of real-world PPI networks of different species with respect to: **(a)** S^3, **(b)** LCCS, and **(c)** Exp-GO. For analogous results for topology-only alignments of real-world PPI networks of different species, see Fig. A.7 in the Appendix.

Topological Alignments. WAVE is always superior to MAGNA, for all noise levels and alignment quality measures (Figs. 7 (c) and A.7 in the Appendix). Only in one out of 18 cases, MAGNA is superior to WAVE: with respect to S^3 for one of the six network pairs. NETAL is overall superior to the other two methods, especially with respect to topological alignment quality measures (S^3 and LCCS) (Figs. 7 (c) and A.7). This could be because NETAL has both different NCF and AS compared to WAVE, and as such, its superiority might be a consequence not of its ASs but rather of its NCF. So, if its NCF was fed into WAVE AS, this could perhaps result in a superior new method. This possibility of designing a novel superior method simply by mixing NCF of one method and AS of another method has already been confirmed on several occasions [7, 42].

Best Alignments. When each method is given the best-case advantage, WAVE remains superior to MAGNA, and moreover, its ranking against NETAL now improves compared to topology-only alignments (Figs. 7 (d) and 9), which confirms real-life relevance of WAVE.

4   Concluding Remarks

We have presented WAVE, a general network alignment strategy for simultaneously optimizing both node conservation and weighted edge conservation, which can be used with any node cost function or combination of multiple node cost functions. We have demonstrated overall superiority of WAVE against existing state-of-the-art alignment strategies under multiple node cost functions, especially with respect to topological alignment quality. Moreover, we have demonstrated that WAVE is comparable or superior even to very recent approaches that became available only close to completion of our study, especially on the synthetic network data. This only further validates the effectiveness of WAVE.

Since WAVE can be combined with any node cost function, doing so for any recent function might improve its alignment quality. Also, WAVE itself can be modified to optimize any other measure of node and edge conservation, which could further improve its accuracy; the measures that we have used are merely a proof of concept that optimizing both node and weighted edge conservation can lead to better alignments compared to optimizing just node conservation (as e.g., MI-GRAAL and GHOST do) or just unweighted edge conservation (as e.g., MAGNA does).

As more biological network data are becoming available, network alignment will only continue to gain importance in the computational biology domain [41,55,56]. For example, network alignment has already redefined the notion of sequence-based orthology to the notion of network-based orthology, as it can identify conserved network (rather than sequence) regions between different species [57]. Then, network alignment can guide the transfer of biological (e.g., aging-related) knowledge from well-studied model species to poorly-studied species such as human. Hence, given WAVE's superiority as demonstrated in our study, WAVE could further our biological insights. Applying WAVE to an interesting biological question is out of the scope of this current method evaluation study and is subject of our future work. Further, network alignment (and thus WAVE) has implications in many domains. For example, it can be used to de-anonymize online social networks and thus impact privacy [37]. Hence, further theoretical improvements that would lead to better network alignments have a potential to lead to important discoveries in different fields.

Acknowledgements. This work was funded by the National Science Foundation CAREER CCF-1452795 and CCF-1319469 grants.

A   Appendix

A.1   Appendix Figures

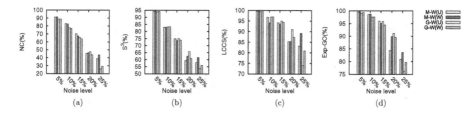

Fig. A.1. Comparison of the edge-weighted and edge-unweighted versions of WAVE on topology-only alignments of real-world PPI networks of different species with respect to **(a)** S^3, **(b)** LCCS, and **(c)** Exp-GO.

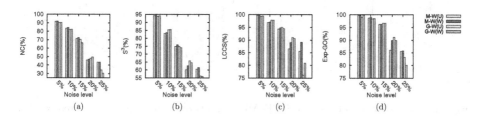

Fig. A.2. Comparison of the edge-weighted and edge-unweighted versions of WAVE on best alignments of real-world PPI networks of different species with respect to **(a)** S^3, **(b)** LCCS, and **(c)** Exp-GO.

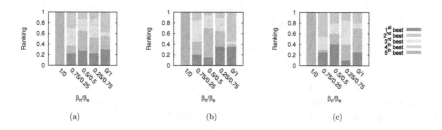

Fig. A.3. Remaining results for overall ranking of each method over all network pairs in a given data set and over all alignment quality measures. The ranking is expressed as a percentage of all cases in which the given method ranks as the kth best method. That is, the more cases in which a given method achieves a higher ranking, the better the method. For example, in panel (b), M-W is the highest scoring of all methods shown on x-axis, since it is ranked the highest (i.e., as the 1st best method) in most of the cases. **(a)** Results for the five NCF-AS methods on best alignments of "synthetic" (noisy yeast) networks. **(b)** Results for the five NCF-AS methods on best alignments of real-world PPI networks of different species. Details (per network pair and alignment quality measure) for panels (a)-(b) are shown in Figs. A.4 and A.5. Recall that M-M and G-G are MI-GRAAL and GHOST.

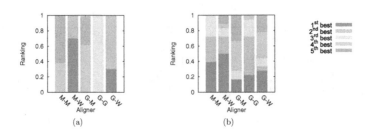

Fig. A.4. Comparison of the five NCF-AS methods on best alignments of "synthetic" (noisy yeast) networks with respect to: **(a)** NC, **(b)** S³, **(c)** LCCS, and **(d)** Exp-GO.

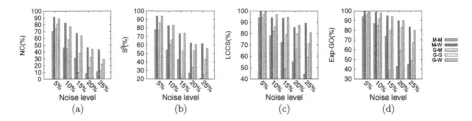

Fig. A.5. Comparison of the five NCF-AS methods on best alignments of real-world PPI networks of different species with respect to: **(a)** S³, **(b)** LCCS, and **(c)** Exp-GO.

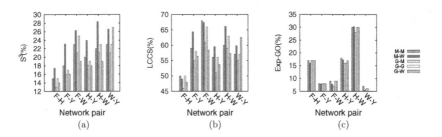

Fig. A.6. Comparison of WAVE (the best of M-W and G-W) with very recent network alignment methods on topology-only alignments of "synthetic" (noisy yeast) networks with respect to: **(a)** NC, **(b)** S³, **(c)** LCCS, and **(d)** Exp-GO.

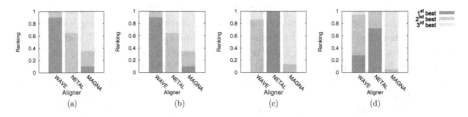

Fig. A.7. Comparison of WAVE (the best of M-W and G-W) with very recent network alignment methods on topology-only alignments of real-world PPI networks of different species with respect to: **(a)** S³, **(b)** LCCS, and **(c)** Exp-GO.

References

1. Kuchaiev, O., Milenković, T., Memišević, V., Hayes, W., Pržulj, N.: Topological network alignment uncovers biological function and phylogeny. J. R. Soc. Interface **7**(50), 1341–1354 (2010)
2. Milenković, T., Leong Ng, W., Hayes, W., Pržulj, N.: Optimal network alignment with graphlet degree vectors. Cancer Inform. **9**, 121–137 (2010)
3. Liao, C.-S., Kanghao, L., Baym, M., Singh, R., Berger, B.: IsoRankN: spectral methods for global alignment of multiple protein networks. Bioinformatics **25**(12), i253–258 (2009)
4. Kuchaiev, O., Pržulj, N.: Integrative network alignment reveals large regions of global network similarity in yeast and human. Bioinformatics **27**(10), 1390–1396 (2011)
5. Patro, R., Kingsford, C.: Global network alignment using multiscale spectral signatures. Bioinformatics **28**(23), 3105–3114 (2012)
6. Saraph, V., Milenković, T.: MAGNA: maximizing accuracy in global network alignment. Bioinformatics **30**(20), 2931–2940 (2014)
7. Faisal, F.E., Zhao, H., Milenković, T.: Global network alignment in the context of aging. IEEE/ACM Trans. Comput. Biol. Bioinf. **12**(1), 40–52 (2014)
8. Li, J., Tang, J., Li, Y., Luo, Q.: Rimom: a dynamic multistrategy ontology alignment framework. IEEE Trans. Knowl. Data Eng. **21**(8), 1218–1232 (2009)
9. Suchanek, F.M., Abiteboul, S., Senellart, P.: Paris: probabilistic alignment of relations, instances, and schema. Proc. VLDB Endowment **5**(3), 157–168 (2011)
10. Lacoste-Julien, S., Palla, K., Davies, A., Kasneci, G., Graepel, T., Ghahramani, Z.: Sigma: simple greedy matching for aligning large knowledge bases. In: Proceedings of the 19th ACM SIGKDD International Conference on Knowledge Discovery and Data Mining, pp. 572–580. ACM (2013)
11. Melnik, S., Garcia-Molina, H., Rahm, E.: Similarity flooding: a versatile graph matching algorithm and its application to schema matching. In: Proceedings of the 18th ICDE Conference (2002)
12. Conte, D., Foggia, P., Sansone, C., Vento, M.: Thirty years of graph matching in pattern recognition. Int. J. Pattern Recognit Artif Intell. **18**(03), 265–298 (2004)
13. Zaslavskiy, M., Bach, Francis, Vert, J.-P.: A path following algorithm for the graph matching problem. IEEE Trans. Pattern Anal. Mach. Intell. **31**(12), 2227–2242 (2009)
14. Koutra, D., Tong, H., Lubensky, D.: Big-align: fast bipartite graph alignment. In: 2013 IEEE 13th International Conference on Data Mining (ICDM), pp. 389–398. IEEE (2013)
15. Zhang, Y., Tang, J.: Social network integration: towards constructing the social graph. CoRR, arXiv:abs/1311.2670 (2013)
16. Bayati, M., Gerritsen, M., Gleich, D.F., Saberi, A., Wang, Y.: Algorithms for large, sparse network alignment problems. In: Ninth IEEE International Conference on Data Mining, ICDM 2009, pp. 705–710. IEEE (2009)
17. Torresani, L., Kolmogorov, V., Rother, C.: Feature correspondence via graph matching: models and global optimization. In: Forsyth, D., Torr, P., Zisserman, A. (eds.) ECCV 2008, Part II. LNCS, vol. 5303, pp. 596–609. Springer, Heidelberg (2008)
18. Noma, A., Cesar, R.M.: Sparse representations for efficient shape matching. In: 2010 23rd SIBGRAPI Conference on Graphics, Patterns and Images (SIBGRAPI), pp. 186–192. IEEE (2010)

19. Duchenne, O., Bach, F., Kweon, I.-S., Ponce, J.: A tensor-based algorithm for high-order graph matching. IEEE Trans. Pattern Anal. Mach. Intell. **33**(12), 2383–2395 (2011)

20. Smalter, A., Huan, J., Lushington, G.: Gpm: a graph pattern matching kernel with diffusion for chemical compound classification. In: 8th IEEE International Conference on BioInformatics and BioEngineering, BIBE 2008, pp. 1–6. IEEE (2008)

21. Faisal, F.E., Milenković, T.: Dynamic networks reveal key players in aging. Bioinformatics **30**(12), 1721–1729 (2014)

22. Cook, S.A.: The complexity of theorem-proving procedures. In: Proceedings of the 3rd Annual ACM Symposium on Theory of Computing, pp. 151–158 (1971)

23. Kelley, B.P., Yuan, B., Lewitter, F., Sharan, R., Stockwell, B.R., Ideker, T.: PathBLAST: a tool for alignment of protein interaction networks. Nucleic Acids Res. **32**, 83–88 (2004)

24. Sharan, R., Suthram, S., Kelley, R.M., Kuhn, T., McCuine, S., Uetz, P., Sittler, T., Karp, R.M., Ideker, T.: Conserved patterns of protein interaction in multiple species. Proc. Natl. Acad. Sci. **102**(6), 1974–1979 (2005)

25. Flannick, J., Novak, A., Srinivasan, B.S., McAdams, H.H., Batzoglou, S.: Graemlin general and robust alignment of multiple large interaction networks. Genome Res. **16**(9), 1169–1181 (2006)

26. Koyutürk, M., Kim, Y., Topkara, U., Subramaniam, S., Szpankowski, W., Grama, A.: Pairwise alignment of protein interaction networks. J. Comput. Biol. **13**(2), 182–199 (2006)

27. Berg, J., Lässig, M.: Local graph alignment and motif search in biological networks. Proc. Natl. Acad. Sci. **101**(41), 14689–14694 (2004)

28. Liang, Z., Meng, X., Teng, M., Niu, L.: NetAlign: a web-based tool for comparison of protein interaction networks. Bioinformatics **22**(17), 2175–2177 (2006)

29. Berg, J., Lässig, M.: Cross-species analysis of biological networks by Bayesian alignment. Proc. Natl. Acad. Sci. **103**(29), 10967–10972 (2006)

30. Mina, M., Guzzi, P.H.: Improving the robustness of local network alignment: design and extensive assessment of a markov clustering-based approach. IEEE/ACM Trans. Comput. Biol. Bioinf. **99**(PrePrints), 1 (2014)

31. Ciriello, G., Mina, M., Guzzi, P.H., Cannataro, M., Guerra, C.: AlignNemo: a local network alignment method to integrate homology and topology. PLOS ONE **7**(6), e38107 (2012)

32. Singh, R., Xu, J., Berger, B.: Pairwise global alignment of protein interaction networks by matching neighborhood topology. In: Speed, T., Huang, H. (eds.) RECOMB 2007. LNCS (LNBI), vol. 4453, pp. 16–31. Springer, Heidelberg (2007)

33. Flannick, J.A., Novak, A.F., Do, C.B., Srinivasan, B.S., Batzoglou, S.: Automatic parameter learning for multiple network alignment. In: Vingron, M., Wong, L. (eds.) RECOMB 2008. LNCS (LNBI), vol. 4955, pp. 214–231. Springer, Heidelberg (2008)

34. Singh, R., Jinbo, X., Berger, B.: Global alignment of multiple protein interaction networks. Proc. Pac. Symp. Biocomputing **13**, 303–314 (2008)

35. Zaslavskiy, M., Bach, F., Vert, J.-P.: Global alignment of protein-protein interaction networks by graph matching methods. Bioinformatics **25**(12), i259–i267 (2009)

36. Neyshabur, B., Khadem, A., Hashemifar, S., Arab, S.S.: NETAL: a new graph-based method for global alignment of protein-protein interaction networks. Bioinformatics **29**(13), 1654–1662 (2013)

37. Narayanan, A., Shi, E., IP Rubinstein, B.: Link prediction by de-anonymization: how we won the Kaggle social network challenge. In: Proceedings of the 2011

International Joint Conference on Neural Networks (IJCNN), pp. 1825–1834. IEEE (2011)

38. Guo, X., Hartemink, A.J.: Domain-oriented edge-based alignment of protein inter-action networks. Bioinformatics **25**(12), i240–1246 (2009)
39. Klau, G.W.: A new graph-based method for pairwise global network alignment. BMC Bioinformatics **10**(Suppl 1), S59 (2009)
40. El-Kebir, M., Heringa, J., Klau, G.W.: Lagrangian relaxation applied to sparse global network alignment. In: Loog, M., Wessels, L., Reinders, M.J.T., de Ridder, D. (eds.) PRIB 2011. LNCS, vol. 7036, pp. 225–236. Springer, Heidelberg (2011)
41. Clark, C., Kalita, J.: A comparison of algorithms for the pairwise alignment of biological networks. Bioinformatics, btu307 (2014)
42. Crawford, J., Sun, Y., Milenković, T.: Fair evaluation of global network aligners. Algorithms Mol. Biol. 10(19) (2015)
43. Milenković, T., Pržulj, N.: Uncovering biological network function via graphlet degree signatures. Cancer Informatics **6**, 257–273 (2008)
44. Solava, R.W., Michaels, R.P., Milenković, T.: Graphlet-based edge clustering reveals pathogen-interacting proteins. Bioinformatics **18**(28), i480–i486 (2012)
45. Memišević, V., Milenković, T., Pržulj, N.: Complementarity of network and sequence information in homologous proteins. J. Integr. Bioinform. **7**(3), 135 (2010)
46. Milenković, T., Memišević, V., Ganesan, A.K., Pržulj, N.: Systems-level can-cer gene identification from protein interaction network topology applied to melanogenesis-related interaction networks. J. R. Soc. Interface **7**(44), 423–437 (2010)
47. Pržulj, N.: Biological network comparison using graphlet degree distribution. Bioin-formatics **23**, e177–e183 (2007)
48. Milenković, T., Lai, J., Pržulj, N.: GraphCrunch: a tool for large network analyses. BMC Bioinformatics 9(70) (2008)
49. Hulovatyy, Y., Solava, R.W., Milenković, T.: Revealing missing parts of the inter-actome via link prediction. PLOS ONE **9**(3), e90073 (2014)
50. Milenković, T., Memišević, V., Bonato, A., Pržulj, N.: Dominating biological net-works. PLOS ONE **6**(8), e23016 (2011)
51. Collins, S.R., Kemmeren, P., Zhao, X.-C., Greenblatt, J.F., Spencer, F., Holstege, F.C.P., Weissman, J.S., Krogan, N.J.: Toward a comprehensive atlas of the phy-isical interactome of Saccharomyces cerevisiae. Molecular Cell. Proteomics **6**(3), 439–450 (2007)
52. Venkatesan, K., Rual, J.-F., Vazquez, A., Stelzl, U., Lemmens, I., Hirozane-Kishikawa, T., Hao, T., Zenkner, M., Xin, X., Goh, K.-I., et al.: An empirical framework for binary interactome mapping. Nat. Methods **6**(1), 83–90 (2009)
53. Altschul, S.F., Gish, W., Miller, W., Myers, E.W., Lipman, D.J.: Basic local align-ment search tool. J. Mol. Biol. **215**(3), 403–410 (1990)
54. De Magalhães, J.P., Budovsky, A., Lehmann, G., Costa, J., Li, Y., Fraifeld, V., Church, G.M.: The human ageing genomic resources: online databases and tools for biogerontologists. Aging Cell **8**(1), 65–72 (2009)
55. Sharan, R., Ideker, T.: Modeling cellular machinery through biological network comparison. Nat. Biotechnol. **24**(4), 427–433 (2006)
56. Vijayan, V., Saraph, V., Milenković, T.: Magna++: maximizing accuracy in global network alignment via both node and edge conservation. Bioinformatics **31**(14), 2409–2411 (2015)
57. Faisal, F.E., Meng, L., Crawford, J., Milenković, T.: The post-genomic era of bio-logical network alignment. EURASIP J. Bioinform. Syst. Biol. 2015(1), (2015)

The Topological Profile of a Model of Protein Network Evolution Can Direct Model Improvement

Todd A. Gibson[1]([✉]) and Debra S. Goldberg[2]

[1] California State University, Chico, USA
tagibson@csuchico.edu
[2] University of Colorado at Boulder, Boulder, USA
debra@colorado.edu

Abstract. Biological networks are an attractive construct for studying evolution. One method for inferring evolutionary mechanics is to construct models which generate networks sharing topological characteristics with their empirical counterparts. It remains a challenge to assess, modify, and improve a model based on the topological values it generates. A large range of parameter values may produce a similar topology, and topological properties may vacillate in unexpected ways, frustrating attempts to determine whether the model is flawed or model parameter values are incorrect. We introduce a new method for evaluating the fidelity of an evolutionary network model with respect to topological characteristics by driving topological characteristics towards empirical values concurrently with network generation. From this we compute a *topological profile* which defines the ability of the network model to produce a desired topology. The topological profile also measures the volatility of characteristics, and the interrelationships among topological characteristics. Our method shows that a top-rated protein interaction network model cannot produce the empirical number of triangles. As triangle count is driven to the empirical value, additional characteristics are propelled towards empirical values. These findings suggest that new model mechanics that increase the number of triangles produced will best enhance the existing model. By providing systematic evaluation of the ability of model mechanics to produce desired topological properties, our framework can help to focus the search for biologically plausible and relevant processes important to network evolution.

1 Introduction

Discovery of the biological processes that influence evolution remains an active area of research. Protein interaction networks are particularly well-suited for undertaking evolutionary studies. Topological properties are readily measured from empirical networks and have been used to guide research into biological processes which may have influenced the formation of these properties. In particular, models of protein interaction network evolution have been built featuring biologically-plausible evolutionary mechanics. When such a model can generate

© Springer-Verlag Berlin Heidelberg 2015
M. Pop and H. Touzet (Eds.): WABI 2015, LNBI 9289, pp. 40–52, 2015.
DOI: 10.1007/978-3-662-48221-6_3

networks sharing topological characteristics with extant networks of biological organisms, it suggests that the modeled evolutionary mechanics played a role in the formation of empirical networks.

A continuing challenge in the search for evolutionary mechanics is associating the mechanics with the topological features generated by the network model. Although analytic and computational results have demonstrated the relevance of specific topological properties to specific evolutionary mechanics [2,5,8], there is little work suggesting the number, type, or relative importance of topological measures appropriate for evolutionary analysis. Different studies have relied on different sets of measures for model validation. Examples of validating sets of topological features include: clustering coefficient, square coefficient, and degree distribution [18], clustering coefficient, degree distribution, and characteristic (average) path length (CPL) [13], clustering coefficient and CPL [1], degree distribution and average connectivity of first neighbors [4], and degree distribution alone [15]. By convention some measures such as the degree distribution are included frequently in model evaluation. Otherwise there is little agreement as to the number or type of topological measures to include in a model assessment. More sophisticated topological measures such as graphlets (small, connected, non-isomorphic subgraphs) [12] have also been used, and subgraphs have been enumerated using machine learning techniques [10]. Thorne and Stumpf [16] used graph spectra as a measure of network similarity, in lieu of traditional topological characteristics.

Compounding this is the determination of parameter values. Parameters which seem to produce a particular topological value can be mistaken if a large range of parameter values produce a similar topology. If a model can produce a large variety of topologies, then choosing an (incorrect) parameter value to produce a valid topology can "validate" a flawed model. Conversely, a model which produces a topology divergent from empirical could indicate either a flaw of the model or incorrect parameters.

Previous methods of evolutionary model validation differ in how model-generated networks are measured against empirical networks. These measures allow different models to be compared and ranked on their ability to generate empirically-observed networks. Absent from previous methods is a mechanism to study a single model, to "stress" it and uncover its strengths and weaknesses. We approach the problem from this different direction by first asking: for which topological properties can empirical values be obtained, and how do multiple topological properties interact during model evolution? By driving various topological characteristics towards empirical values *during network generation* we are able to observe the relationships between multiple topological measures and assess their influence on each other and their contribution to the network topology. We are not finding the "best", or "optimized" topology a model can produce. We are exercising a model to explore its *topological profile*–the limits of a model to approximate an empirical network topology and the interrelationships among topological characteristics of a model. We coerce topological characteristics towards empirical values, revealing the model's (in)ability to achieve empirical values, and measured each topological property's impact on other topological properties.

Though the optimization strategy shares some qualities of genetic algorithms, it must be stressed that the optimization strategy is not being used as a "natural selection" proxy for optimizing the networks generated by the model, nor are we suggesting that natural selection has driven empirical values of topological properties. The optimization strategy is used only to measure the potential of a model to achieve various empirical topological properties.

2   Methods

To drive individual topological properties produced by a model of network evolution, we generated populations of networks via an optimization process. Each topological property we examined was "optimized" in a population. Starting with a population of 100 seed graphs, each with 100 nodes, a network model was used to grow each of the 100 networks by 10 proteins. The target property was then measured across all of the networks. The 10 networks whose property most-closely approached that of an empirical network were culled and duplicated ten times to restore the population to 100 networks. The networks were iteratively grown 10 nodes at a time, retaining the optimal networks at each iteration until the number of nodes in the networks matched that of the empirical network. Topological values were measured on the final best 10 networks from the population and averaged. Although this method is inspired by genetic or evolutionary algorithms, it is used only to define the topological profile of the model-generated networks. The method we use is not intended to reflect how networks evolve in nature.

2.1   The Evolutionary Model

The duplication and divergence [18] model was selected for this study. The model offers simple and biologically-plausible evolutionary mechanics of duplication and divergence (see Fig. 1). Specifically, duplication is modeled by adding a progeny protein to the network which interacts with the same protein "neighbors" as its progenitor. Divergence is modeled by removing, with some probability, one of each pair of redundant interactions the progeny/progenitor pair has with each neighbor. Subsequent to its initial publication, additional studies have further validated the Vázquez et al. model [5,8,10]. To this model we have added the asymmetric loss of interactions during subfunctionalization due to its better performance [7].

We begin our populations of networks with a 100-node scale-free network (i.e., with a power-law degree distribution that follows a power-law). Because seed graphs can affect the topology of generated networks [1], we also analyzed populations seeded with a 100-node Erdős-Renyí random graph (i.e., with Poisson degree distribution), with comparable results.

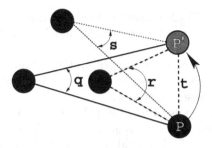

Fig. 1. Duplication and divergence model of evolution [18]. **Duplication:** protein P' is created as the progeny of P. Progeny protein P' interacts with the same neighbors as its progenitor P. With some probability interaction t is added, representing the paralogous interaction formed when a homomeric protein duplicates. **Divergence:** one interaction is lost from each redundant pair of interactions q, r and s, with some probability. In the original model, when an interaction is lost from a redundant pair, the lost interaction is selected equiprobably. We select the lost interaction asymmetrically due to its better performance [7].

2.2 The Empirical Network

Protein interaction data was assembled from two high-quality *Saccharomyces cerevisiae* (yeast) data sets: a meta study that produced a "second-generation" high quality interaction data set by Yu et al. [19], and high-confidence interactions from an in vivo protein-fragment complementation assay (PCA) by Tarassov et al. [14]. The Yu et al. data represents directly-measured binary interactions such as from yeast two-hybrid assays and interactions curated from the literature. The data set includes high-confidence interactions from earlier studies including [17] and [9]. The Tarassov data set was published subsequent to the Yu et al. study. It is the first high-throughput study to identify binary interactions in vivo. The combined empirical data sets have 2647 proteins and 5449 interactions.

Model parameters were estimated from this empirical network via an evolutionary reconstruction method as described in [6] (probability of edge loss = 0.887, asymmetry of loss = 0.82, probability of a paralogous interaction = 0.46).

2.3 Topological Measures Assessed

Although the method allows for any arbitrary set of network measures to be used, such as graphlet degree distribution [12] or graph spectra [16], here we confine ourselves to eight, simply computed topological characteristics.

- **Connected triples** ("Trips") A node connected to an unordered pair of nodes (i.e., a path of length 2).
- **Triangles** ("Tris") Three fully-connected nodes.

- **Transitivity** ("Trans") Also known as the *clustering coefficient*. It is given by $C = \frac{3T}{\Gamma}$, where T is the number of triangles and Γ is the number of connected triples [11].
- **Size** The number of interactions (edges) in the network.
- **Diameter** ("Diam") The maximum of the shortest paths between all pairs of proteins in the largest component.
- **Characteristic (Average) Path Length** ("CPL") The mean of the shortest paths between all pairs of proteins in the network's largest component.
- **Assortativity** ("Assort") The tendency of proteins to interact with proteins having similar degree. The assortativity is the equivalent of the Pearson Correlation Coefficient between the degrees of pairs of nodes which interact.
- **Degree distribution** ("Deg Dist") A histogram of the number of interacting partners for each protein.

A population was evolved for each of the eight measures. Each population had a topological measure driven towards the empirically observed value. In addition, one population was grown without any constraints (that is, the normal, "neutral" model), for a total of 9 populations to be compared.

The topological measures were averaged from the ten best-performing individuals in each population. Each average was then converted to a proportion by dividing it by the equivalent empirical value.

The difference between the model's degree distribution and the empirical distribution was converted to a single value by first summing up the absolute value of differences between degrees:

$$1 - \frac{\sum_i |k_{ei} - k_{mi}|}{n_e}$$

where k_{ei} is the number of proteins in the empirical network with degree i, k_{mi} is the number of proteins in the model network with degree i, and n_e is the number of nodes in the empirical network. Because the model network has the same number of nodes as the empirical network, dividing by n_e normalizes the summation, and subtracting it from 1 ensures that a distribution identical to empirical will have a value of 1, and a maximally disparate model distribution (i.e., with no nodes sharing a degree with empirical), will have a value of 0.

A desktop workstation can easily handle the computational demands of the method and a combination of C++ and Python code generated complete results in a few minutes. The limiting factor of the computation is determined by the cost of running the model and the topological measure calculations.

3   Results

Table 1 shows the topological profile of the model, presented as proportions of empirical topological values. A value of 1.0 represents perfect agreement with the empirical value. Each row in the table is a population with selective pressure applied to the topological property indicated by the row-label. Each column is a

Table 1. Topological values observed as a proportion of empirical. Each row contains the topological values for one population driven towards empirical by the characteristic specified by the row label. Values are shown as proportions of empirical. A value of 1.0 represents perfect agreement with the empirical value. Each column is a topological property measured.

Driving property	Scale-free seed graph							
	Trips	Tris	Trans	Size	Diam	CPL	Assort	DD
Neutral	0.22	0.11	0.51	0.66	1.58	1.40	0.57	0.66
Trips	**1.00**	0.15	0.15	0.72	1.27	1.02	-1.72	0.64
Tris	0.59	**0.47**	0.80	0.89	1.13	1.09	1.72	0.60
Trans	0.33	0.39	**1.19**	0.78	1.73	1.40	5.67	0.58
Size	0.45	0.31	0.69	**1.00**	1.05	1.12	0.63	0.42
Diam	0.22	0.11	0.50	0.66	**1.00**	1.37	0.31	0.67
CPL	0.41	0.14	0.35	0.68	0.80	**1.01**	-0.44	0.64
Assort	0.25	0.12	0.51	0.64	1.69	1.37	**1.00**	0.68
DD	0.20	0.10	0.53	0.64	1.60	1.43	1.09	**0.83**

Driving property	Erdős-Renyí seed graph							
	Trips	Tris	Trans	Size	Diam	CPL	Assort	DD
Neutral	0.20	0.11	0.55	0.66	1.61	1.45	0.03	0.66
Trips	**0.71**	0.27	0.38	0.88	1.27	1.01	-1.83	0.59
Tris	0.44	**0.45**	1.00	0.89	1.27	1.19	1.97	0.58
Trans	0.31	0.38	**1.22**	0.82	1.43	1.38	3.28	0.56
Size	0.43	0.33	0.75	**0.99**	1.20	1.16	0.22	0.43
Diam	0.20	0.12	0.60	0.67	**1.00**	1.38	0.05	0.64
CPL	0.31	0.15	0.50	0.69	0.83	**1.06**	-0.21	0.68
Assort	0.20	0.10	0.51	0.65	1.72	1.45	**1.00**	0.66
DD	0.19	0.10	0.52	0.64	1.49	1.47	-0.13	**0.83**

Legend:

Neutral	Normal model run	CPL	Characteristic Path Length	Size	Size (# edges)
Trips	Triples	Tris	Triangles	Trans	Transitivity
Diam	Diameter	Assort	Assortativity	DD	Degree Distribution

topological property measured on the optimal members of that population. For an alternative visualization, Fig. 2 presents bar plots of each row for networks generated from the scale-free seed graph.

3.1 Success of Driving Properties Towards the Empirical Topology

Five topological measures (triples, size, diameter, CPL, and assortativity) achieved parity with the empirical value in the population towards which they were driven. Driving some parameters towards empirical has expected effects. For example, when triples are increased towards empirical values, the transitivity (a measure whose calculation includes triples in the denominator) decreases. Three measures produced suboptimal results:

1. Triangles. Driving triangles towards empirical values produced fewer than 50 % of the empirical triangles (Table 1, Fig. 2). This is the most discrepant

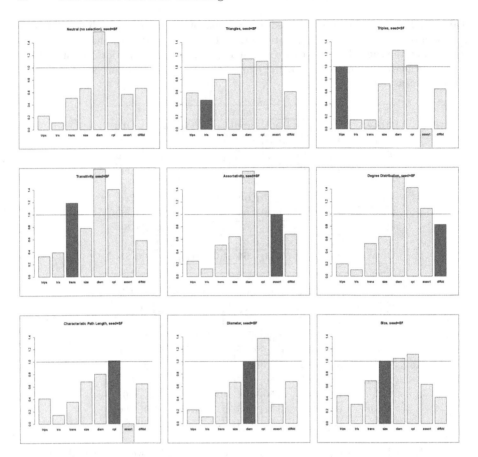

Fig. 2. From scale-free seed graphs, the proportion of model-generated to empirical values when driving various measures towards empirical values. The bars in each plot represent (from left to right): triples, triangles, transitivity, size, diameter, characteristic path length, assortativity, and degree distribution difference. The red bar in each of the nine plots indicate the topoological property that was driven towards the empirical value. From left to right, top to bottom they are: neutral model, triangles, triples, transitivity, assortativity, degree distribution difference, characteristic path length, diameter, and size (Color figure online).

of the measures and reflects a limitation of the model [7]. This suggests that a mechanism should be modified or included in the model that increases the number of triangles.

2. Transitivity. Optimizing for empirical transitivity (clustering coefficient) results in a final network population having a ≈20 % larger transitivity than empirical. This is a deviation in the opposite direction as the neutral model, which achieves only about half of the empirical transitivity. Note that in the populations driven towards empirical values, the triangles and triples comprising the transitivity were still much lower than empirical, though higher than in the neutral (nonevolved) networks.

3. Degree Distribution. Driving the degree distribution towards empirical achieves ≈80 % of empirical. This may be an artifact of the measure, as we must summarize many different values (one for each observed degree) while all other topological measures considered are a single value. Although one might think that the size (number of edges), which is much lower in the neutral model than in the empirical network, would be related to degree distribution (as we've measured it), driving size achieves the empirical size but does not significantly impact the distance from the empirical degree distribution, while driving the degree distribution does not achieve the empirical degree distribution nor does it significantly impact size.

3.2 Non-Optimized Topological Properties that Trend Towards Empirical Values

Topological values driven towards empirical had varying tendencies to draw neutral measures towards empirical values. We calculated an overall distance from the population's topological values to empirical values (d_p) by:

$$d_p = \sqrt{\sum_{i=1}^{8} (\ln |v_{pi}|)^2}$$

where v_{pi} is the distance from empirical of topological property i in population p (the values are shown in Table 1). The d_p values are summarized in Table 2. Each row represents a population as in Table 1, and "Distance from Empirical" is d_p.

As Table 2 summarizes, there are several topological characteristics which, when driven towards empirical values, show better fits to the empirical model than the neutral model. Driving the population towards triangles has the closest

Table 2. Aggregate logarithmic distance of normalized topological measures from empirical

Directed Measure	Scale-free seed	Erdős-Renyí seed
Triangles	1.22	1.48
Size	1.78	2.27
Transitivity	2.44	2.08
Char. Path Len.	2.59	2.84
Assortativity	2.74	3.00
Triples	2.83	1.87
neutral	2.90	4.61
Degree Dist	2.95	3.64
Diameter	3.06	4.05

fit to empirical, as illustrated in Fig. 2 (top, center) by the several bars which are within close proximity of the 1.0 line.

3.3 Variability of Topological Measures

We also measured the extent to which each topological property varied across the different optimization regimes (Fig. 3). Specifically, Fig. 3 shows the standard deviations of each column in Table 1. Properties having low standard deviations suggest stable measures which are more robust against the varying conditions impacting the evolutionary model. Basing model evaluation on these low-variance characteristics increases confidence that the model mechanics are indeed relevant to the evolution of the network and not just a coincidental match with more variable measures for the given model.

Reassuringly, the degree distribution–putatively the most-studied topological property among biological networks–showed the least variability across the different populations. At the other extreme, the assortativity is extremely sensitive. Its instability can be seen in Table 1 where the value vacillates between positive and negative values.

Triangles merit additional attention. As noted earlier, the model is unable to approach the number of triangles found in the empirical network. At the same time driving triangles towards empirical has the greatest success in coercing other properties towards the empirical topology. If the stability of the measure as shown in Fig. 3 remains, a relevant and specific modification to the model would be one that generates a greater number of triangles.

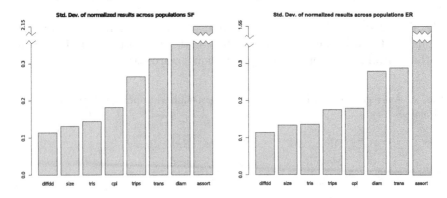

Fig. 3. The standard deviation of each normalized topological value across all populations.

3.4 Correlation Among Topological Characteristics

Table 3 shows the correlation between pairs of topological characteristics for the generated networks. Only one measure from each highly-correlated pair is informative with regard to evaluating model-generated networks against an empirical

Table 3. Pearson correlation. Values above 0.5 and below −0.5 in bold

Scale-free seed graph

	Trips	Tris	Trans	Size	Diam	CPL	Assort	DD
Trips	**1.00**	0.19	-0.39	0.30	-0.25	**-0.76**	-0.44	-0.28
Tris	0.19	**1.00**	**0.76**	**0.76**	0.03	-0.19	**0.61**	**-0.60**
Trans	-0.39	**0.76**	**1.00**	0.43	0.40	0.42	**0.96**	-0.35
Size	0.30	**0.76**	0.43	**1.00**	-0.27	-0.43	0.18	**-0.86**
Diam	-0.25	0.03	0.40	-0.27	**1.00**	**0.70**	**0.58**	0.34
CPL	**-0.76**	-0.19	0.42	-0.43	**0.70**	**1.00**	**0.56**	0.45
Assort	-0.44	**0.61**	**0.96**	0.18	**0.58**	**0.56**	**1.00**	-0.15
DD	-0.28	**-0.60**	-0.35	**-0.86**	0.34	0.45	-0.15	**1.00**

Erdős-Renyí seed graph

	Trips	Tris	Trans	Size	Diam	CPL	Assort	DD
Trips	**1.00**	**0.57**	-0.06	**0.75**	-0.16	**-0.81**	-0.37	**-0.52**
Tris	**0.57**	**1.00**	**0.76**	**0.84**	-0.02	-0.38	**0.50**	**-0.70**
Trans	-0.06	**0.76**	**1.00**	0.42	0.13	0.20	**0.91**	-0.45
Size	**0.75**	**0.84**	0.42	**1.00**	-0.12	**-0.59**	0.08	**-0.86**
Diam	-0.16	-0.02	0.13	-0.12	**1.00**	**0.61**	0.33	0.17
CPL	**-0.81**	-0.38	0.20	**-0.59**	**0.61**	**1.00**	0.46	0.44
Assort	-0.37	**0.50**	**0.91**	0.08	0.33	0.46	**1.00**	-0.22
DD	**-0.52**	**-0.70**	-0.45	**-0.86**	0.17	0.44	-0.22	**1.00**

network. By contrast, pairs showing little correlation would each provide independent evidence of the plausibility of the model mechanics. Highly-correlated properties in an accepted model may also suggest that these topological properties are closely-related in evolution.

Table 3 indicates that the number of triangles and the diameter are essentially uncorrelated.

It is also notable that transitivity and assortativity are highly correlated despite their high variability. Size and degree distribution on the other hand have the least variability and are the most highly anticorrelated.

3.5   Improving the Evolutionary Model

The relative dearth of triangles, even when driven towards the empirical number of triangles, reflects a limitation of the mechanics of this particular model. While optimizing the number of triangles fails to produce enough triangles, it simultaneously brings other topological properties more closely to empirical values than other topological properties. Additionally, triangles were among the most stable measures. This suggests that incorporating mechanics that will increase the number of triangles will significantly improve the model.

Evolutionarily, a greater number of triangles are produced when a homomeric (self-interacting) protein is duplicated [5]. The current model considers

Table 4. Topological values observed as a proportion of empirical. Two rows are shown. The first is the original model's proportions. This row is the same row as the *Neutral* row in Table 1. The second row are the proportions generated from neutral (non-optimized) runs of the iSite model. Values are shown as proportions of empirical. A value of 1.0 represents perfect agreement with the empirical value. Each column is a topological property measured.

Driving property	Trips	Tris	Trans	Size	Diam	CPL	Assort	DD
Original	0.22	0.11	0.51	0.66	1.58	1.40	0.57	0.66
iSite	0.50	0.52	1.04	1.02	1.60	1.92	0.62	0.26

Legend:

Neutral	Normal model run	CPL	Characteristic Path Length
Trips	Triples	Assort	Assortativity
Tris	Triangles	DD	Degree Distribution
Trans	Transitivity	Size	Size (# edges)
Diam	Diameter		

each redundant interaction pair independently when determining interaction loss, and uses a simple probability to add paralogous interactions (which simulates the duplication of a homomeric protein). A more biologically faithful interpretation acknowledges that interactions are formed at specific sites on the protein's surface, and that these sites are heritable through duplication. The iSite evolutionary model adds precisely this enhancement to the original model, including support for homomeric interactions [7]. By eschewing independent loss of interactions in favor of grouping interactions (including self-interactions) into heritable sites when calculating interaction loss, the iSite model generates significantly more triangles. Table 4 shows the topological characteristics of generated by both the original model and the iSite model as a proportion of the empirical network. Nearly every topological characteristic of networks generated by the iSite model is closer than the original model to the empirical network.

4 Conclusion

As empirical networks become more complete and correct, evolutionary network models have a greater potential to enhance our understanding of the mechanisms affecting evolution [3,7]. Driving topological features towards empirical values, and feature interrelationship analysis provide network model builders with additional insights into model behavior leading to more informed model construction.

There are many strategies for improving an evolutionary model that fails to achieve topological parity with empirical networks. These include adjusting model parameter values, changing model mechanics, and finding evolutionary phenomena to explain model discrepancies. These techniques can be informed by the systematic evaluation our approach provides of the capabilities and limitations of a model to produce desired topological properties.

Unsurprisingly, driving individual topological properties towards empirical values generates networks that generally do well in that property when compared to empirical networks. More notable is the ability to identify:

1. the limitations of existing mechanics to achieve an empirical topological value (for example, the original model is unable to attain the number of triangles found in the empirical network),
2. individual topological characteristics which beneficially affect many other characteristics,
3. the volatility of individual characteristics with regard to the model being used, and
4. correlated changes among properties.

With this additional topological information, we can make more informed assessments of the biological processes which might have the most relevant impact on the evolution of biological networks.

A mathematical analysis of the interrelationships among a cohort of topological characteristics is intractable in models striving for biological realism. A more tractable approach is to identify biological processes to incorporate into models, but this requires much trial and error. An alternative approach is needed. Our method provides a simple mechanism for observing these topological interrelationships which influence the evolution of protein interactions.

Acknowledgements. *Funding:* National Science Foundation grant DGE-0841423; National Institutes of Health training grant T15LM009451.

References

1. Bhan, A., Galas, D.J., Dewey, T.G.: A duplication growth model of gene expression networks. Bioinformatics **18**(11), 1486–1493 (2002). http://bioinformatics.oxfordjournals.org/cgi/content/abstract/18/11/1486
2. Chung, F., Lu, L., Dewey, T.G., Galas, D.J.: Duplication models for biological networks. J. Comput. Biol. **10**(5), 677–687 (2003). http://dx.doi.org/10.1089/106652703322539024
3. de Silva, E., Stumpf, M.P.H.: Complex networks and simple models in biology. J. R. Soc. Interface **2**, 419–430 (2005)
4. Evlampiev, K., Isambert, H.: Modeling protein network evolution under genome duplication and domain shuffling. BMC Syst. Biol. **1**, 49 (2007). http://dx.doi.org/10.1186/1752-0509-1-49
5. Gibson, T.A., Goldberg, D.S.: Questioning the ubiquity of neofunctionalization. PLoS Comput. Biol. **5**(1), e1000252 (2009). http://dx.doi.org/10.1371/journal.pcbi.1000252
6. Gibson, T.A., Goldberg, D.S.: Reverse engineering the evolution of protein interaction networks. In: Altman, R.B., Dunker, A.K., Hunter, L., Murray, T., Klein, T.E. (eds.) Pacific Symposium on Biocomputing, pp. 190–202 (2009)

7. Gibson, T.A., Goldberg, D.S.: Improving evolutionary models of protein interaction networks. Bioinformatics **27**(3), 376–382 (2011). http://dx.doi.org/10.1093/bioinformatics/btq623
8. Ispolatov, I., Krapivsky, P.L., Mazo, I., Yuryev, A.: Cliques and duplication-divergence network growth. New J. Phys. **7**, 145 (2005). http://stacks.iop.org/1367-2630/7/145
9. Ito, T., Chiba, T., Ozawa, R., Yoshida, M., Hattori, M., Sakaki, Y.: A comprehensive two-hybrid analysis to explore the yeast protein interactome. Proc. Natl. Acad. Sci. USA **98**(8), 4569–4574 (2001). http://dx.doi.org/10.1073/pnas.061034498
10. Middendorf, M., Ziv, E., Wiggins, C.H.: Inferring network mechanisms: the Drosophila melanogaster protein interaction network. Proc. Natl. Acad. Sci. USA **102**(9), 3192–3197 (2005). http://dx.doi.org/10.1073/pnas.0409515102
11. Newman, M.E.: The structure of scientific collaboration networks. Proc. Natl. Acad. Sci. USA **98**(2), 404–409 (2001). http://dx.doi.org/10.1073/pnas.021544898
12. Pržulj, N.: Biological network comparison using graphlet degree distribution. Bioinformatics **23**(2), e177–e183 (2007). http://dx.doi.org/10.1093/bioinformatics/btl301
13. Solé, R.V., Pastor-Satorras, R., Smith, E., Kepler, T.B.: A model of large-scale proteome evolution. Advs. Complex Syst. **5**, 43–54 (2002). http://www.citebase.org/cgi-bin/citations?id=oai:arXiv.org:cond-mat/0207311
14. Tarassov, K., Messier, V., Landry, C.R., Radinovic, S., Molina, M.M.S., Shames, I., Malitskaya, Y., Vogel, J., Bussey, H., Michnick, S.W.: An in vivo map of the yeast protein interactome. Science **320**, 1465–1470 (2008). http://dx.doi.org/10.1126/science.1153878
15. Thomas, A., Cannings, R., Monk, N.A.M., Cannings, C.: On the structure of protein-protein interaction networks. Biochem. Soc. Trans. **31**(Pt 6), 1491–1496 (2003). http://dx.doi.org/10.1042/
16. Thorne, T., Stumpf, M.P.H.: Graph spectral analysis of protein interaction network evolution. J. R. Soc. Interface **12**(108), 1–14 (2012)
17. Uetz, P., Giot, L., Cagney, G., Mansfield, T.A., Judson, R.S., Knight, J.R., Lockshon, D., Narayan, V., Srinivasan, M., Pochart, P., Qureshi-Emili, A., Li, Y., Godwin, B., Conover, D., Kalbfleisch, T., Vijayadamodar, G., Yang, M., Johnston, M., Fields, S., Rothberg, J.M.: A comprehensive analysis of protein-protein interactions in Saccharomyces cerevisiae. Nature **403**(6770), 623–627 (2000). http://dx.doi.org/10.1038/35001009
18. Vázquez, A., Flammini, A., Maritan, A., Vespignani, A.: Modeling of protein interaction networks. ComPlexUs **1**, 38–44 (2003)
19. Yu, H., Braun, P., Yildirim, M.A., Lemmens, I., Venkatesan, K., Sahalie, J., Hirozane-Kishikawa, T., Gebreab, F., Li, N., Simonis, N., Hao, T., Rual, J.F., Dricot, A., Vazquez, A., Murray, R.R., Simon, C., Tardivo, L., Tam, S., Svrzikapa, N., Fan, C., de Smet, A.S., Motyl, A., Hudson, M.E., Park, J., Xin, X., Cusick, M.E., Moore, T., Boone, C., Snyder, M., Roth, F.P., Barabási, A.L., Tavernier, J., Hill, D.E., Vidal, M.: High-quality binary protein interaction map of the yeast interactome network. Science **322**, 104–110 (2008). http://dx.doi.org/10.1126/science.1158684

Algorithms for Regular Tree Grammar Network Search and Their Application to Mining Human-Viral Infection Patterns

Ilan Smoly[1], Amir Carmel[1], Yonat Shemer-Avni[2], Esti Yeger-Lotem[3](✉), and Michal Ziv-Ukelson[1](✉)

[1] Department of Computer Science, Ben-Gurion University of the Negev, Beer Sheva, Israel
{smolyi,karmela,michaluz}@cs.bgu.ac.il
[2] Department of Virology, Ben-Gurion University of the Negev, Beer Sheva, Israel
yonat@bgu.ac.il
[3] Department of Clinical Biochemistry and Pharmacology, Ben-Gurion University of the Negev, Beer Sheva, Israel
estiyl@bgu.ac.il

Abstract. Network querying is a powerful approach to mine molecular interaction networks. Most network querying tools support queries in the form of a template sub-network, in case of topology-constrained queries, or a set of colored vertices in case of topology-free queries. A third approach is grammar-based queries, which are more flexible and expressive as they allow the addition of logic rules to the query. Previous grammar-based querying tools defined queries via string grammars and identified paths in graphs. In this paper, we extend the scope of grammar-based queries to regular tree grammar (RTG), and the scope of the identified sub-graphs from paths to trees. We introduce a new problem and propose a novel algorithm to search a given graph for the k highest scoring sub-graphs matching a tree accepted by an RTG. Our algorithm is based on dynamic programming and combines an extension to k-best parsing optimization with color coding. We implement the new algorithm and exemplify its application to mining the human-viral interaction network. Our code is available at http://www.cs.bgu.ac.il/~smolyi/RTGnet/.

1 Introduction

Molecular interaction networks have become a prominent resource for inferring protein functions, detecting cellular processes, predicting disease pathways, and more [5]. A powerful approach to mine these networks is via network querying: Given a query sub-network, network querying tools identify instances within the network that match the query. Previous network querying tools mostly support queries in the form of a "template" sub-network that is confined to a specific topology, such as paths [33], trees [28,30] or graphs with bounded tree-width [10,30].

These authors contributed equally to the paper.

© Springer-Verlag Berlin Heidelberg 2015
M. Pop and H. Touzet (Eds.): WABI 2015, LNBI 9289, pp. 53–65, 2015.
DOI: 10.1007/978-3-662-48221-6_4

Other approaches, denoted topology free, support queries in the form of a set of colors, and seek in the network vertices with matching colors that can be connected by a spanning tree [7,23].

Here we focus on a third approach to network querying, denoted grammar-based queries. Such queries are given a grammar as input, and seek in the network sub-graphs accepted by the grammar. On one hand, grammar-based queries are more flexible than other approaches because neither the exact topology nor the exact set of colors are specified in advance. On the other hand, they are more expressive as they allow the addition of logic rules to the query. Previous works in this area defined queries via string grammars, including regular expressions [12,22] and context-free grammars [31], and identified paths in graphs. Here, we extend the scope of the grammars to regular tree grammars (RTGs), and the scope of the identified sub-graphs from paths to trees.

Note that even the most basic problem variant, that of finding a path in a network based on a regular grammar query descriptor, is generally intractable [26]. Thus, previous grammar-based search approaches either bounded the size of the sought paths [12] or applied inadmissible heuristics such as seed-and-extend [22]. In our solution, we also bound the size of the sought subtrees.

We introduce a new problem and propose a novel algorithm to search a given graph for the k highest scoring sub-graphs, of size bounded by m, matching a tree accepted by a regular tree grammar (RTG). Our proposed approach is based on a two-stage dynamic programming algorithm combining RTG k-best parsing with network search. One challenge encountered here is that RTGs recognize ordered trees, while the subtrees in the graph to be searched are unordered. We handle this by incorporating redundancy-avoidance logic and show how to extend an efficient k-best parsing algorithm [19] to support it. Another challenge is that in order to support the bottom-up parsing employed in this paper, we need to use a normalized form of the grammar, similarly to the way the CYK algorithm parses context-free string grammars in Chomsky normal form (CNF) [24]. For this purpose, we harness a previously known binarization approach to RTG parsing [25]. To handle the exponentially growing search space, and to avoid cycles, we employ color-coding [2], which was also used by other network querying tools [7,10,30,33], and show how to probabilistically bound the number of color-coding iterations needed to compute the k-best subtrees for a given error threshold.

We implement the proposed algorithm in a software package, denoted RTGnet, and demonstrate its usage in two separate applications which demonstrate the expressive power of RTGs and the topological flexibility of the identified patterns. Due to space constraints, supplementary materials including some of the figures and proofs to all theorems, are deferred to an appendix that is available at http://www.cs.bgu.ac.il/~smolyi/RTGnet/

2   Regular Tree Grammars and Curry-Encoding

The patterns we seek are ordered rooted trees, to be recognized by Regular Tree Grammars [9].

Definition 1. *A Regular Tree Grammar (RTG) is a tuple $A = \{N, \Sigma, P, S\}$, where N is a finite set of non-terminal symbols, Σ is an alphabet of terminal symbols, S is the initial non-terminal, and P is a finite set of productions of type $X \to a(R)$, where R is a regular expression over N and $a \in \Sigma$.*

Each production rule $X \to a(R)$ defines a vertex x labelled a, while R expresses the non-terminal symbols that are used to recursively generate the child subtrees of x. If $R = \epsilon$, then x is a leaf. Let $T(\Sigma)$ denote the set of all ordered, rooted trees with vertex labels in Σ. An ordered rooted vertex-labelled tree $\tau \in T(\Sigma)$ is accepted by A if there is a chain of consecutive derivations of production rules from P that starts in S and generates τ. The language $L(A) \subseteq T(\Sigma)$ is the set of all trees accepted by A. Regular Tree Languages (RTL) is the class of tree languages that are generated by RTGs.

RTGs are classically categorized as either ranked or unranked. Ranked RTGs generate trees for which there is a global bound on the number of children each vertex may have. Unranked RTGs, which are the more general case supported by our framework, have no such bound. For our bottom-up RTG parsing, we harness a recursive deterministic binarization approach, denoted *curryfication* [8] (see Figure S1). This approach encodes an unranked tree τ into a ranked binary tree, denoted $curry(\tau)$, as follows. Given a tree $\tau = a(\tau_1, \ldots, \tau_n) \in T(\Sigma)$, $curry(\tau) = @(curry(\tau'), curry(\tau_n))$ with $\tau' = a(\tau_1, \ldots, \tau_{n-1})$. If $n = 0$ then $curry(\tau) = a$. The *extension operator* @ is used to denote the labels of the internal vertices of the binarized trees, while their leaves correspond to the vertices of the original tree. Correspondingly, the given RTG A is also converted to a Curry-encoded RTG (defined below), $curry(A)$, that accepts the curry encodings of the trees in $L(A)$, such that the bijection $curry(L(A)) = L(curry(A))$ is obeyed [8,9].

Definition 2. *A Curry-Encoded Regular Tree Grammar is a tuple $A = \{N, \Sigma, P, S\}$ where N is a finite set of non-terminal symbols, Σ is an alphabet of terminal symbols, S is the initial non-terminal, and P is a finite set of productions of type $X \to @(Y, Z)$ or $X \to a$, where $X, Y, Z \in N$ and $a \in \Sigma$ and $@ \notin \Sigma$.*

In [8,9] it is shown that any unranked RTG A can be translated in linear time to $curry(A)$ via the construction of the corresponding deterministic Stepwise Automaton. Furthermore, it is shown that for any recognizable RTL $L \subseteq T(\Sigma)$ there is a unique minimal deterministic Stepwise Automaton accepting L [25].

Based on this, in the rest of this paper we assume that the input grammar A is in Curry-Encoded form, and denote by g the number of its production-rules. The binarization of the candidate trees will be done during the parsing process. Also, for convenience, we write $X \to Y@Z$ when $X \to @(Y, Z)$.

3   The RTG Network Parse-and-Search Problem

Given an RTG A of size g, a directed vertex-labeled graph $G = (V, E)$, and an integer parameter m. Let $T[G, A, m]$ be a set of labeled trees, such that for any tree $\tau \in T[G, A, m]$: (1) $\tau \in L(A)$, (2) $|\tau| \le m$, and (3) τ is a subtree in G. Let $s(\tau) \in \Re$ denote a predefined monotonic scoring scheme on τ, where $\Re$ denotes the real numbers. Let $Max_s^k(T[G, A, m])$ denote the subset of k-best scoring subtrees in $T[G, A, m]$ according to s. Our search is based on the following optimization problem.

Problem 1 (RTG Network Parse-and-Search Problem). Given an RTG $A = \{N, \Sigma, P, S\}$, a monotonic scoring scheme s, a graph $G = (V, E)$, and two integer parameters m and k. The RTG Network Parse-and-Search Problem is to compute $Max_s^k(T[G, A, m])$.

The problem of deciding whether a given ordered rooted tree is accepted by a given RTG can be solved in polynomial time. However, solving Problem 1 entails extending RTG parsing to handle, as input, a *directed graph* rather than an *ordered rooted tree* and to support a local RTG derivation-tree search rather than just tree parsing.

To handle the exponentially growing search space, and to avoid cycles, we apply iterative randomized color-coding to our search space [2]. Color-coding is a probabilistic approach that bounds the error expectancy by executing multiple graph coloring iterations. In each iteration, every vertex in the queried network is assigned a color chosen randomly out of m colors. The colored network is then searched for colorful subgraphs in which each color appears exactly once. Thus, instead of maintaining an enumeration space of size $\binom{n}{m}$, due to all possible selections of subsets of size m from among n vertices, one can maintain a search space of size 2^m enumerating sets of m distinct colors in considerably lower complexity. In each color-coding trial, the probability to obtain a certain tree of the sought k-best trees is $\frac{m!}{m^m} \ge e^{-m}$ (i.e. the probability that a tree of m vertices is colorful). The following theorem, which is proven in the Appendix, bounds the number of required color-coding iterations.

Theorem 1. *For any $\epsilon > 0$, after $e^m \ln(\frac{k}{\epsilon})$ iterations, the output list contains all the k-best subtrees with error probability $\le \epsilon$.*

We next formalize the work per color-coding iteration, which will be the subject of the next section. Let $C : V \to \{1, \ldots, m\}$ be a coloring function. We define $T^C[G, A, m]$ to be a subset of $T[G, A, m]$, such that for any tree $\tau \in T^C[G, A, m]$, each vertex of τ is distinctly colored according to C (i.e. τ is colorful). Let $Max_s^k(T^C[G, A, m])$ denote the subset of k-best scoring subtrees in $T^C[G, A, m]$ according to s.

Problem 2 (RTG Colored Network Parse-and-Search Problem). Given an RTG $A = \{N, \Sigma, P, S\}$, a coloring function C, a monotonic scoring scheme s, a vertex-labelled graph $G = (V, E)$, and two integer parameters m and k. The RTG Colored Network Parse-and-Search Problem is to compute $Max_s^k(T^C[G, A, m])$.

An overall illustration of our framework for solving Problem 1, by solving Problem 2 within color-coding iterations, is shown in Fig. S2.

Across color-coding trials, we maintain a global sorted list of size k, which stores the trees corresponding to the k-best derivations found so far. After each color-coding iteration, this list is updated with the k-best derivations yielded by that iteration. This is done in $O(km \log k)$ time and $O(m \cdot k)$ space, i.e. without increasing the overall time and space complexities of the algorithm we propose for solving Problem 2, which will be described and analyzed in the next section.

4   Algorithms for Solving Problem 2

In this section we describe a two-stage algorithm for solving Problem 2. We formulate our dynamic programming algorithm as an instance of optimal derivation in the ordered hypergraph framework [13,19,27,29]. In the first stage of our algorithm (Sect. 4.1), the curryfied representations of all the candidate subtrees within the graph G are constructed and encoded in an ordered acyclic hypergraph. Then, in the second stage (Sect. 4.2), the hypergraph is processed again to compute the k-best scoring subtrees in the hypergraph.

4.1   Stage 1: Hypergraph Construction and Its Optimal Derivation

A directed ordered hypergraph is a pair $H = (V_H, E_H)$, where V_H is the set of hypernodes, and E_H is the set of ordered, directed hyperedges. Each hyperedge e is a pair $\langle head(e), tail(e) \rangle$, where $head(e)$ is a hypernode called the *head* of the hyperedge and $tail(e)$ is an ordered pair of hypernodes, called the *tail* of the hyperedge. Denote by $t_i(e)$ the i'th element of the tail of e. Note that the hyperedges in our approach are all binary. Thus, henceforth we write a hyperedge with head h and tail t_1, t_2 as $(h \leftarrow \langle t_1, t_2 \rangle)$.

Given an RTG $A = \{N, \Sigma, P, S\}$, a directed graph $G = (V, E)$, a bound m on the size of the sought trees, and a coloring function C of vertices in the input graph. We define the hypergraph in our framework as follows:

Definition 3. *A hypernode $x \in V_H$ is of the form (W, v, q), where $W \in N$, $v \in V$, $q \subseteq \{1, \ldots, m\}$ and $C(v) \in q$.*

A hypernode $x = (W, v, q)$ represents a class of subtrees T_x from G, such that for each $\tau \in T_x$, (1) τ is rooted in v, (2) τ is composed of vertices with distinct colors from q, and (3) τ matches (in terms of both topology and vertex labels) a tree obtained by consecutive derivations of production-rules from P, that start with the non-terminal W.

Definition 4. *A hyperedge $e \in E_H$ is of the form $(z \leftarrow \langle x, y \rangle)$, such that $x = (X, u, q')$, $y = (Y, v, q'')$, $z = (Z, u, q' \cup q'')$, and the following conditions hold:*

1. $Z \rightarrow X@Y \in P$
2. $(u, v) \in E$
3. $q' \cap q'' = \emptyset$

Figure 1 exemplifies the construction of hyperedges incoming to a specific hypernode (x_1), for a given colored vertex-labelled graph $G = (V, E)$ (Fig. 1.A), and a given RTG $A = \{N, \Sigma, P, S\}$. Two hyperedges (Fig. 1.B, black arrows) are incoming to hypernode x_1, satisfying all three conditions: (1) $(u, v) \in E$, (2) $W \to Y@Z \in P$, and (3) the tail hypernodes contain disjoint sets of colors. The hyperedge $(x_1 \leftarrow \langle x_2, x_5 \rangle)$ is not constructed, since both x_2 and x_5 contain the color pink, in contradiction to condition 3.

We call the set of hyperedges, with x as their *head*, the *backward star* of x, and denote it by $BS(x) = \{e \in E_H | x = head(e)\}$. Let $L_H \subseteq V_H$ denote the set of leaf nodes of H. Each hyperedge $e \in E_H$ and each leaf hypernode $\ell \in L_H$, is associated with weights $c(e)$, $c(\ell)$. The function $c : E_H \cup L_H \to \Re$ could be grammar-based, reflecting a derivation-rule specific scoring, or network-based, reflecting the confidence level of the corresponding edges or vertices in the network, or alternatively some combination of the two strategies. For $e \in E_H$, let $s_c(e)$ denote a monotonic scoring scheme according to c. For each hyperedge $e \in E_H$, $s_c(e)$ computes its score based on the score of the optimal derivation of its tails. The score of the optimal derivation of a hypernode x in an acyclic, directed, binary hypergraph H is denoted $D^*(x)$, defined as:

$$D^*(x) = \begin{cases} c(x) & v \text{ is a leaf} \\ \max_{e \in BS(x)} s_c(e) & otherwise \end{cases}$$

In our applications, s_c is an addition function, defined as follows: $s_c(e) = c(e) + D^*(t_1(e)) + D^*(t_2(e))$.

Based on the above definitions, Stage 1 of our algorithm applies an agglomerative bottom-up construction of the hypergraph, while simultaneously computing $D^*(x)$ for all $x \in H$.

The following theorem gives the time and space complexities for Stage 1 of our algorithm. The proof is given in the Appendix, and is based on some principles from [7,30].

Theorem 2. *The time complexity of Stage 1 of the proposed algorithm for solving Problem 2 is $O(mg|E|3^m)$, and the space complexity is $O(|E_H|) = O(g|E|3^m)$.*

4.2   Stage 2: Computing K-Best Scoring Trees

The hypernodes V_H of the hypergraph H, constructed in stage 1, encode all trees that are accepted by the grammar A, together with the subtrees used to construct them. To extract the k-best scoring trees from H, we extend the k-best optimization algorithm, *FindKBest*, developed by Huang and Chiang [19].

For exemplifications of our next definitions, we refer the reader again to Fig. 1.B. Let $D(x)$ denote the list of the k-best derivations of a hypernode x (gray arrows) and $D^i(x)$ denote the i'th best derivation of x. (For example, in the figure, the leftmost hypernode (x_2) shows two derivations, each one consisting of pink and yellow vertices, while the next hypernode (x_3) has one derivation,

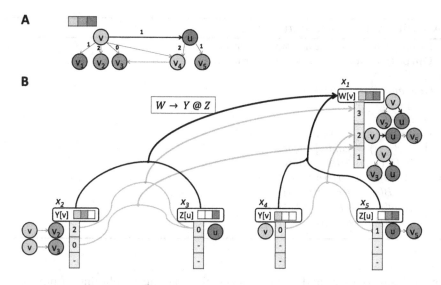

Fig. 1. Example of a step in the construction of the hypergraph H.

consisting of a blue vertex). Each derivation in $D(x)$ is a tuple, $(e, (i, j))$ where $e = x \leftarrow \langle t_1, t_2 \rangle$ and (i, j) is a pair of indices specifying two sub-derivations, $D^i(t_1)$ and $D^j(t_2)$, respectively. Each derivation corresponds to a specific tree in T_x. For example, the third derivation of hypernode x_1 in the figure is marked $(x_1 \leftarrow \langle x_2, x_3 \rangle, (2, 1))$.

To compute $D(x)$ for all hypernodes, Algorithm *FindKBest* (see pseudocode in Algorithm 1) is executed for each hypernode $x \in V_H$ in topological order, starting from the leaves and moving up the hypergraph. The algorithm maintains a priority queue $cand[x]$ of potential candidates to be added to $D(x)$, sorted in decreasing score order. Initially, $cand[x]$ is populated with the k top-scoring derivations among all hyperedges in $BS(x)$. This is achieved by calling the procedure *GetCandidates* (line 2 of the pseudocode). If $|D(x)| < k$, then $cand[x]$ is further populated iteratively. In each such iteration, the top derivation $(e, (i, j))$ in $cand[x]$ is removed from the priority queue and added to $D(x)$. Then, its neighboring derivations, $(e, (i, j + 1))$ and $(e, (i + 1, j))$ are inserted into $cand[x]$. This procedure continues until $|D(x)| = k$ or until all candidate derivations are exhausted.

The correctness of Algorithm *FindKBest* is based on the monotonicity of s_c, by which for any derivation $(e, (i, j))$, the next-best scoring candidate derivation following $(e, (i, j))$ is either $(e, (i, j + 1))$ or $(e, (i + 1, j))$. This is the essential observation behind the *cube pruning* and *cube growing* approaches [16,19] that is crucial to correctly, yet efficiently, enumerate the k-best derivations.

One obstacle encountered in our application is that the trees in G are unordered while the trees defined by RTGs are ordered. This means that any given subtree τ from G could have a one-to-many mapping to several top-scoring derivations in H. These derivations correspond to ordered trees which are isomorphic under Rooted Unordered Isomorphism. (In Rooted Unordered

---

Algorithm 1. $FindKBest(x, k)$

Input: A hypernode x and an integer k.
Output: Vector $D(x)$ of k-best scoring derivations rooted in x.

1 $cand[x] \leftarrow \emptyset$
2 $GetCandidates(x, \alpha mk)$
3 $D(x) \leftarrow \emptyset$
4 $B(x) \leftarrow \emptyset$
5 **while** $|cand[x]| > 0$ **and** $|D(x)| < k$ **do**
6 $\quad$ $(e, (i, j)) = cand[x].dequeue()$
7 $\quad$ **if** $(e, (i, j)) \notin B(x)$ **then**
8 $\quad\quad$ $D(x).insert((e, (i, j)))$
9 $\quad\quad$ $B(x).insert((e, (i, j)))$
10 $\quad$ $cand[x].insert(e, (i + 1, j))$
11 $\quad$ $cand[x].insert(e, (i, j + 1))$

Isomorphism, trees τ_1 and τ_2 are isomorphic if τ_1 can be obtained from τ_2 by subtree reordering operations). As some or all of these isomorphic derivations could be reported among the k-best derivations, this undermines the correctness of the algorithm as a solution to Problem 2.

To prevent this from happening, we assign to each vertex $v \in G$ a unique integer id, and define a unique canonical encoding for each candidate derivation, representing its corresponding subtree as a parentheses-annotated sequence over lexicographic orderings of its vertex ids. This property ensures that any two isomorphic trees have the same canonical encoding (see Fig. S3).

The canonical encoding logic is integrated into Algorithm *FindKBest* (line 7 of the pseudocode), ensuring that, for each hypernode x and for any two isomorphic trees τ_1 and τ_2 that are induced by two distinct derivations and are competing for $D(x)$, only a single, highest-scoring representative is kept. It is implemented as follows. For each hypernode x, we maintain an additional binary-search tree $B(x)$ containing pointers to all elements in $D(x)$, sorted by lexicographic order of their canonical codes. When a new candidate derivation is about to be added to $D(x)$, we first search for its canonical encoding in $B(x)$. If it is found, the new derivation is not added to $D(x)$.

The work per iteration of Algorithm *FindKBest* is computed as follows. Constructing the canonical representation of a subtree can be naively implemented in $O(m)$ time, while operations on $B(x)$ take $O(m \log k)$ time. Moreover, since the candidate subtree is not necessarily added to $D(x)$ (due to the fact that an isomorphic tree could already be in $D(x)$), the number of iterations of the *while* loop in Algorithm *FindKBest* (line 5 of the pseudocode) is no longer limited to k, as was the case in the original algorithm. However, the following observation helps to bound the number of redundant iterations:

Lemma 1. *For hypernode* $x = (X, v, q_x)$, *the number of distinct incoming derivations that may produce the same tree (up to isomorphism) is bounded by* $m \cdot \alpha_X$, *where* α_X *denotes the number of different production rules derived by* X.

This leads to the following theorem which bounds the time and space complexities of Stage 2 of the algorithm.

Theorem 3. *The time complexity of Stage 2 of the proposed algorithm for solving Problem 2 is $O(g \cdot |E| \cdot 3^m + g \cdot |V| \cdot 2^m \cdot km(m \log k + \log \alpha))$, where $\alpha = \max_{X \in N} \alpha_X$. The space complexity is $O(m \cdot k \cdot |V_H|) = O(m \cdot k \cdot |N| \cdot |V| \cdot 2^m)$.*

To obtain the k-best trees which are the final solution to Problem 2, we add to the hypergraph H a pseudo-root hypernode, denoted v^p, such that all hypernodes in H that are marked with the initial symbol S are in the backward star of v^p. After the computation of Stage 2, the k-best list of v^p will contain the k-best trees solving Problem 2.

5   Results

We demonstrate the power of RTG queries in two applications, both of which mine the human protein interaction network for viral infection patterns. The first application exemplifies the expressive power of grammar-based queries. The second application exemplifies the ability to score patterns as part of the grammar.

Our first query aimed to identify temporal responses of human cells to infection by the flu virus Influenza Type A. This is a single-stranded, negative-sense segmented RNA virus that infects hundreds of millions of people and results in numerous hospitalizations and deaths world-wide [6]. The transcriptional response to infection by the virus or its components was measured across time in human primary bronchial epithelial cells [32]. To gain a mechanistic insight into the cellular response to infection, we mapped these data onto a network representing physical interactions between human proteins. We defined a query that searches this network for temporal pathway cascades: Starting with a protein that was up-regulated at the earliest time-point following infection, the cascade proceeds iteratively with one or more proteins that were up-regulated at the same or the subsequent time point, while requiring that the two proteins interact with each other (Fig. 1.A). To prioritize cascades that are biologically relevant, we scored instances based on the sum of transcript levels of the proteins in the cascade and the reliability of their interactions. Using RTGnet we identified the top 1,000 up-regulated cascades ($\epsilon = 0.05$) involving at most nine genes, which required 80,248 iterations of the color-coding procedure.

We demonstrate the mechanistic insight that can be gained from this query by discussing the top-scoring up-regulated cascade (Fig. 2.A). This top-most cascade delineated important defence steps in the host response to influenza infection, starting from detection of viral particles, inhibition of their production, and ending with cell death. The cascade starts with up-regulation of ISG15, an immune-related ubiquitin-like protein that recruits its interacting partners to fight viral infection [35], and two interferon-induced proteins, IFIT1 and IFIT2, which target viral RNA and the synthesis of viral proteins [1]. It follows with up-regulation of MX1, an interferon-induced GTPase with a known antiviral activity, and DDX85 (also known as RIG-I), a receptor activating a cascade of

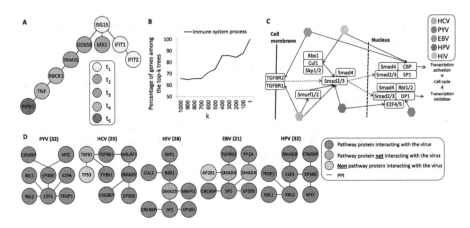

Fig. 2. (A) The maximal-scoring instance of the temporal cascade following influenza infection. Protein colors denote the first time point at which they were up-regulated. (B) The percentage of proteins annotated to 'Immune system process' in the k top-scoring subtrees increases as k is decreased from 1,000 to 1. (C) A schematic view of the TGF-β signaling pathway showing that distinct viruses interact with different proteins in the pathway. (D) Maximal-scoring subtree per virus (and score) in the TGF-β demonstrating the power of an RTG to define various patterns in a single grammar (Color figure online).

antiviral responses, whose conjugation with ISG15 is essential for the initiation of this pathway [3]. Up-regulated at the consecutive time point was TRIM25, forming the DDX58-TRIM25 complex that mediates the DDX58-ISG15 conjugation [14]. Up-regulated at the fourth time point was RBCK1, which regulates the activity of the DDX58-TRIM25 complex [20], and TNF, that together with RBCK1 initiates TNF-mediated gene induction [18]. This is followed by up-regulation of RIPK1, that initiates inflammatory and cell-death cascades upon TNF-receptor signaling [11]. This cascade is especially interesting since several of its proteins interact with the virus: ISG15 and MX1 interact with influenza's NS1 and NP proteins, correspondingly, to inhibit viral replication [17,34], and TRIM25 was found be to be targeted by influenza's NS1 protein to evade recognition of viral RNA by RIG-I [15]. Lastly, the silencing of ISG15, IFIT2, TRIM25 or RBCK1 was found to promote influenza life-cycle and replication [32]. To assess our prioritization scheme we computed the fraction of proteins annotated to the immune system (GO:0002376) [4] across different values of k (Fig. 2.B). This fraction increased as we limited k, and was enriched in the top-ranking subtree relative to other subtrees (p = 0.0125, Fisher's exact test)

Our second query aimed to identify human pathways that a specific virus targets in multiple, related points, as these pathways are likely to be of high importance to its survival and replication. Therefore, we designed a query that accepted subtrees in which several proteins in the pathway interact with each other and with the virus, either directly or via a mediator protein. The accepted subtrees were scored as part of the grammar according to the viral interactions of pathway proteins (see Appendix for details): Proteins with direct interactions scored high (+4), while proteins with indirect interactions scored according to the number of mediator proteins (+1 for each mediator protein). Note that the topology of the subtrees was not predefined. To implement this query we augmented the human protein interaction network described above with data of protein interactions between human and viral proteins for six viruses: Epstein-Barr virus (EBV), hepatitis C virus (HCV), human papillomavirus (HPV), polyoma virus (PYV), human immunodeficiency virus (HIV1), and Influenza A virus subtype H1N1 (H1N1). These viruses were selected because they are evolutionarily remote and their interactions with human proteins were relatively mapped. Associations of human proteins to cellular pathways were extracted from KEGG [21]. As proof-of-concept we focus on the transforming growth factor-β (TGF-β) signaling pathway that regulates cell growth and differentiation, which is crucial for replication of viruses that establish persistent (chronic or latent) infections. Indeed, the only virus without a high-scoring instance was H1N1, a virus that does not establish such persistent infections (Fig. 2.C). Top-scoring instances were identified for all other viruses and had distinct topologies, demonstrating the ability of RTGs to define flexible patterns with a single query (Fig. 2.D). Results pertaining to other pathways are described in the Appendix for lack of space.

The instances we found could have been identified by other types of network querying approaches, though they would require hard-tailoring of the application to ensure that all specifications are followed. RTGnet on the other hand offers a general framework, in which various such queries can be expressed and scored.

6   Discussion

We introduce a novel framework, RTGnet, that extends the class of grammar-based queries that can be efficiently searched within a network to include all languages defined by RTGs, and produces the list of k-best results. We demonstrated the capabilities of RTGnet in two applications that highlight the generality of the tool and the topological flexibility of the identified instances. The RTG search modelling that we suggest is particularly handful in rich networks with many vertex-labels, enabling the design of more expressive grammars. Another advantage of our framework is the ability of k-best optimization to identify a space of near-optimal solutions, making them suitable for learning stochastic production scores, given a reliable source of training data. Future extensions could include the derivation and usage of more informed scoring schemes, more complex grammars to express the queries, and the extension of tree-based instances to general sub-networks.

Acknowledgments. We thank the anonymous WABI referees for their many helpful comments.The work of Ilan Smoly, Amir Carmel and Michal Ziv-Ukelson was partially supported by the Frankel Center for Computer Science at Ben Gurion University of the Negev and by the Israel Science Foundation (ISF 179/14).

References

1. Abbas, Y.M., Pichlmair, A., Górna, M.W., Superti-Furga, G., Nagar, B.: Structural basis for viral 5 [prime]-PPP-RNA recognition by human IFIT proteins. Nature **494**, 60–64 (2013)
2. Alon, N., Yuster, R., Zwick, U.: Color-coding. J. ACM (JACM) **42**(4), 844–856 (1995)
3. Arimoto, K.I., Konishi, H., Shimotohno, K.: Ubch8 regulates ubiquitin and ISG15 conjugation to RIG-I. Mol. Immunol. **45**(4), 1078–1084 (2008)
4. Ashburner, M., Ball, C., Blake, J., et al.: Gene ontology: tool for the unification of biology. The gene ontology consortium database resources of the national center for biotechnology information. Nucleic Acids Res. 34 (2006)
5. Barabási, A.L., Gulbahce, N., Loscalzo, J.: Network medicine: a network-based approach to human disease. Nat. Rev. Genet. **12**(1), 56–68 (2011)
6. Belshe, R.B.: Implications of the emergence of a novel H1 influenza virus. N. Engl. J. Med. **360**(25), 2667–2668 (2009)
7. Bruckner, S., Hüffner, F., Karp, R.M., Shamir, R., Sharan, R.: Topology-free querying of protein interaction networks. J. Comput. Biol. **17**(3), 237–252 (2010)
8. Carme, J., Niehren, J., Tommasi, M.: Querying unranked trees with stepwise tree automata. In: van Oostrom, V. (ed.) RTA 2004. LNCS, vol. 3091, pp. 105–118. Springer, Heidelberg (2004)
9. Comon, H., Dauchet, M., Gilleron, R., Löding, C., Jacquemard, F., Lugiez, D., Tison, S., Tommasi, M.: Tree automata techniques and applications (2007) (release 12 October 2007)
10. Dost, B., Shlomi, T., Gupta, N., Ruppin, E., Bafna, V., Sharan, R.: Qnet: a tool for querying protein interaction networks. J. Comput. Biol. **15**(7), 913–925 (2008)
11. Ea, C.K., Deng, L., Xia, Z.P., Pineda, G., Chen, Z.J.: Activation of IKK by TNFα requires site-specific ubiquitination of RIP1 and polyubiquitin binding by NEMO. Mol. Cell **22**(2), 245–257 (2006)
12. Fan, W., Li, J., Ma, S., Tang, N., Wu, Y.: Adding regular expressions to graph reachability and pattern queries. In: 2011 IEEE 27th International Conference on Data Engineering (ICDE), pp. 39–50. IEEE (2011)
13. Finkelstein, A., Roytberg, M.: Computation of biopolymers: a general approach to different problems. BioSystems **30**(1), 1–19 (1993)
14. Gack, M.U., Shin, Y.C., Joo, C.H., Urano, T., Liang, C., Sun, L., Takeuchi, O., Akira, S., Chen, Z., Inoue, S., et al.: TRIM25 RING-finger E3 ubiquitin ligase is essential for RIG-I-mediated antiviral activity. Nature **446**(7138), 916–920 (2007)
15. Gack, M.U., Albrecht, R.A., Urano, T., Inn, K.S., Huang, I.C., Carnero, E., Farzan, M., Inoue, S., Jung, J.U., García-Sastre, A.: Influenza a virus NS1 targets the ubiquitin ligase TRIM25 to evade recognition by the host viral RNA sensor RIG-I. Cell Host Microbe **5**(5), 439–449 (2009)
16. Gesmundo, A., Henderson, J.: Faster cube pruning. In: IWSLT, pp. 267–274. Citeseer (2010)
17. Guan, R., Ma, L.C., Leonard, P.G., Amer, B.R., Sridharan, H., Zhao, C., Krug, R.M., Montelione, G.T.: Structural basis for the sequence-specific recognition of human ISG15 by the NS1 protein of influenza B virus. Proc. Natl. Acad. Sci. **108**(33), 13468–13473 (2011)
18. Haas, T.L., Emmerich, C.H., Gerlach, B., Schmukle, A.C., Cordier, S.M., Rieser, E., Feltham, R., Vince, J., Warnken, U., Wenger, T., et al.: Recruitment of the linear ubiquitin chain assembly complex stabilizes the TNF-R1 signaling complex and is required for TNF-mediated gene induction. Mol. Cell **36**(5), 831–844 (2009)

19. Huang, L., Chiang, D.: Better k-best parsing. In: Proceedings of the Ninth International Workshop on Parsing Technology, pp. 53–64. Association for Computational Linguistics (2005)

20. Inn, K.S., Gack, M.U., Tokunaga, F., Shi, M., Wong, L.Y., Iwai, K., Jung, J.U.: Linear ubiquitin assembly complex negatively regulates RIG-I-and TRIM25-mediated type I interferon induction. Mol. cell **41**(3), 354–365 (2011)

21. Kanehisa, M., Goto, S., Sato, Y., Kawashima, M., Furumichi, M., Tanabe, M.: Data, information, knowledge and principle: back to metabolism in KEGG. Nucleic Acids Res. **42**(D1), D199–D205 (2014)

22. Koschmieder, A., Leser, U.: Regular path queries on large graphs. In: Ailamaki, A., Bowers, S. (eds.) SSDBM 2012. LNCS, vol. 7338, pp. 177–194. Springer, Heidelberg (2012)

23. Lacroix, V., Fernandes, C.G., Sagot, M.F.: Motif search in graphs: application to metabolic networks. IEEE/ACM Trans. Comput. Biol. Bioinform. **3**(4), 360–368 (2006)

24. Lange, M., Leiß, H.: To CNF or not to CNF? An efficient yet presentable version of the CYK algorithm. Informatica Didactica **8**, 2008–2010 (2009)

25. Martens, W., Niehren, J.: Minimizing tree automata for unranked trees. In: Bierman, G., Koch, C. (eds.) DBPL 2005. LNCS, vol. 3774, pp. 232–246. Springer, Heidelberg (2005)

26. Mendelzon, A.O., Wood, P.T.: Finding regular simple paths in graph databases. SIAM J. Comput. **24**(6), 1235–1258 (1995)

27. Patro, R., Kingsford, C.: Predicting protein interactions via parsimonious network history inference. Bioinformatics **29**(13), i237–i246 (2013)

28. Pinter, R.Y., Rokhlenko, O., Yeger-Lotem, E., Ziv-Ukelson, M.: Alignment of metabolic pathways. Bioinformatics **21**(16), 3401–3408 (2005)

29. Ponty, Y., Saule, C.: A combinatorial framework for designing (pseudoknotted) RNA algorithms. In: Przytycka, T.M., Sagot, M.-F. (eds.) WABI 2011. LNCS, vol. 6833, pp. 250–269. Springer, Heidelberg (2011)

30. Scott, J., Ideker, T., Karp, R.M., Sharan, R.: Efficient algorithms for detecting signaling pathways in protein interaction networks. J. Comput. Biol. **13**(2), 133–144 (2006)

31. Sevon, P., Eronen, L.: Subgraph queries by context-free grammars. J. Integr. Bioinform. **5**(2), 100 (2008)

32. Shapira, S.D., Gat-Viks, I., Shum, B.O., Dricot, A., de Grace, M.M., Wu, L., Gupta, P.B., Hao, T., Silver, S.J., Root, D.E., et al.: A physical and regulatory map of host-influenza interactions reveals pathways in H1N1 infection. Cell **139**(7), 1255–1267 (2009)

33. Shlomi, T., Segal, D., Ruppin, E., Sharan, R.: Qpath: a method for querying pathways in a protein-protein interaction network. BMC Bioinform. **7**(1), 199 (2006)

34. Verhelst, J., Parthoens, E., Schepens, B., Fiers, W., Saelens, X.: Interferon-inducible protein MX1 inhibits influenza virus by interfering with functional viral ribonucleoprotein complex assembly. J. Virol. **86**(24), 13445–13455 (2012)

35. Zhao, C., Denison, C., Huibregtse, J.M., Gygi, S., Krug, R.M.: Human ISG15 conjugation targets both ifn-induced and constitutively expressed proteins functioning in diverse cellular pathways. Proc. Natl. Acad. Sci. USA **102**(29), 10200–10205 (2005)

Orthology Relation and Gene Tree Correction: Complexity Results

Manuel Lafond$^{(\boxtimes)}$ and Nadia El-Mabrouk

Department of Computer Science, Université de Montréal, Montréal, QC, Canada
`lafonman@iro.umontreal.ca`

Abstract. Tree-oriented methods for inferring orthology and paralogy relations between genes are based on reconciling a gene tree with a species tree. On the other hand, many tree-free methods, mainly based on sequence similarity, are also available. The link between orthology relations and gene trees has been formally considered recently from the angle of reconstructing phylogenies from orthology relations. Here, we rather consider this link from a correction point of view. While a gene tree induces a set of relations, the converse is not always true, as a set of relations is not necessarily in agreement with any gene tree. How can we minimally correct an infeasible set of relations? On the other hand, given a gene tree and a set of relations, how to minimally correct a gene tree in order to fit the set of relations? In this paper, various objective functions are considered for the minimality criterion, among them the Robinson-Foulds distance between the initial and corrected gene tree. All considered problem variants are shown to be NP-complete.

1   Introduction

Genes are the molecular units of heredity, holding the information to build and maintain cells. In the course of evolution, they are duplicated, lost, and passed to organisms through speciation. Genes originating from the same ancestral copy are called *homologs*. They are usually inferred from sequence similarity and grouped into *Gene Families*. Two homologous genes are *orthologous* if their parental origin is a speciation, and *paralogous* if it is a duplication. From the orthology conjecture, orthologs tend to be more similar in function than paralogs [29]. This is a major motivation for inferring gene evolution, as it is a prerequisite for functional prediction purposes.

The tree-based method requires to build, classically from a DNA or protein sequence alignment, a phylogenetic tree for the considered gene family. Reconciliation [12] with the species tree then allows to label internal nodes as duplications and speciations, inducing a full orthology and paralogy set of relations between gene pairs. On the other hand, tree-free orthology detection methods are also available. They are based on gene clustering according to sequence similarity, (cf. e.g. the COG database [34], OrthoMCL [24], InParanoid [3], Proteinortho [22]), synteny [20,21] or functional annotation of genes [7]. Only partial sets of relations are usually inferred from these methods.

© Springer-Verlag Berlin Heidelberg 2015
M. Pop and H. Touzet (Eds.): WABI 2015, LNBI 9289, pp. 66–79, 2015.
DOI: 10.1007/978-3-662-48221-6_5

Recent papers have been dedicated to the formal study of the link between trees and orthology/paralogy relations (we just say "relations" in the following) [15,16]. Given a gene family Γ and a set C of pairwise relations, can we reconstruct a labeled gene tree for Γ inducing C? The question can be subdivided into two parts: 1. Is C *satisfiable*, i.e. is there an event-labeled gene tree G in agreement with C? However satisfiability is not sufficient to ensure the possibility for the relation set to reflect a true history, as nodes of G labeled as speciations can be contradictory. This raises the second question; 2. Is there an event-labeled gene tree G which is *S-consistent*, i.e. obtained from reconciliation, with a species tree S? A simple characterization of satisfiability is given in [15] in the case of C being a full set of relations (i.e. each pair of genes of Γ is in C). Moreover, a polynomial-time algorithm can be devised to check for S-consistency [1,17]. In [19], we generalized these results to partial relations.

In this paper we explore the link between relations and trees for the purpose of relation and tree correction. Several gene tree databases from whole genomes are available, including for instance Ensembl Compara [36], Hogenom [30], Phog [8], MetaPHOrs [31], PhylomeDB [18], Panther [26]. However, due to various limitations such as alignment errors, systematic artifacts of inference methods or unsufficient differentiation between sequences, trees are known to contain errors and uncertainties. Consequently, a great deal of effort has been put towards tools for gene tree editing [5,6,9,13,14,33,35]. Most of them are based on selecting, in a neighborhood of an input tree, one best fitting the species tree.

Recently, we developed the first algorithm for gene tree correction using orthology relations [20]. Here we address, from a complexity point of view, the more general problem of correcting a gene tree according to a set of orthology and paralogy relations. Two objective functions are considered: the number of unchanged relations and the number of unchanged clades (the Robinson-Foulds distance [32]). Conversely, we also address the problem of correcting a set of relations so that it represents a valid history in terms of an S-consistent gene tree. Two criteria are considered: maximize the number of unchanged relations, and minimize the number of genes that should be removed for the relation set to be S-consistent. These problems are all shown to be NP-complete.

We introduce the notations and known results in Sect. 2, and show the NP-completeness of two relation correction problems in Sect. 3, namely the Minimum Edge-Removal Consistency and Minimum Node-Removal Consistency problems. In Sect. 4, we then provide analogous complexity results for two gene tree correction problem: the Maximum Homology Correction and the Maximum Clade Correction problems. Algorithmic avenues are discussed in Sect. 5. Due to space constraints, some of the proof have been relegated to Supplementary materials, which can be accessed at http://www-ens.iro.umontreal.ca/~lafonman/en/publications.php.

2   Trees and Orthology Relations

All trees considered in this paper are assumed to be rooted. We also assume that trees have no nodes of degree 2, except possibly the root. Given a set X, a *tree T for X* is a tree whose leafset $\mathcal{L}(T)$ is in bijection with X. We denote by $V(T)$ the set of nodes and by $r(T)$ the root of T. Given an internal node u of T, the subtree rooted at u is denoted T_u and we call the leafset $\mathcal{L}(T_u)$ the *clade of u*. A node u is an *ancestor* of v if u is on the (inclusive) path between v and the root, and we then call v a *descendant* of u. If u and v are connected by an edge of T, then v is a *direct descendant* of u. We denote by $ch(u)$ the set of direct descendants (children) of u. The *lowest common ancestor* (lca) of u and v, denoted $lca_T(u, v)$, is the ancestor common to both nodes that is the most distant from the root. We say that u and v are *separated* iff $lca_T(u, v) \notin \{u, v\}$ (i.e. none is an ancestor of the other). We define $lca_T(U)$ analogously for a set U of nodes. Let L' be a subset of $\mathcal{L}(T)$. The *restriction $T|_{L'}$ of T to L'* is the tree with leaf set L' obtained from the subtree of T rooted as $lca_T(L')$ by removing all leaves that are not in L', and all internal nodes of degree 2, except the root. Let T' be a tree such that $\mathcal{L}(T') = L' \subseteq \mathcal{L}(T)$. We say that T *displays T'* iff $T|_{L'}$ is label-isomorphic to T'.

2.1   Evolution of a Gene Family

Species evolve through *speciation*, which is the separation of one species into distinct ones. A species tree S for a species set Σ represents an ordered set of speciation events that have led to Σ: an internal node is an ancestral species at the moment of a speciation event, and its children are the new descendant species. Inside the species' genomes, genes undergo speciation when the species to which they belong do, but also duplications, and losses (other events such as transfers can happen, but we ignore them here). A *gene family* is a set of genes Γ accompanied by a *mapping function* $s : \Gamma \to \Sigma$ mapping each gene to its corresponding species. The evolutionary history of Γ can be represented as a node-labeled *gene tree* for Γ, where each internal node refers to an ancestral gene at the moment of an event (either speciation or duplication), and is labeled as a speciation (*Spec*) or duplication (*Dup*) accordingly.

   Formally, we call a *DS-tree for Γ* a pair (G, ev_G), where G is a tree with $\mathcal{L}(G) = \Gamma$, and $ev_G : V(G) \setminus \mathcal{L}(G) \to \{Dup, Spec\}$ is a function labeling each internal node of G as a duplication or a speciation node (we drop the G subscript from ev_G when it is clear from the context). Given a species tree S, the *LCA-mapping* function s_G maps each gene, ancestral or extant, to a species as follows: if $g \in \mathcal{L}(G)$, then $s_G(g) = s(g)$; otherwise, $s_G(g) = lca_S(\{s(g') : g' \in \mathcal{L}(G_g)\})$. An example is given in Fig. 1, where the label of each node of G represents its LCA-mapping with respect to S.

   According to the Fitch [11] terminology, we say that two genes x, y of Γ are *orthologous in G* if $ev(lca_G(x, y)) = Spec$, and *paralogous in G* if $ev(lca_G(x, y)) = Dup$. We denote by $\mathcal{O}(G)$, respectively $\mathcal{P}(G)$, the set of all gene pairs that are orthologous, respectively paralogous in G. By $xy \in \mathcal{O}(G)$ we mean $\{x, y\} \in \mathcal{O}(G)$ (the same applies for $\mathcal{P}(G)$). In Fig. 1, $a_1 c_1 \in \mathcal{O}(G)$ while $a_1 b_1 \in \mathcal{P}(G)$. We say that $a_1 c_1$ (respec. $a_1 b_1$) is an orthology (respec. paralogy) relation *induced* by G.

   While a history for Γ can be represented as a *DS-tree*, the converse is not always true, as a *DS-tree G* for Γ does not necessarily represent a valid history.

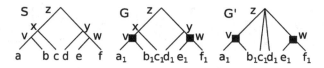

Fig. 1. A species tree S, a binary DS-tree G and a non-binary DS-tree G'. In DS-trees, Dup nodes are indicated by squares, and each leaf α_i denotes a gene belonging to the genome α. G is a refinement of G' such that $\mathcal{O}(G) = \mathcal{O}(G')$ and $\mathcal{P}(G) = \mathcal{P}(G')$.

For this to hold, any speciation node of G should reflect a clustering of species in agreement with S [19]. Formally G should be S-*consistent*, as defined below.

Definition 1. *Let S be a species tree and G be a DS-tree. Let v be an internal node of G such that $ev(v) = Spec$. Then the speciation node v is S-consistent iff for any $v_1, v_2 \in ch(v)$, $s_G(v_1)$ and $s_G(v_2)$ are separated in S.*
We say that G is S-consistent iff every speciation node of G is S-consistent.

Notice that G and S are not required to be binary. In particular, the definition of S-consistency for a speciation node v of G does not require v to be binary, even if S is binary. In this case, one can "refine" v into a set of binary S-consistent speciation nodes based on the topology of S. This operation does not affect the orthology and paralogy relations of by G (see Fig. 1). Duplication nodes can be refined as well. Lemma 1 formalizes this intuition - we leave the proof to the Supplementary materials.

Lemma 1. *Let G be an S-consistent DS-tree for some binary species tree S. Then there is a binary DS-tree G' such that G' is S-consistent, $\mathcal{O}(G) = \mathcal{O}(G')$ and $\mathcal{P}(G) = \mathcal{P}(G')$.*

We can verify that both DS-trees in Fig. 1 are S-consistent. For example, the speciation node in G' has children from species v, c, d and w, which are pairwise separated in S. Notice that, from Definition 1, if G is a DS-tree, then the lca of two leaves of G belonging to the same species must be a duplication node. The converse is not true. For example, in the S-consistent gene tree G of Fig. 1, the parental node of e_1 and f_1 is a duplication node even though e_1 and f_1 belong to two different species.

2.2 Relation Graph

A set of orthology/paralogy relations on Γ (or simply a relation set) is a pair $C = (C_O, C_P)$ of subsets $C_O, C_P \subseteq \binom{\Gamma}{2}$ such that $C_O \cap C_P = \emptyset$ and if $s(x) = s(y)$, then $\{x, y\} \in C_P$. The relation set is said *full* if $C_O \cup C_P = \binom{\Gamma}{2}$. A DS-tree G induces a full set $(\mathcal{O}(G), \mathcal{P}(G))$ of relations.

We adopt the graph representation considered in [19] for full relation sets. A *relation graph* R on a gene family Γ is a graph with vertex set $V(R) = \Gamma$, in which we interpret each edge uv of the edge set $E(R)$ of R as an orthology relation between u and v, and each missing edge (non-edge) $uv \notin E(R)$ as a

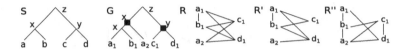

Fig. 2. A species tree S and a DS-tree G. The full orthology set induced by G is represented by the relation graph R. The following graph R' is an example of a not satisfiable graph, as $\{c_1, b_1, d_1, a_2\}$ induces a P_4, while R'' is an example of a satisfiable (it has no induced P_4), but not S-consistent graph.

paralogy relation [1]. Note that if $s(u) = s(v)$, then $uv \notin E(R)$. The relation graph of a DS-tree G, denoted by $R(G)$, is the graph with vertex set $\mathcal{L}(G)$ and edge set $\mathcal{O}(G)$ (for example, see the relation graph R in Fig. 2).

A DS-tree for a gene family Γ leads to a relation graph, but the converse is not always true. A relation graph R is *satisfiable* if there exists a DS-tree G such that $R(G) = R$. The problem of relation graph satisfiability has been addressed in [15]. The following theorem is a reformulation of one of the main results of this paper.

Theorem 1 ([15]). *A relation graph R is satisfiable if and only if R is P_4-free, meaning that no four vertices of R induce a path of length 4.*

For example, in Fig. 2, the relation graphs R and R'' are satisfiable, while the graph R' is not. As a DS-tree does not necessarily represent a true history for Γ (see previous section and Definition 1), satisfiability of a relation graph does not ensure a possible translation in terms of a history for Γ. For this to hold, R should be *consistent* with the species tree, according to the following definition.

Definition 2. *A relation graph R for Γ is S-consistent if and only if R is satisfiable by a DS-tree G which is itself S-consistent.*

For example the graph R in Fig. 2 is S-consistent. Note that S-consistency implies satisfiability. Results from [19] complete the characterization of S-consistent graphs through Theorem 2. A triplet is a binary tree with leaf-set L of size three. For $L = \{x, y, z\}$, we denote by $xy|z$ the unique triplet T on L for which $lca_T(x, y) \neq r(T)$ holds. Now $P_3(R)$ is the subset of triplets of species induced by paths of length 3 in $R = (V, E)$:

$$P_3(R) = \{s(x)s(y)|s(z) : zx, zy \in E \text{ and } xy \notin E \text{ and } s(x) \neq s(y)\}$$

Theorem 2. *Let $R = (V, E)$ be a satisfiable relation graph. Then R is S-consistent if and only if S displays all the triplets of $P_3(R)$.*

Theorem 2 is an immediate consequence of Theorem 5 in [19]. For the sake of completeness, we include the full proof in the Supplementary materials.

---

[1] It has been pointed out to us that the term 'relation graph' is also used in phylogenetics in the form of a generalization of a median network to a set of partitions. To make it clear, relation graphs in this paper have nothing to do with this notion.

As an example, the graph R'' in Fig. 2 is satisfiable but not S-consistent as the path of length 3 containing $\{a_1, b_1, c_1\}$ induces the triplet $ac|b$, while the triplet displayed by S is $ab|c$.

We end this section with additional notations that will be of use later. A *subgraph* H' of H is a graph with $V(H') \subseteq V(H)$ and $E(H') \subseteq E(H)$. For a graph H and some $V' \subseteq V(H)$, the *subgraph of H induced by V'*, denoted $H[V']$, is the subgraph of H with vertex-set V' having the maximum number of edges. We say that H' is an *induced subgraph of H* if there is a subset $V' \subseteq V(H)$ such that $H' = H[V']$. If I is another graph, we say H is I-free if there is no $V' \subseteq V(H)$ such that $H[V']$ is isomorphic to I. Finally, for some edge set $E' \subseteq E(H)$, $H - E'$ is the subgraph H' with $V(H') = V(H)$ and $E(H') = E(H) \setminus E'$.

3 Relation Correction Problems

We raise the issue of leaving out a minimum of information from a relation graph R in order to reach satisfiability or S-consistency. The problem limited to satisfiability reduces to modifying, i.e. adding or removing, a minimum number of edges of R in order to make it P_4-free, which is known to be NP-Hard [25]. In [16], an integer linear programming formulation is used to correct relation graphs of reasonable size.

We first extend the above problem to S-consistency: given a relation graph R and a species tree S, what is the minimum number of edges that need to be modified in order to reach S-consistency? Then, we study the problem of removing as few genes as possible from the gene family in order for the set of relations to be consistent.

3.1 The Minimum Edge-Removal Consistency Problem

Based on the same construction used in paper [10], we show that adding the information on the species tree S does not make the problem of removing the minimum number of edges leading to a P_4-free graph simpler. Although a similar reduction is likely to hold in the general case of edge-modification (removal or insertion) [25], here we focus on edge removal, as this formulation is needed in subsequent developments (Sect. 4). We show the NP-Completeness of this problem, even when every gene from the family Γ comes from a distinct species.

Minimum Edge-Removal Consistency Problem:
Input: A relation graph R for a gene family Γ, a species tree S and an integer k;
Output: An S-consistent subgraph R' of R with $V(R) = V(R')$ such that $|E(R) \setminus E(R')| \leq k$.

Theorem 3. *The* Minimum Edge-Removal Consistency Problem *is NP-Complete, even if for any distinct $g_1, g_2 \in \Gamma$, $s(g_1) \neq s(g_2)$.*

Proof. Given R', Theorem 2 easily translates into a polynomial-time algorithm to verify that R' is S-consistent. It is also clear that verifying if $|E(R) \setminus E(R')| \leq k$

can be done quickly. The problem is therefore in NP. As for the NP-Hardness, the reduction is from the exact 3-cover problem, a classic NP-Hard problem [27]: given a set $W = \{w_1, \ldots, w_{3t}\}$ and a collection $Z = \{Z_1, \ldots, Z_r\}$ of 3-elements of W, does there exists $Z' \subseteq Z$ such that $|Z'| = t$ and Z' is a partition of W ?

Given arbitrary W and Z, we construct R and S by first defining the species set Σ. Let $\alpha = \binom{3t}{2}$ and let $X = \{X_1, \ldots, X_r\}$ and $Y = \{Y_1, \ldots, Y_r\}$ be two collections of all disjoint sets (i.e. for any distinct set $A, B \in X \cup Y$, $A \cap B = \emptyset$), with $|X_i| = \alpha$ and $|Y_i| = r^2\alpha$, for all $1 \le i \le r$. Let $X_\Sigma = \bigcup_{1 \le i \le r} X_i$ and $Y_\Sigma = \bigcup_{1 \le i \le r} Y_i$ be the species in X and Y. Then the species set is $\Sigma = W \cup X_\Sigma \cup Y_\Sigma$. Let S_W, S_X and S_Y be three trees such that $\mathcal{L}(S_W) = W, \mathcal{L}(S_X) = X_\Sigma$ and $\mathcal{L}(S_Y) = Y_\Sigma$. Then S is obtained by first connecting $r(S_Y)$ with $r(S_W)$ to obtain a new tree S_{WY}, then connecting $r(S_{WY})$ with $r(S_X)$ (see Fig. 3). Therefore S has exactly $|\Sigma| = 3t + r(\alpha + r^2\alpha)$ leaves. The gene family Γ is then constructed so that it contains exactly one gene per species, as mentioned in the Theorem statement. In other words the mapping $s : \Gamma \to \Sigma$ is one-to-one. Since s is a bijection, we make no distinction between a gene g and its species $s(g)$. We then define R with $V(R) = \Sigma$ such that each of the sets $W, X_1, \ldots, X_r, Y_1, \ldots, Y_r$ forms an individual clique. Finally we add two edge-sets E_1 and E_2 to R, where $E_1 = \{g_1 g_2 : g_1 \in X_i, g_2 \in Z_i, \text{ for a given } 1 \le i \le r\}$ and $E_2 = \{g_1 g_2 : g_1 \in X_i, g_2 \in Y_i, \text{ for a given } 1 \le i \le r\}$. Then R has $2r + 1$ cliques, namely $W, X_1, \ldots, X_r, Y_1, \ldots, Y_r$. Also, for $1 \le i \le r$, all edges between X_i and Y_i are present, as well as all edges between X_i and Z_i. Figure 3 gives an example with $t = 2$ and $W = \{1, 2, 3, 4, 5, 6\}$.

We show that W and Z admit an exact 3-cover if and only if R admits an S-consistent DS-tree after the deletion of at most $3\alpha(r - t) + (\alpha - 3t)$ edges. Notice that the construction of R described above can clearly be done in polynomial time.

$(\Rightarrow)$: let $Z' \subseteq Z$ be a partition of W, $|Z'| = t$. Let R' be the subgraph of R in which all edges between Z_i and X_i are removed iff $Z_i \notin Z'$ (which removes $3\alpha(r - t)$ edges), and the only edges not removed from the W-clique are those belonging to a Z_i triangle with $Z_i \in Z'$ (which removes $\alpha - 3t$ edges). An example of R' is given in Fig. 3. Thus there are exactly $3\alpha(r - t) + (\alpha - 3t)$ edges of R missing from R', as desired. Clearly, R' is P_4-free and thus satisfiable. To see that R' is S-consistent, we use Theorem 2. Notice that any path of length 3 in R' has the form wx_iy_i with $w \in W, x_i \in X_i$ and $y_i \in Y_i$ for some i, inducing the $wy_i|x_i$ speciation triplet, which is in agreement with S. Therefore there exists an S-consistent gene tree G' satisfying R'.

$(\Leftarrow)$: The construction of R is exactly the same as in Theorem 3 in [10], and the proof is directly applicable to our case. Still, we have included a complete proof in the Supplementary materials.   □

3.2   The Minimum Node-Removal Consistency Problem

Minimum Node-Removal Consistency Problem:
Input: A relation graph R for a gene family Γ, a species tree S and an integer k;
Output: An S-consistent induced subgraph R' of R with $|V(R')| \ge k$.

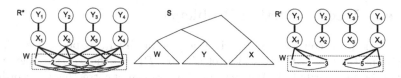

Fig. 3. S represents the species tree and R^* the relation graph constructed from the sets W, Z, X and Y. The illustration is given for $W = \{1, 2, 3, 4, 5, 6\}$ and $Z = \{\{1, 2, 3\}, \{2, 3, 4\}, \{3, 5, 6\}, \{4, 5, 6\}\}$. $Z' = \{\{1, 2, 3\}, \{4, 5, 6\}\}$ is a subset of Z which is a partition of W. R' is the "corrected" relation graph corresponding to Z'.

We use a reduction similar to that in [23], where it was remarkably shown that finding a maximum induced subgraph of some graph H having some property Π is NP-Hard whenever Π is a hereditary property, i.e. applies to any induced subgraph of H. Though it can be shown that S-consistency is indeed hereditary, the reduction assumes H is unlabeled and unconstrained, which is not the case of R .

Theorem 4. *The* Minimum Node-Removal Consistency Problem *is NP-Complete.*

Proof. Again by Theorem 2, verifying that R' is indeed a solution can be done in polynomial time and the problem is thus in NP. The reduction is from the maximum independent set problem. That is, given a graph H, is there an induced subgraph H' of H having at least k nodes such that H' has no edge. Let $n = |V(H)|$. We construct R and S from H as follows: R starts as a copy of H, and for each node x of R, we add a single neighbor x^* (i.e. xx^* is an edge of R and x^* is of degree one). Denote by X the nodes of R originally from H, and by X^* the newly added nodes. Each gene in R is assigned to a distinct species. To construct S, first let S_X be a tree with leafset $s(X)$, and S_{X^*} be a tree with leafset $s(X^*)$. Then S is obtained by connecting $r(S_X)$ and $r(S_{X^*})$ under a common parent. We show that H has an independent set of size at least k if and only if R admits an induced subgraph of size at least $n + k$ that is S-consistent.

Let H' be a solution to the independent set problem with $|V(H')| \geq k$, and let X' be the nodes of X corresponding to $V(H')$. Let $R' = R[X' \cup X^*]$. Now, no two nodes of X' share an edge, and thus the only edges left in R' are of the form xx^*. Therefore, R' is P_3-free and thus, by Theorem 2, is S-consistent. Moreover, $|V(R')| = |X' \cup X^*| \geq k + n$.

Conversely, let R' be an S-consistent induced subgraph of R with $|V(R')| \geq n + k$. Let $W = \{x \in X : x \in V(R')$ and $x^* \in V(R')\}$. We first claim that no two nodes $x, y \in W$ share an edge in R'. For otherwise, x^*xy induce a P_3 with x in the center, inducing the $s(x^*)s(y)|s(x)$ speciation triplet. This contradicts the triplet $s(x)s(y)|s(x^*)$ found in S, and R' is not S-consistent. Therefore, by letting W' denote the nodes of H corresponding to W, we get that $H[W']$ is an independent set. Our final claim is that $|W| \geq k$. Indeed if $|W| < k$, then there are strictly more than $n - k$ node pairs $\{x, x^*\}$ from which at least one of x or x^* is missing in R'. This implies that $|V(R')| < 2n - (n - k) = n + k$, contradicting our initial assumption. □

4   Gene Tree Correction Problems

In this section, we consider we are given a gene family Γ, a species tree S, an S-consistent DS-tree G for Γ, and a set $C = (O, P)$ of orthology/paralogy constraints (not necessarily full). We focus on the problem of correcting G according to C in a minimal way. The goal is thus to find a DS-tree G' inducing C such that the difference between G and G' is minimum. We consider two ways of measuring the difference (or symetrically the similarity) between gene trees, one based on conserved orthology/paralogy relations induced by the two trees, and one based on the number of conserved clades between the two trees, which is the Robinson-Foulds in the case that G, G' and S are all binary trees.

4.1   The Maximum Homology Correction Problem

Maximum Homology Correction Problem :
Input: A species tree S, an S-consistent DS-tree G for a gene family Γ, an integer k, a set O of orthology and a set P of paralogy relations;
Output: An S-consistent DS-tree G' for Γ with $O \subseteq \mathcal{O}(G')$, $P \subseteq \mathcal{P}(G')$ such that $|\mathcal{O}(G) \cap \mathcal{O}(G')| + |\mathcal{P}(G) \cap \mathcal{P}(G')| \geq k$.

Theorem 5. *The* Maximum Homology Correction Problem *is NP-Complete, even if S, G and G' are required to be binary.*

Proof. The problem is clearly in NP, as verifying S-consistency can be done in polynomial time, as well as counting the common orthologs/paralogs relations (the set of relations is quadratic in size). For our reduction, we use the Minimum Edge-Removal Consistency Problem for the case of a gene family with at most one gene per genome, which is NP-Hard by Theorem 3. Given a species tree S, a relation graph R with $V(R)$ in bijection with $\mathcal{L}(S)$ and an integer k, we construct an instance of the Maximum Homology Correction Problem, i.e. a species tree S', a DS-tree G, an orthologous set O and paralogous set P. We show that there is an S-consistent subgraph R' of R obtained by removing at most k edges iff there is an S'-consistent DS-tree G' satisfying O and P with at most $|P| + k$ relations that are not induced by G.

Let $S' = S$ and construct G by mimicking S - that is by first copying S and its leaf labels, then replacing each leaf ℓ of G by the gene $s^{-1}(\ell)$. Note that if S is binary, then so is G. All internal nodes of G are labeled as speciations, so all genes of Γ are pairwise orthologous. Thus $R(G)$ is a clique. Finally, let $O = \emptyset$ and $P = \{g_1 g_2 : g_1 g_2 \notin E(R)\}$. Notice that $R = R(G) - P$.

$\Rightarrow$: Let R' be a solution to the Minimum Edge-Removal Consistency Problem for R and S. Then there exists a S-consistent DS-tree G' satisfying R', which is obtained by deleting at most k edges from R. By Lemma 1, we may assume that if S is binary, then so is G'. Now, since R' has at most $|P| + k$ non-edges, G' has at most $k + |P|$ paralogs and is therefore a solution to the constructed instance of the Maximum Homology Correction Problem that breaks at most $k + |P|$ orthologies.

$\Leftarrow$: Let G' be a solution, binary or not, to the constructed Maximum Homology Correction Problem instance and let $R' = R(G')$. Since G' satisfies P and breaks at most $|P| + k$ orthologies, R' must have P as non-edges, plus at most k other non-edges. Thus R' can be obtained by removing at most k edges from $R(G) - P = R$, as desired. $\square$

4.2   The Maximum Clade Correction Problem

Maximum Clade Correction Problem:
Input: A gene tree G, a species tree S, a set O of orthology and a set P of paralogy relations and an integer k;
Output: An S-consistent DS-tree G' satisfying O and P such that G and G' have at least k clades in common.

Notice that if S, G and G' are required to be binary, the effective measure between G and G' is the Robinson-Foulds distance. This special case is handled as part of the general proof.

Theorem 6. *The* Maximum Clade Correction Problem *is NP-Complete, even if* S, G *and* G' *are required to be binary.*

The proof of Theorem 6 is a bit involving, and due to space constraints we only provide the construction and intuition of the NP-Hardness reduction. The complete proof can be accessed in the supplementary materials.

We use the Minimum Node-Removal Consistency Problem for our reduction. Let R be the input relation graph with $V(R) = \{v_1, \ldots, v_n\}$, S be the species tree and k be an integer. Let $\alpha = n(n-1-k) + 2k$. The constructed instance of the Maximum Clade Correction Problem uses the same species tree S. Construct G as follows: first consider G as a binary tree with n leaves $l_1, \ldots, l_n$, where each leaf l_i is mapped to v_i. Then, replace each leaf l_i by a subtree T_i constructed as follows: T_i is a caterpillar tree with $n - 1 + \alpha$ leaves, and each leaf ℓ of T_i is such that $s(\ell) = s(v_i)$ (a caterpillar tree is a path to which we add a leaf child to each internal node). Let L_i denote the set of the $n - 1$ deepest leaves of T_i (the depth of a leaf ℓ being the number of nodes on the path between ℓ and the root). Each leaf of L_i is mapped to a distinct node of $V(R) \setminus \{v_i\}$. Denote by $\ell_{i,j}$ the leaf of T_i mapped to v_j. Then G has exactly $n(n - 1 + \alpha)$ leaves and $n(n - 1 + \alpha) - 1$ clades (since it is binary). Finally define $O = \{\{\ell_{i,j}, \ell_{j,i}\} : v_i v_j \in E(R)\}$ the set of orthology relations and $P = \{\{\ell_{i,j}, \ell_{j,i}\} : v_i v_j \notin E(R)\}$ the set of paralogy relations. Note that each $\ell_{i,j}$ is present in exactly one relation.

It can be shown that R admits an S-consistent induced subgraph R' with at least k nodes if and only if G, O and P admit an S-consistent DS-tree G' satisfying O and P such that G and G' share at least $k(\alpha + n - 2)$ clades. The idea is that given R', we can construct an S-consistent gene tree H satisfying R. To each leaf v_i of H corresponds a subtree T_i in G. We obtain G^* by replacing each such leaf v_i by its corresponding T_i, which guarantees that the required number of clades were preserved (as there were k such leaves in H). Noting that G^* does not include every gene of G, the difficulty of the proof consists in including every such missing gene whilst satisfying the relations of O and P.

In the other direction, i.e. if we are given a solution G' that preserves enough clades, it can be shown that G' must preserve at least k of the T_i subtrees intact, and restricting G' to these k subtrees, then replacing each such T_i by its corresponding vertex v_i in R, we obtain a gene tree G^* whose relation graph R' is the solution we are looking for.

5   Algorithmic Avenues

As the problems presented in this work are NP-complete, non-polynomial exact algorithms or approximation algorithms avenues should be explored. Let us generalize the Minimum Edge-Removal Consistency Problem to the minimum *editing* problem (i.e. minimzing edge removals and insertions). It is not hard to imagine a branch-and-bound algorithm that solves the problem. Call an induced subgraph H of a relation graph R *bad* if it is either a P_4, or a P_3 in contradiction with S. Each P_4 can be solved by 6 possible edge editings, and each contradictory P_3 can be solved by 3 possible editings. Therefore, in a branch-and-bound process, one would verify if a given graph R' contains a bad subgraph and if so, proceed recursively on each graph obtained by an editing that removes it. If no bad subgraph exists, then R' is a possible solution and its number of editings is retained. If, at any point, R' has had more editings than the best solution encountered so far, the algorithm can stop the recursion. Notice however that an edge should not be edited more than once in order to avoid infinite loops. The idea of this branch-and-bound algorithm can also be applied to the Minimum Node-Removal Consistency problem. It is known that a P_4, if one exists, can be found in linear time [4]. It remains to see if we can find a contradictory P_3 in time better than $O(n^3)$.

As for approximations, an algorithm proposed in [28] can be directly applied to the Minimum Edge-Removal Consistency Problem and guarantees that we do not remove more than $4\Delta(R)$ times more edges than the optimal solution, where $\Delta(R)$ is the maximum degree of R. The idea is simple : as long as R has a bad subgraph H, remove every edge incident to a vertex of H and continue. Even though this is the best known approximation algorithm so far, it has the undesirable effect of isolating many vertices, motivating the exploration of alternative algorithms. One direction would be to consider existing ideas on the problem of satisfiability, i.e. what is the minimum number of editings required to make a graph P_4-free, and adapt them to the consistency problem - for instance the Min-Cut algorithm proposed in [2].

For gene tree correction, we have developed in [19] a polynomial-time algorithm which, given a species tree S and a partial set of relations O and P, verifies if there exists an S-consistent gene tree G' satisfying O and P and if so, constructs one among the set of all possible solutions. It would be interesting to explore the possibility of providing an input gene tree G to the algorithm in order to pick a solution that is close to G (either in terms of common homology relations or clades).

It is also worth mentioning that relations are not always fully known, and instead of a yes or no orthology assignment between two genes, existing methods for orthology prediction can rather motivate a way of assigning a probabilistic score to a given relation [19]. A natural extension to the edge removal/editing problems is therefore to add a weight to each edge and non-edge, so that each insertion/removal has its own weight. The objective then becomes to minimize the total weight of a set of edited edges. Notice that the branch-and-bound algorithm given above can easily be adapted to support weights on editings. This generalization actually encompasses the Maximum Homology Correction problem. Indeed, given a gene tree G and relations O and P to satisfy, one can create a weighted relation graph R in this way: each relation in O (resp. P) is an edge (resp. non-edge) with infinite weight, and each relation in $\mathcal{O}(G) \setminus O$ (resp. $\mathcal{P}(G) \setminus P$) is an edge (resp. non-edge) with a weight of one. Therefore a minimum S-consistent edge-editing of R corresponds to a gene tree G' that satisfies O and P and has a maximum number of common homologies with G.

6   Conclusion

A gene tree induces a set of orthology and paralogy relations between members of a gene family, but the converse is not always true. In this paper we show that attempting to modify a set of relations as least as possible in order to ensure consistency with a species tree leads to the formulation of NP-Complete problems. Moreover, even assuming that the given relations are error-free, it remains computationally difficult to correct a gene tree in order to fit the given set of relations. As various model-free methods are available to infer orthology and paralogy, these correction problems are of practical biological interest. A future direction would be to explore fast approximation algorithms for the relation graph and gene tree editing.

References

1. Aho, A.V., Sagiv, Y., Szymanski, T.G., Ullman, J.D.: Inferring a tree from lowest common ancestors with an application to the optimization of relational expressions. SIAM J. Comp. **10**, 405–421 (1981)
2. Altenhoff, A.M., Gil, M., Gonnet, G.H., Dessimoz, C.: Inferring hierarchical orthologous groups from orthologous gene pairs. PLoS One **8**(1), e53786 (2013)
3. Berglund, A.C., Sjolund, E., Ostlund, G., Sonnhammer, E.L.: InParanoid 6: eukaryotic ortholog clusters with inparalogs. Nucl. Acids Res. **36**, 263–266 (2008)
4. Bretscher, A., Corneil, D.G., Habib, M., Paul, C.: A simple linear time LexBFS cograph recognition algorithm. In: Bodlaender, H.L. (ed.) WG 2003. LNCS, vol. 2880, pp. 119–130. Springer, Heidelberg (2003)
5. Chaudhary, R., Burleigh, J.G., Eulenstein, O.: Efficient error correction algorithms for gene tree reconciliation based on duplication, duplication and loss, and deep coalescence. BMC Bioinf. **13**(Supp. 10), S11 (2011)
6. Chen, K., Durand, D., Farach-Colton, M.: Notung: dating gene duplications using gene family trees. J. Comp. Biol. **7**, 429–447 (2000)

7. The Gene Ontology Consortium: Gene ontology: tool for the unification of biology. Nat. Genet. **25**(1), 25–29 (2000)
8. Datta, R.S., Meacham, C., Samad, B., Neyer, C., Sjölander, K.: Berkeley PHOG: PhyloFacts orthology group prediction web server. Nucl. Acids Res. **37**, W84–W89 (2009)
9. Doroftei, A., El-Mabrouk, N.: Removing noise from gene trees. In: Przytycka, T.M., Sagot, M.-F. (eds.) WABI 2011. LNCS, vol. 6833, pp. 76–91. Springer, Heidelberg (2011)
10. El-Mallah, E.S., Colbourn, C.J.: The complexity of some edge deletion problems. IEEE Trans. Circ. Syst. **35**(3), 354–362 (1988)
11. Fitch, W.M.: Homology a personal view on some of the problems. TIG **16**(5), 227–231 (2000)
12. Goodman, M., Czelusniak, J., Moore, G.W., Romero-Herrera, A.E., Matsuda, G.: Fitting the gene lineage into its species lineage, a parsimony strategy illustrated by cladograms constructed from globin sequences. Syst. Zool. **28**, 132–163 (1979)
13. Gorecki, P., Eulenstein, O.: Algorithms: simultaneous error-correction and rooting for gene tree reconciliation and the gene duplication problem. BMC Bioinf. **13**(Supp 10), S14 (2011)
14. Górecki, P., Eulenstein, O.: A linear time algorithm for error-corrected reconciliation of unrooted gene trees. In: Chen, J., Wang, J., Zelikovsky, A. (eds.) ISBRA 2011. LNCS, vol. 6674, pp. 148–159. Springer, Heidelberg (2011)
15. Hellmuth, M., Hernandez-Rosales, M., Huber, K., Moulton, V., Stadler, P., Wieseke, N.: Orthology relations, symbolic ultrametrics, and cographs. J. Math. Biol. **66**(1–2), 399–420 (2013)
16. Hellmuth, M., Wieseke, N., Lechner, M., Middendorf, M., Stadler, P.F., Lenhof, H-P.: Phylogenomics with paralogs. In: PNAS (2014)
17. Hernandez-Rosales, M., Hellmuth, M., Wieseke, N., Huber, K.T., Moulton, V., Stadler, P.: From event-labeled gene trees to species trees. BMC Bioinf. **13**(Suppl. 19), 56 (2012)
18. Huerta-Cepas, J., Capella-Gutierrez, S., Pryszcz, L.P., Denisov, I., Kormes, D., Marcet-Houben, M., Gabaldón, T.: Phylomedb v3.0: an expanding repository of genome-wide collections of trees, alignments and phylogeny-based orthology and paralogy predictions. Nucl. Acids Res. **39**, D556–D560 (2011)
19. Lafond, M., El-Mabrouk, N.: Orthology and paralogy constraints: satisfiability and consistency. BMC Genom. **15**(Suppl. 6), S12 (2014)
20. Lafond, M., Semeria, M., Swenson, K.M., Tannier, E., El-Mabrouk, N.: Gene tree correction guided by orthology. BMC Bioinf. **14**(Suppl. 15), S5 (2013)
21. Lafond, M., Swenson, K.M., El-Mabrouk, N.: Models and algorithms for genome evolution. In: Chauve, C., El-Mabrouk, N., Tannier, E. (eds.) Error detection and correction of gene trees, pp. 261–285. Springer, Heidelberg (2013)
22. Lechner, M., Findeib, S.S., Steiner, L., Marz, M., Stadler, P.F., Prohaska, S.J.: Proteinortho: detection of (co-)orthologs in large-scale analysis. BMC Bioinf. **12**, 124 (2011)
23. Lewis, J.M., Yannakakis, M.: The node-deletion problem for hereditary properties is np-complete. J. Comput. Syst. Sci. **20**(2), 219–230 (1980)
24. Li Jr, L., Stoeckert, C.J., Roos, D.S.: OrthoMCL: identification of ortholog groups for eukaryotic genomes. Genome Res. **13**, 2178–2189 (2003)
25. Liu, Y., Wang, J., Guo, J., Chen, J.: Cograph editing: complexity and parameterized algorithms. In: Fu, B., Du, D.-Z. (eds.) COCOON 2011. LNCS, vol. 6842, pp. 110–121. Springer, Heidelberg (2011)

26. Mi, H., Muruganujan, A., Thomas, P.D.: Panther in 2013: modeling the evolution of gene function, and other gene attributes, in the context of phylogenetic trees. Nucl. Acids Res. **41**, D377–D386 (2012)

27. Michael, R.G., David, S.J.: Computers and intractability: a guide to the theory of np-completeness. WH Freeman & Co., San Francisco (1979)

28. Natanzon, A., Shamir, R., Sharan, R.: Complexity classification of some edge modification problems. Discrete Appl. Math. **113**(1), 109–128 (2001)

29. Ohno, S.: Evolution by Gene Duplication. Springer, Berlin (1970)

30. Penel, S., Arigon, A.M., Dufayard, J.F., Sertier, A.S., Daubin, V., Duret, L., Gouy, M., Perrière, G.: Databases of homologous gene families for comparative genomics. BMC Bioinf. **10**, S3–S6 (2009)

31. Pryszcz, L.P., Huerta-Cepas, J., Gabaldón, T.: MetaPhOrs: orthology and paralogy predictions from multiple phylogenetic evidence using a consistency-based confidence score. Nucl. Acids Res. **39**, e32 (2011)

32. Robinson, D., Foulds, L.: Comparison of phylogenetic trees. Math. Biosc. **53**, 131–147 (1981)

33. Swenson, K.M., Doroftei, A., El-Mabrouk, N.: Gene tree correction for reconciliation and species tree inference. Alg. Mol. Biol. **7**(1), 31 (2012)

34. Tatusov, R.L., Galperin, M.Y., Natale, D.A., Koonin, E.V.: The COG database: a tool for genome-scale analysis of protein functions and evolution. Nucl. Acids Res. **28**, 33–36 (2000)

35. Nguyen, T.H., Ranwez, V., Pointet, S., Chifolleau, A.M., Doyon, J.P., Berry, V.: Reconciliation and local gene tree rearrangement can be of mutual profit. Alg. Mol. Biol. **8**(8), 12 (2013)

36. Vilella, A.J., Severin, J., Ureta-Vidal, A., Heng, L., Durbin, R., Birney, E.: EnsemblCompara gene trees: complete, duplication-aware phylogenetic trees in vertebrates. Gen. Res. **19**, 327–335 (2009)

Finding a Perfect Phylogeny from Mixed Tumor Samples

Ademir Hujdurović[1,2], Urša Kačar[2], Martin Milanič[1,2(✉)], Bernard Ries[3,4], and Alexandru I. Tomescu[5]

[1] UP IAM, University of Primorska, Koper, Slovenia
[2] UP FAMNIT, University of Primorska, Koper, Slovenia
{ademir.hujdurovic,martin.milanic}@upr.si, ursa.kacar@student.upr.si
[3] PSL, Université Paris-Dauphine, Paris, France
[4] CNRS, LAMSADE UMR 7243, Paris, France
bernard.ries@dauphine.fr
[5] Helsinki Institute for Information Technology HIIT,
Department of Computer Science, University of Helsinki, Helsinki, Finland
tomescu@cs.helsinki.fi

Abstract. Recently, Hajirasouliha and Raphael (WABI 2014) proposed a model for deconvoluting mixed tumor samples measured from a collection of high-throughput sequencing reads. This is related to understanding tumor evolution and critical cancer mutations. In short, their formulation asks to split each row of a binary matrix so that the resulting matrix corresponds to a perfect phylogeny and has the minimum number of rows among all matrices with this property. In this paper we disprove several claims about this problem, including an NP-hardness proof of it. However, we show that the problem is indeed NP-hard, by providing a different proof. We also prove NP-completeness of a variant of this problem proposed in the same paper. On the positive side, we obtain a polynomial time algorithm for matrix instances in which no column is contained in both columns of a pair of conflicting columns.

1 Introduction

Tumor progression is assumed to follow a phylogenetic evolution in which each tumor cell passes its somatic mutations to its daughter cells as it divides, with new mutations being accumulated over time. It is important to discover what tumor types are present in the sample, at what evolutionary stage the tumor is in, or what are the "founder" mutations of the tumor, mutations that trigger an uncontrollable growth of the tumor. These can lead to better understanding of cancer [2,17], better diagnosis, and more targeted therapies [16].

This work was supported in part by the Slovenian Research Agency (I0-0035, research program P1-0285, research projects N1-0032, J1-5433, J1-6720, and J1-6743), by the bilateral project BI-FR/15–16–PROTEUS–003, and by the Academy of Finland, grant 274977.

© Springer-Verlag Berlin Heidelberg 2015
M. Pop and H. Touzet (Eds.): WABI 2015, LNBI 9289, pp. 80–92, 2015.
DOI: 10.1007/978-3-662-48221-6_6

DNA sequencing is one method for discovering the somatic mutations present in each tumor sample. The most accurate possible observation would come from sampling and sequencing every single cell. However, because of single-cell sequencing limitations, and the sheer number of tumor cells, one usually samples populations of cells. Even though the samples are taken spatially and morphologically apart, they can still contain millions of different cancer cells. Moreover, this mixing is not consistent across different collections of samples. Therefore, studying only these mixed samples poses a serious challenge to understanding tumors, their evolution, or their founding mutations.

Solutions for overcoming this limitation can come from a computational approach, as one could deconvolute each sample by exploiting some properties of the tumor progression. One common assumption is that all mutations in the parent cells are passed to the descendants. Another one, called the "infinite sites assumption", postulates that once a mutation occurs at a particular site, it does not occur again at that site. These two assumptions give rise to the so-called *perfect phylogeny* evolutionary model. Hajirasouliha and Raphael proposed in [8] a model for deconvoluting each sample into a set of tumor types so that the *multi-set* of all resulting tumor types forms a perfect phylogeny, and is minimum with this property. Even though this model has some limitations, for example it assumes no errors, and only single nucleotide variant mutations, it is a fundamental problem whose understanding can lead to more practical extensions.

Other major approaches for deconvoluting tumor heterogeneity include methods based on somatic point mutations, such as PyClone [20], SciClone [15], PhyloSub [11], and methods based on somatic copy number alterations, such as THetA [18], TITAN [7] and MixClone [14].

In this paper we show that several claims from [8] about this problem are incorrect, including an NP-hardness proof of it. However, we show that the problem is indeed intractable, by providing a different proof. We also adapt this NP-completeness proof to a variant of the problem also proposed in [8] but whose complexity was left open. This problem asks to minimize the *set* (instead of multi-set) of all tumor types of the perfect phylogeny.

Moreover, we obtain a polynomial time algorithm for a collection of instances of the former problem, which can be biologically characterized as follows. Say that two mutations i and j are *exclusive* if i is present in a sample in which j is absent, and j is present in a sample in which i is absent. Observe that exclusive mutations cannot both be present in the same vertex of a perfect phylogeny. Thus, we say that a sample is a *mixture* at exclusive mutations i and j if both i and j are present in that sample. The instances for which we can solve the problem in polynomial time are such that for any two exclusive mutations i and j, no mutation is present only in the samples mixed at i and j.

Paper outline. In Sect. 2 we give all formal definitions and review the approach of [8]. In Sect. 3 we give a complete characterization of a class of graphs considered in [8]. The complexity results are presented in Sect. 4, and the above-mentioned polynomial time algorithm is given in Sect. 5. Some proofs are omitted due to space limitations.

2   Problem Formulation

As mentioned in the introduction, we assume that we have a set of sequencing reads from each tumor sample, and that based on these reads we have discovered the sample variants with respect to a reference (e.g., by using a somatic mutation caller such as VarScan 2 [13]). This gives rise to an $m \times n$ matrix M whose m rows are the different samples, and whose n columns are the genome loci where a mutation was observed with respect to the reference. The entries of M are either 0 or 1, with 0 indicating the absence of a mutation, and 1 indicating the presence of the mutation. We assume that the matrix has no row whose all entries are 0.

Under ideal conditions, e.g., each mutation was called without errors, and the samples do not contain reads from several leaves of the perfect phylogeny, M corresponds to a perfect phylogeny matrix. Such matrices are characterizable by a simple property, called *conflict-freeness*.

Definition 1. *Two columns i and j of a binary matrix M are said to be in conflict if there exist three rows r, r', r'' of M such that $M_{r,i} = M_{r,j} = 1$, $M_{r',i} = M_{r'',j} = 0$, and $M_{r',j} = M_{r'',i} = 1$. A binary matrix M is said to be conflict-free if no two columns of M are in conflict.*

It is well known that the rows of M are leaves of a perfect phylogenetic tree if and only if M is conflict-free (see [3,6]). Moreover, if this is the case, then the corresponding phylogenetic tree can be retrieved from M in time linear in the size of M [5].

However, in practice, each tumor sample is a mixture of reads from several tumor types, and thus possibly M is not conflict-free. If we are not allowed to edit the entries of M as done e.g. by methods such as [19], [21], Hajirasouliha and Raphael proposed in [8] to turn M into a conflict-free matrix M' by splitting each row r of M into some rows $r_1, \ldots, r_k$ such that r is the bitwise OR of $r_1, \ldots, r_k$; that is, for every column c, $M_{r,c} = 1$ if and only if $M_{r_i,c} = 1$ for at least one r_i. The rows $r_1, \ldots, r_k$ can be seen as the deconvolution of the mixed sample r into samples from single vertices of a perfect phylogeny. One can then build the perfect phylogeny corresponding to M' and carry further downstream analysis. Let us make this row split operation precise.

Definition 2. *Given a binary matrix $M \in \{0,1\}^{m \times n}$ with rows labeled $r_1, r_2, \ldots, r_m$, we say that a binary matrix $M' \in \{0,1\}^{m' \times n}$ is a row split of M if there exists a partition of the set of rows of M' into m sets $R'_1, R'_2, \ldots, R'_m$ such that for all $i \in \{1, 2, \ldots, m\}$, r_i is the bitwise OR of the binary vectors given by the rows of R'_i. The set R'_i of rows of M' is said to be a set of split rows of row r_i.*

Observe that a simple strategy for obtaining a conflict-free row split of M is to split every row r into as many rows as there are 1s in r, with a single 1 per row. While this might be an informative solution for some instances (cf. also Corollary 2 on p. 9), Hajirasouliha and Raphael proposed in [8] as criterion for obtaining a meaningful conflict-free row split M' the requirement that the number of rows of M' is minimum among all conflict-free row splits of M.

In this paper we consider the following problem, which we call MINIMUM CONFLICT-FREE ROW SPLIT problem. For a binary matrix M, we denote by $\overline{\gamma}(M)$ the minimum number of rows in a conflict-free row split M' of M. This notation is in line with notation $\gamma(M)$ used in [8] to denote the minimum number of *additional* rows in a conflict-free row split M' of M, that is, $\gamma(M) = \overline{\gamma}(M) - m$, where m is the number of rows of M.

MINIMUM CONFLICT-FREE ROW SPLIT:
Input: Binary matrix M, an integer k.
Question: Is it true that $\overline{\gamma}(M) \leq k$?

The optimization version of the above problem (in which only a given subset of rows needs to be split) was called the Minimum-Split-Row problem in [8], however, all results from [8] deal with the variant of the problem in which all rows need to be split (some perhaps trivially by setting $R'_i = \{r_i\}$), which is equivalent to the MINIMUM CONFLICT-FREE ROW SPLIT problem.

Given a binary matrix M and a row r of M, the *conflict graph of* (M, r) is the graph $G_{M,r}$ defined as follows: with each entry 1 in r, we associate a vertex in $G_{M,r}$, and two vertices in $G_{M,r}$ are connected by an edge if and only if their corresponding columns in M are in conflict. Denoting by $\chi(G)$ the chromatic number of a graph G, Hajirasouliha and Raphael proved in [8] the following lower bound on the value of $\overline{\gamma}(M)$:

Lemma 1. *[8] Let M be a binary matrix with a conflict-free row split M'. Then, for every row r_i of M with a set R'_i of split rows of M', we have $|R'_i| \geq \chi(G_{M,r_i})$.*

Corollary 1. *For every binary matrix M, we have $\overline{\gamma}(M) \geq \sum_r \chi(G_{M,r})$.*

Hajirasouliha and Raphael also claimed in [8] the following hardness result.

Theorem 1. *[8] The* MINIMUM CONFLICT-FREE ROW SPLIT *problem is NP-hard.*

To recall their approach for proving Theorem 1, we need one more definition. We denote the fact that two graphs G and H are isomorphic by $G \cong H$.

Definition 3. *A graph G is a* row-conflict graph *if there exists a binary matrix M and a row r of M such that $G \cong G_{M,r}$.*

The proof of Theorem 1 was based on a reduction from the chromatic number problem in graphs and relied on three ingredients: the lower bound given by Corollary 1, Theorem 4 from [8] stating that every graph is a row-conflict graph, and an algorithm based on graph coloring, also proposed in [8], for optimally solving the MINIMUM CONFLICT-FREE ROW SPLIT problem by constructing a conflict-free row split of M with exactly $\sum_r \chi(G_{M,r})$ rows. In particular, their results would imply that the lower bound on $\overline{\gamma}(M)$ given by Corollary 1 is always attained with equality.

Contrary to what was claimed in [8], we show that there exist graphs that are not row-conflict graphs. In fact, we give a complete characterization of row-conflict graphs, showing that a graph is a row-conflict graph if and only if its complement is transitively orientable (see Theorem 2). Using a reduction from 3-edge-colorability of cubic graphs, we show that it is NP-complete to test whether a given binary matrix M has a conflict-free row split M' with a number of rows achieving the lower bound given by Corollary 1 (see Theorem 3). This implies that there exist infinitely many matrices for which this bound is not achieved.

A corollary of our characterization of row-conflict graphs is that the chromatic number is polynomially computable for this class of graphs. This fact with the assumption that $P \neq NP$, as well as the existence of matrices M with $\overline{\gamma}(M) > \sum_r \chi(G_{M,r})$, each individually imply that the claimed NP-hardness proof of the MINIMUM CONFLICT-FREE ROW SPLIT problem given in [8] is flawed. Nevertheless, our NP-completeness proof (see Theorem 3) implies that Theorem 1 is correct.

On the positive side, we give a polynomial time algorithm for the MINIMUM CONFLICT-FREE ROW SPLIT problem on input matrices M in which no column is contained in both columns of a pair of conflicting columns (see Theorem 5).

We also consider a variant of the problem, also proposed in [8], in which we are only interested in minimizing the number of *distinct* rows in a conflict-free row split of M. This problem is similar to the Minimum Perfect Phylogeny Haplotyping problem [1], in which we need to explain a set of genotypes with a minimum number of haplotypes admitting a perfect phylogeny. For a binary matrix M, we denote by $\overline{\eta}(M)$ the minimum number of *distinct* rows in a conflict-free row split M' of M. We establish NP-completeness of the following problem (see Theorem 4), which was left open in [8].

MINIMUM DISTINCT CONFLICT-FREE ROW SPLIT:
Input: Binary matrix M, an integer k.
Question: Is it true that $\overline{\eta}(M) \leq k$?

3   A Characterization of Row-Conflict Graphs

Definition 4. *Given a binary matrix M and two columns i and j of M, column i is said to be* contained in *column j if $M_{k,i} \leq M_{k,j}$ holds for every k. The* undirected containment graph H_M *is the undirected graph whose vertices correspond to the columns of M and in which two vertices i and j, $i \neq j$, are adjacent if and only if the column corresponding to vertex i is contained in the column corresponding to vertex j or vice-versa.*

Recall that an *orientation* of an undirected graph $G = (V, E)$ is a directed graph $D = (V, A)$ such that for every edge $uv \in E$, either $(u, v) \in A$ or $(v, u) \in A$, but not both. An orientation is said to be *transitive* if the presence of the directed edges (u, v) and (v, w) implies the presence of the directed edge (u, w). A graph is said to be *transitively orientable* if it has a transitive orientation.

The *complement* of a graph G is a graph $\overline{G}$ with the same vertex set as G in which two distinct vertices are adjacent if and only if they are non-adjacent in G. Transitively orientable graphs appeared in the literature under the name of *comparability graphs* (and their complements under the name of *co-comparability graphs*). Transitively orientable graphs and their complements form a subclass of the well known class of perfect graphs [4]. Therefore, odd cycles of length at least 5 and their complements are examples of graphs that are not transitively orientable.

Observation 1. *For every binary matrix M, the graph H_M is transitively orientable.*

Proof. We say that column i is *properly contained in* column j if i is contained in j and $M_{k,i} < M_{k,j}$ for some k. Fix an ordering $\{c_1, \ldots, c_n\}$ of the columns of M. Let us define a binary relation $\sqsubset$ on the set of columns on M by setting, for every two columns c_i and c_j of M, $c_i \sqsubset c_j$ if and only if either c_i is properly contained in c_j, or $i < j$ and each of c_i and c_j is contained in the other one (that is, as binary vectors they are the same). Observe that for a pair of columns c_i and c_j with $c_i c_j \in E(H_M)$ we have either $c_i \sqsubset c_j$ or $c_j \sqsubset c_i$ but not both. The binary relation $\sqsubset$ defines an orientation of H_M, by orienting each edge $c_i c_j$ as going from c_i to c_j if and only if $c_i \sqsubset c_j$. This orientation can be easily verified to be transitive. $\qquad\square$

In the next theorem, we characterize row-conflict graphs (cf. Definition 3).

Theorem 2. *A graph G is a row-conflict graph if and only if $\overline{G}$ is transitively orientable.*

Proof. ($\Rightarrow$) Let M be an arbitrary binary matrix, r an arbitrary row of M, and let $G = G_{M,r}$. Let N be the submatrix of M consisting of the columns of M that have 1 in row r. It is now easy to see that $G_{M,r} \cong G_{N,r}$. Moreover, any two columns of N are either in conflict or their corresponding vertices are adjacent in H_N. Therefore, $H_N \cong \overline{G_{N,r}}$. Since H_N is transitively orientable (by Observation 1), it follows that $\overline{G}$ is transitively orientable as well.

($\Leftarrow$) The reverse implication follows from the proof of Theorem 4 in [8] (which works for complements of transitively orientable graphs). $\qquad\square$

Theorem 2 implies that odd cycles of length at least 5 and their complements are not row-conflict graphs. The reader not familiar with transitively orientable graphs might find it useful to verify that the cycle of length 5 cannot be transitively oriented.

4   Complexity Results

Theorem 3. *The following two problems are NP-complete:*

- *The* MINIMUM CONFLICT-FREE ROW SPLIT *problem.*
- *Given a binary matrix M, is it true that $\overline{\gamma}(M) = \sum_r \chi(G_{M,r})$?*

Proof. The MINIMUM CONFLICT-FREE ROW SPLIT problem is in NP, since testing if a given binary matrix M' with at most k rows, equipped with a partition of its rows into m sets, satisfies the condition in the definition of a row split, as well as the conflict-freeness, can be done in polynomial time. To argue that the second problem is in NP, we proceed similarly as above, performing an additional test checking that the number of rows of M' equals $\sum_r \chi(G_{M,r})$. (In this case, we will have $\overline{\gamma}(M) \leq \sum_r \chi(G_{M,r})$ and equality will follow from Corollary 1.) The value of $\sum_r \chi(G_{M,r})$ can be computed in polynomial time, since each graph $G_{M,r}$ is the complement of a transitively orientable graph (by Theorem 2), and the chromatic number of complements of transitively orientable graphs can be computed in polynomial time (see, e.g., [4]).

   We prove hardness of both problems at once, making a reduction from the following NP-complete problem [9]: Given a simple cubic graph $G = (V, E)$, is G 3-edge-colorable? (A graph is *cubic*, or *3-regular*, if every vertex is incident with precisely three edges. A *matching* in a graph is a set of pairwise disjoint edges. A graph is *3-edge-colorable* if its edge set can be partitioned into 3 matchings.)

   Given a simple cubic graph $G = (V, E)$, we construct an instance (M, k) of the MINIMUM CONFLICT-FREE ROW SPLIT problem as follows:

- M is a $(|V|+3) \times (|E|+3)$ binary matrix, with rows indexed by $V \cup \{r_1, r_2, r_3\}$, columns indexed by $E \cup \{c_1, c_2, c_3\}$, and entries defined as follows (see Fig. 1 for an example):
  - For every row indexed by a vertex $v \in V$ and every column indexed by an edge e, we have

$$M_{v,e} = \begin{cases} 1, \text{ if } v \text{ is an endpoint of } e; \\ 0, \text{ otherwise.} \end{cases}$$

  - For every row indexed by a vertex $v \in V$ and every column indexed by some $c \in \{c_1, c_2, c_3\}$, we have $M_{v,c} = 1$.
  - For every row indexed by some $r \in \{r_1, r_2, r_3\}$ and every column indexed by an edge $e \in E$, we have $M_{r,e} = 0$.
  - For every row indexed by some $r_i \in \{r_1, r_2, r_3\}$ and every column indexed by some $c_j \in \{c_1, c_2, c_3\}$, we have

$$M_{r_i,c_j} = \begin{cases} 1, \text{ if } i = j \\ 0, \text{ otherwise.} \end{cases}$$

- $k = 3|V| + 3$.

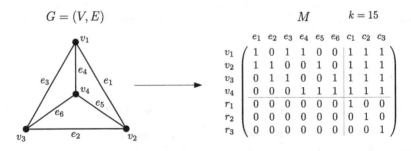

Fig. 1. An example construction of (M, k) from G.

Note that for each row indexed by a vertex $v \in V$, the graph $G_{M,v}$ is isomorphic to the disjoint union of two complete graphs with three vertices each, hence $\chi(G_{M,v}) = 3$. For each row indexed by some $r \in \{r_1, r_2, r_3\}$, the graph $G_{M,r}$ consists in a single vertex, thus $\chi(G_{M,r}) = 1$. It follows that $k = \sum_r \chi(G_{M,r})$ and therefore M is a yes instance to the second problem ("Given a binary matrix M, is $\overline{\gamma}(M) = \sum_r \chi(G_{M,r})$?") if and only if (M, k) is a yes instance for the MINIMUM CONFLICT-FREE ROW SPLIT problem. Hardness of both problems follows from the following claim (the proof of which is omitted due to space limitations): G is 3-edge-colorable if and only if $\overline{\gamma}(M) \leq k$.   □

Hajirasouliha and Raphael proposed in [8] an algorithm based on graph coloring for optimally solving the MINIMUM CONFLICT-FREE ROW SPLIT problem by constructing a conflict-free row split of M with exactly $\sum_r \chi(G_{M,r})$ rows. Since there are infinitely many cubic graphs that are not 3-edge-colorable (see, e.g., [10]), the proof of Theorem 3 implies that there exist infinitely many matrices M such that $\overline{\gamma}(M) > \sum_r \chi(G_{M,r})$. On such instances, the algorithm from [8] will not produce a valid (that is, conflict-free) solution.

Since the smallest cubic 4-edge-chromatic graph is the Petersen graph, the smallest matrix M with $\overline{\gamma}(M) > \sum_r \chi(G_{M,r})$ that can be obtained using the construction given in the proof of Theorem 3 is of order 13×18. A smaller matrix M for which the bound from Corollary 1 is not tight can be obtained by applying a similar construction starting from the complete graph of order 3 (which is a 2-regular 3-edge-chromatic graph):

$$M = \left(\begin{array}{ccc|cc} 1 & 1 & 0 & 1 & 1 \\ 1 & 0 & 1 & 1 & 1 \\ 0 & 1 & 1 & 1 & 1 \\ \hline 0 & 0 & 0 & 1 & 0 \\ 0 & 0 & 0 & 0 & 1 \end{array} \right) .$$

We leave it as an exercise for the reader to verify that $\sum_r \chi(G_{M,r}) = 8$ and $\overline{\gamma}(M) \geq 9$ (in fact, $\overline{\gamma}(M) = 9$). Let us also remark that in [12, Section 4.2.1] a binary matrix M is given with $\overline{\gamma}(M) = \sum_r \chi(G_{M,r})$, on which the algorithm from [8] fails to produce a conflict-free solution.

We conclude this section with another hardness result.

Theorem 4. *The* MINIMUM DISTINCT CONFLICT-FREE ROW SPLIT *problem is NP-complete.*

Proof. Membership in NP of the MINIMUM DISTINCT CONFLICT-FREE ROW SPLIT problem can be argued similarly as for the MINIMUM CONFLICT-FREE ROW SPLIT problem. It suffices to argue that there is a polynomially-sized conflict-free matrix M' such that M' is a row split of M with at most k distinct rows. We may assume that for a partition $R'_1, \ldots, R'_m$ of rows of M' into m sets satisfying the condition in the definition of a row split, the rows within each R'_i are pairwise distinct. Recall from e.g. [6] that a conflict-free matrix N with d distinct rows and n columns corresponds to a perfect phylogenetic rooted tree T such that: T has d leaves (the rows of the matrix), all internal vertices of T

are branching, and all edges from a vertex to its children are injectively labeled with a column of N, with the exception of at most one edge per node that is unlabeled. Thus T has and at most $2n$ edges, and we infer that $d \leq 2n$. Therefore, the total number of rows of M' does not exceed $2nm$, where m and n are the numbers of rows and columns of M, respectively.

The hardness proof is based on a slight modification of the reduction used in the proof of Theorem 3. (See Fig. 2 for an example.) Given a cubic graph $G = (V, E)$, we map it to $(\overline{M}, \overline{k})$ where

- $\overline{M}$ is the binary matrix obtained from the binary matrix M described in the proof of Theorem 3 by adding to it three columns d_1, d_2, d_3, which on the rows indexed by V equal 0, and on the rows indexed by r_1, r_2, r_3, each d_i equals c_i, $i \in \{1, 2, 3\}$.
- $\overline{k} = |E| + 3$.

We claim that $(\overline{M}, \overline{k})$ is an instance of the MINIMUM DISTINCT CONFLICT-FREE ROW SPLIT problem such that G is 3-edge-colorable if and only if $\overline{\eta}(\overline{M}) \leq \overline{k}$. The proof of the claim is omitted due to space limitations. □

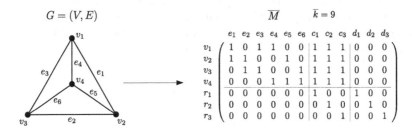

Fig. 2. An example construction of $(\overline{M}, \overline{k})$ from G.

5 An Algorithm for the Case When No Column Is Contained in both Columns of a Pair of Conflicting Columns

In this section we consider the binary matrices in which no column is contained in both columns of a pair of conflicting columns, and derive a polynomial time algorithm for the MINIMUM CONFLICT-FREE ROW SPLIT problem on such matrices. The main idea behind the algorithm is the fact that on such matrices the lower bound from Corollary 1 is achieved, and the bound can be expressed in terms of parameters of a set of derived digraphs, the so-called directed containment graph (see Definition 5 below).

Let M be a binary matrix such that no column of M is contained in two or more conflicting columns. If there are duplicated columns in M, then we form a new matrix where we take just one copy of the columns that are duplicated. Since an optimal solution of the reduced instance can be mapped to an optimal solution of the original instance (by duplicating the columns corresponding to the copies of the duplicated columns in M kept by the reduction), we may assume that there are no duplicated columns in M.

Definition 5. *Given a binary matrix M with distinct columns $c_1, \ldots, c_n$ and a row r of M, the* directed containment graph of (M, r) *is the graph $\overrightarrow{H}_{M,r}$ whose vertex set is the set of columns of M having a 1 in row r, in which there is a directed edge from c_i to c_j if and only if $i \neq j$ and c_i is contained in c_j.*

We will use the notation $c_i \sqsubset_r c_j$ as a shorthand for the fact that (c_i, c_j) is a directed edge of $\overrightarrow{H}_{M,r}$. We say that c_i is a *source* of $\overrightarrow{H}_{M,r}$ if $c_i \in V(\overrightarrow{H}_{M,r})$ and there is no c_j with $c_j \sqsubset_r c_i$. Let $\sigma(M, r)$ denote the number of sources in $\overrightarrow{H}_{M,r}$.

Lemma 2. *If there are no duplicated columns in M, then $\sigma(M, r) \leq \chi(G_{M,r})$ holds for any row r of M.*

Proof. Observe that the underlying undirected graph of $\overrightarrow{H}_{M,r}$ is equal to the complement of $G_{M,r}$. The set of all sources of $\overrightarrow{H}_{M,r}$ forms an independent set in its underlying undirected graph. This set corresponds to a clique in $G_{M,r}$. Therefore $\sigma(M, r) \leq \omega(G_{M,r}) \leq \chi(G_{M,r})$ (where $\omega(G_{M,r})$ denotes the maximum size of a clique in $G_{M,r}$). □

$$M = \begin{array}{c} \\ \\ r \\ \\ \\ \\ \end{array}\!\!\begin{array}{c} \begin{array}{ccccc} c_1 & c_2 & c_3 & c_4 & c_5 \end{array} \\ \begin{pmatrix} 1 & 0 & 0 & 0 & 0 \\ 1 & 1 & 1 & 1 & 0 \\ 0 & 1 & 0 & 1 & 0 \\ 0 & 1 & 1 & 0 & 1 \\ 0 & 1 & 0 & 0 & 0 \end{pmatrix} \end{array} \qquad M' \quad \begin{array}{c|ccccc} & c_1' & c_2' & c_3' & c_4' & c_5' \\ r_1' & 1 & 0 & 0 & 0 & 0 \\ r_2' & 0 & 1 & 1 & 0 & 0 \\ r_3' & 0 & 1 & 0 & 1 & 0 \end{array}$$

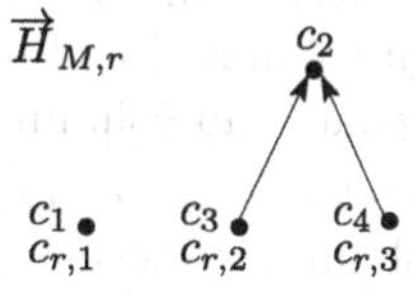

Fig. 3. An example of a matrix M in which no column is contained in both columns of a pair of conflicting columns (c_1, c_2 and c_3, c_4 are conflicting). The rows r_1', r_2', r_3' constructed by the algorithm corresponding to row r of M are shown in the center. On the right, the directed containment graph of (M, r).

Our algorithm is the following one (see also Fig. 3 for an example).

Input: An $m \times n$ binary matrix M with columns $c_1, c_2, \ldots, c_n$, without duplicated columns, and such that no column of M is contained in a pair of conflicting columns.

Algorithm:

Define a new matrix M' with columns $c_1', c_2', \ldots, c_n'$;

For each row r of M, add the rows $r_1', \ldots, r_{\sigma(M,r)}'$ to M', defined as:

let $c_{r,1}, \ldots, c_{r,\sigma(M,r)}$ denote the sources of $\overrightarrow{H}_{M,r}$;

$$M'_{r_i', c_j'} = \begin{cases} 1, \text{ if } c_{r,i} = c_j \text{ or } c_{r,i} \sqsubset_r c_j; \\ 0, \text{ otherwise}; \end{cases}$$

Return M'.

Theorem 5. *For any $m \times n$ binary matrix M such that no column of M is contained in a pair of conflicting columns, it holds that $\overline{\gamma}(M) = \sum_r \chi(G_{M,r}) = \sum_r \sigma(M, r)$. Moreover, a conflict-free row split M' of M with $\overline{\gamma}(M)$ rows can be constructed in polynomial time.*

Proof. As argued above, we may assume that M has no duplicated columns. We claim that the matrix M' produced by the above algorithm is a conflict-free row split of M with a number of rows equal to $\overline{\gamma}(M)$.

It is clear that M' is a row split of M. Let us prove that M' is conflict-free. Suppose the contrary, that is, let c'_i and c'_j be two columns of M' which are in conflict. Then, there exists a row r'_k of M' (obtained by splitting a row r of M) which has 1 in columns c'_i and c'_j.

We will first show that c_i is contained in c_j or viceversa. Assume for a contradiction that this is not the case, and suppose first that $c_{r,k} \notin \{c_i, c_j\}$. Since r'_k has 1 in columns c'_i and c'_j it follows that $c_{r,k} \sqsubset_r c_i$ and $c_{r,k} \sqsubset_r c_j$. This implies that column $c_{r,k}$ is contained in both column c_i and column c_j. By the assumption on M, c_i and c_j cannot be in conflict, hence, one of them is contained in the other one. This violates our first assumption, and thus $c_{r,k} \in \{c_i, c_j\}$. If $c_{r,k} = c_i$ (resp. $c_{r,k} = c_j$) then $c_i \sqsubset_r c_j$ (resp. $c_j \sqsubset_r c_i$) and therefore column c_i is contained in column c_j (resp. c_j is contained in c_i).

Thus, we may assume without loss of generality that c_i is contained in c_j. Since c'_i and c'_j are in conflict it follows that there exists a row w'_ℓ of M' which has 1 in column c'_i and 0 in column c'_j. This implies that the corresponding row w of M has 1 in column c_i, and consequently also in c_j, since c_i is contained in c_j. Therefore, both c_i and c_j are vertices of $\overrightarrow{H}_{M,w}$. If $c_i = c_{w,\ell}$, then w'_ℓ has value 1 in column c'_j (since c_i is contained in c_j), which contradicts the choice of w'_ℓ. Thus, $c_i \neq c_{w,\ell}$ and $c_{w,\ell} \sqsubset_w c_i$. However, since c_i is contained in c_j and $\overrightarrow{H}_{M,w}$ is transitive, it follows that $c_{w,\ell} \sqsubset_w c_j$. This implies that row w'_ℓ has value 1 in column c'_j, which again contradicts the choice of w'_ℓ. This finally shows that M' is conflict-free.

Since the number of rows in M' is $\sum_r \sigma(M, r)$ and M' is conflict free, we have $\overline{\gamma}(M) \leq \sum_r \sigma(M, r)$. By Corollary 1 and Lemma 2 we have $\sum_r \sigma(M, r) \leq \sum_r \chi(G_{M,r}) \leq \overline{\gamma}(M)$. This implies equality. Clearly, the algorithm can be implemented to run in polynomial time. $\qquad\square$

Observe that when the input matrix satisfies the stronger property that no column is contained in another one, Theorem 5 implies that the naive solution obtained by splitting each row r into as many 1s as it contains always produces an optimal solution. This is true since all vertices of $\overrightarrow{H}_{M,r}$ are sources. We obtain:

Corollary 2. *For any binary $m \times n$ matrix M such that no column of M is contained in another one, it holds that $\overline{\gamma}(M) = m'$, where m' equals the number of 1s in M. Moreover, a conflict-free row split M' of M of size $m' \times n$ can be constructed in time $O(m'n)$.*

6   Discussion

In this paper we gave a polynomial time algorithm for instances of the MIN-IMUM CONFLICT-FREE ROW SPLIT problem where no column is contained in both columns of a pair of conflicting columns. It remains to be verified if real instances satisfy this property. More general tractable instances could be found by inspecting further dependencies between column containment and conflict-ness. For example, it remains open whether the MINIMUM CONFLICT-FREE ROW SPLIT problem is tractable on matrices in which no pair of conflicting columns is contained in both columns of a pair of conflicting columns. It is also interesting to identify polynomially solvable cases of the MINIMUM DISTINCT CONFLICT-FREE ROW SPLIT problem and to explore variations of the problems in which we are also allowed to edit the entries of the input matrix. Finally, observe that in [8] it was assumed that the matrices have no duplicated columns, which was not necessary in this paper.

Note added in proof. After the final version of this paper was due, it has come to the authors' knowledge that Hajirasouliha and Raphael mentioned during their WABI 2014 talk that their claim about every graph being a row-conflict graph (Theorem 4 in [8]) contained a flaw and proposed a correction stating that for every binary matrix M with an all-zeros row and an all-ones row, the complement of $G_{M,r}$ (for any row r of M) is transitively orientable (cf. Theorem 2 above).

References

1. Bafna, V., et al.: A note on efficient computation of haplotypes via perfect phylogeny. J. Comput. Biol. **11**(5), 858–866 (2004). http://dx.doi.org/10.1089/cmb.2004.11.858
2. Campbell, P.J., et al.: Subclonal phylogenetic structures in cancer revealed by ultra-deep sequencing. Proc. Natl. Acad. Sci. **105**(35), 13081–13086 (2008). http://dx.doi.org/10.1073/pnas.0801523105
3. Estabrook, G.F., et al.: An idealized concept of the true cladistic character. Math. Biosci. **23**(3–4), 263–272 (1975)
4. Golumbic, M.C.: Algorithmic Graph Theory and Perfect Graphs. Annals of Discrete Mathematics, vol. 57, 2nd edn. Elsevier Science BV, Amsterdam (2004)
5. Gusfield, D.: Efficient algorithms for inferring evolutionary trees. Networks **21**(1), 19–28 (1991)
6. Gusfield, D.: Algorithms on Strings, Trees and Sequences: Computer Science and Computational Biology. Cambridge University Press, New York (1997)
7. Ha, G., et al.: Titan: inference of copy number architectures in clonal cell populations from tumor whole-genome sequence data. Genome Res. **24**(11), 1881–1893 (2014). http://genome.cshlp.org/content/24/11/1881.abstract
8. Hajirasouliha, I., Raphael, B.J.: Reconstructing mutational history in multiply sampled tumors using perfect phylogeny mixtures. In: Brown, D., Morgenstern, B. (eds.) WABI 2014. LNCS, vol. 8701, pp. 354–367. Springer, Heidelberg (2014). http://dx.doi.org/10.1007/978-3-662-44753-6_27

9. Holyer, I.: The NP-completeness of edge-coloring. SIAM J. Comput. **10**(4), 718–720 (1981). http://dx.doi.org/10.1137/0210055

10. Isaacs, R.: Infinite families of nontrivial trivalent graphs which are not Tait colorable. Amer. Math. Monthly **82**, 221–239 (1975)

11. Jiao, W., et al.: Inferring clonal evolution of tumors from single nucleotide somatic mutations. BMC Bioinform. **15**, 35 (2014)

12. Kačar, U.: Problemi popolne filogenije (Perfect Phylogeny Problems). Final project paper. University of Primorska, Faculty of Mathematics, Natural Sciences and Information Technologies, Koper, Slovenia (2015). http://www.famnit.upr.si/sl/izobrazevanje/zakljucna_dela/view/276

13. Koboldt, D.C., et al.: VarScan 2: somatic mutation and copy number alteration discovery in cancer by exome sequencing. Genome Res. **22**, 568–576 (2012). http://genome.cshlp.org/content/early/2012/02/02/gr.129684.111.abstract

14. Li, Y., Xie, X.: Mixclone: a mixture model for inferring tumor subclonal populations. BMC Genomics **16**(S–2), S1 (2015). http://dx.doi.org/10.1186/1471-2164-16-S2-S1

15. Miller, C.A., et al.: SciClone: inferring clonal architecture and tracking the spatial and temporal patterns of tumor evolution. PLoS Comput. Biol. **10**(8), e1003665+ (2014). http://dx.doi.org/10.1371/journal.pcbi.1003665

16. Newburger, D.E., et al.: Genome evolution during progression to breast cancer. Genome Res. **23**(7), 1097–1108 (2013). http://dx.doi.org/10.1101/gr.151670.112

17. Nik-Zainal, S., et al.: The life history of 21 breast cancers. Cell **149**(5), 994–1007 (2012). http://dx.doi.org/10.1016/j.cell.2012.04.023

18. Oesper, L., et al.: THetA: inferring intra-tumor heterogeneity from high-throughput DNA sequencing data. Genome Biol. **14**(7), R80 (2013). http://dx.doi.org/10.1186/gb-2013-14-7-r80

19. van Rens, K.E., et al.: SNV-PPILP: refined SNV calling for tumor data using perfect phylogenies and ILP. Bioinformatics **31**(7), 1133–1135 (2015). http://bioinformatics.oxfordjournals.org/cgi/content/abstract/btu755?ijkey=XNg7zdRpqjrCkRUI&keytype=ref

20. Roth, A., et al.: PyClone: statistical inference of clonal population structure in cancer. Nat. Methods **11**(4), 396–398 (2014). http://view.ncbi.nlm.nih.gov/pubmed/24633410

21. Salari, R., et al.: Inference of tumor phylogenies with improved somatic mutation discovery. J. Comput. Biol. **20**(11), 933–944 (2013). http://dx.doi.org/10.1089/cmb.2013.0106

A Sub-quadratic Time and Space Complexity Solution for the Dated Tree Reconciliation Problem for Select Tree Topologies

Benjamin Drinkwater[1][(✉)] and Michael A. Charleston[2]

[1] School of Information Technologies, University Of Sydney,
Sydney, NSW 2006, Australia
`benjamin.drinkwater@sydney.edu.au`
[2] School of Physical Sciences, University Of Tasmania,
Hobart, TAS 7005, Australia

Abstract. Recently coevolutionary analysis has turned to tree topology, specifically the unbalanced nature of evolutionary trees, as a means to reduce the asymptotic complexity associated with inferring coevolutionary interrelationships that exist between organismal trees. The leveraging of tree topology for coevolutionary analysis has been shown to be highly successful, with one recent result demonstrating a logarithmic space complexity reduction for the dated tree reconciliation problem. In this work we build on this prior result providing a reduced complexity bound by applying a new model to construct the dynamic programming table. The new complexity bound is the first sub quadratic running time solution for the dated tree reconciliation problem for selected tree topologies and is shown to be, in practice, the fastest method for solving the dated tree reconciliation problem for expected evolutionary trees. Our theoretical results are then validated using a combination of synthetic and biological data with our proposed model shown to save almost $O(\sqrt{n})$ space while finishing in half the time compared to existing methodologies.

Keywords: Coevolution · Phylogeny · Tree reconciliation · Tree shape · NP-Hard

1 Background

Phylogenetics considers the evolutionary drivers that have given rise to the diversity present within the tree of life [1]. A long standing area of interest in the field of phylogenetics is the topology of evolutionary trees, specifically their unbalanced nature. Analysis of this imbalance dates back to at least Yule's modelling for the evolutionary process [2]. While to date no model has captured the topological variation of all evolutionary trees, it has been shown that this variation is bounded by two synthetic models; the *Yule* and *Uniform* models [3,4]. This result is quite powerful as it allows for algorithmic development to target this set of topologies when developing new phylogenetic analysis techniques [5].

© Springer-Verlag Berlin Heidelberg 2015
M. Pop and H. Touzet (Eds.): WABI 2015, LNBI 9289, pp. 93–107, 2015.
DOI: 10.1007/978-3-662-48221-6_7

The Yule model is a bifurcating tree generation model which applies a continuous pure birth process to create a bifurcating tree. Under this model branch lengths have no bearing on the rate of speciation [3], rather, each leaf exists for a random period of time before it diverges, with this process repeating until the tree has n taxa. Trees produced under this model represent the most balanced evolutionary trees within the tree of life [6].

The Uniform model, also known as proportional-to-distinguishable arrangements (PDA), is a bifurcating tree generation model which selects trees through uniform sampling from all trees with n taxa [4]. This approach does not model the evolutionary process, however, trees produced under this model are representative of the most unbalanced evolutionary trees found within the tree of life. As a result, Yule and Uniform trees provide a tight topological bound for expected evolutionary data [6].

Until recently little focus has been given to how the unbalanced nature of phylogenetic trees may be applied to the analysis of coevolving systems. The topology of phylogenetic trees, however, has the potential to be leveraged within this context, as such analysis often considers the coevolutionary relationships between a pair of phylogenetic trees, based on the known associations between their extant taxa. These relationships are often inferred using a technique known as *cophylogeny mapping*, the complexity of which is highly dependent on the shape of the phylogenetic trees in question [5].

Cophylogeny mapping is the process of mapping a dependent (parasite) phylogeny into an independent (host) phylogeny based on the associations between their extant taxa [7]. This technique is able to determine the congruence between a pair of phylogenetic trees, along with inferring their shared evolutionary history using four evolutionary events: codivergence, duplication, host switch and loss [8]; as can be seen in Fig. 1.

The development of cophylogeny mapping algorithms has gained significant interest due to its close association with the gene–species tree reconciliation problem [9], where the evolutionary events within this context are cospeciation, gene duplication, horizontal transfer or lateral gene transfer and loss [10].

Cophylogeny mapping aims to reconcile the minimum cost map, where each evolutionary event is assigned a penalty score, such that the minimum cost map is representative of the most likely shared evolutionary history [11]. The inference of such a map, the cophylogeny reconstruction problem, is NP-Hard [12]. Therefore, techniques aiming to reconcile the most likely evolutionary history between a pair of phylogenetic trees are often forced to rely on heuristics [7,13]. There are currently two popular heuristics applied to this computationally intractable problem, the first ignores the order of evolutionary events defined by the parasite phylogeny, and the second constrains the order of evolutionary events within the host phylogeny, reducing this problem to the simpler dated tree reconciliation (DTR) problem.

In the case where the order of the evolutionary events in the parasite's phylogeny are not considered, it is possible to reconstruct a map where the order of evolutionary events as defined in the reconciled map contradict the order defined

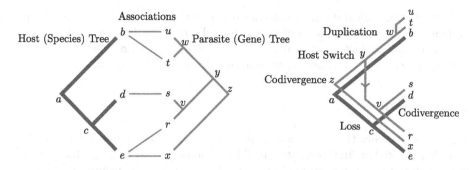

Fig. 1. The resultant map (right) of a pair of phylogenetic trees based on their known associations (left). What is unique about this particular map is that it is composed of all four evolutionary events, codivergence (at nodes z and v), duplication (at node w), host switch (at node y) and loss (edge (z, x) at host node c).

by the parasite's phylogeny, that is, the parasite phylogeny and resultant map describe a different order of evolutionary events. Such a map is often referred to as time-inconsistent or biologically infeasible [14]. While it has been shown that such solutions are uncommon [9], in cases where they do occur this technique may be unable to recover a biologically feasible alternative [14,15].

On the other hand, fixing the order of evolutionary events in the host tree gives rise to a worst case exponential number of internal node orderings that need to be considered to guarantee optimality. To overcome this limitation approaches which leverage the DTR problem often use a metaheuristic to traverse only a subset of the exponential number of possible node orderings. This approach has been shown to be highly accurate and therefore recent algorithmic analysis has tried to minimise the time complexity of the DTR problem to allow for further exploration using a metaheuristic within a fixed period of time.

To date the fastest known approach for solving the DTR problem is Bansal et al.'s reconciliation algorithm, implemented within RANGER-DTL [10]. This algorithm is asymptotically bounded by $O(n^2 \log n)$ and requires $O(n^2)$ space, which is a significant reduction from the initial method proposed in 2009 [7] ($O(n^7)$). Any further reduction to this computational complexity, however, would allow further exploration of the complex search space using metaheuristics, where even a space complexity reduction would allow additional threads as systems will quickly become bound by the quadratic space requirement of this problem as data sets continue to grow [15].

In this work we re-examine the recent complexity analysis which aims to leverage tree topology as a means to reduce the asymptotic complexity of the DTR problem [5]. We build on these past results to provide a new worst case complexity bound which is asymptotically faster than Bansal et al.'s reconciliation method for a select set of topologies, where these selected tree topologies represent the bounds for expected evolutionary data. As a result, our model represents the most time and space efficient approach for solving the DTR problem for expected biological data. The asymptotic bounds which we present are then evaluated using a combination of synthetic and biological data to validate that our algorithm is in fact superior to Bansal et al.'s reconciliation algorithm, which until now was the best known approach for solving the DTR problem.

2   Methodology

This work presents the construction of a new asymptotic bound for the node mapping algorithm first presented in [7] and improved in [5,11], which relies heavily on the tree topology analysis presented in [1,5]. The asymptotic bound presented herein has the potential to be improved further, although appears to provide a relatively tight bound, as observed in Sect. 3.

The new bound represents the lowest worst case time and space complexity bound for solving the DTR problem for coevolving systems composed of trees produced under the Yule model, where it is shown that it is possible to solve the DTR problem in sub-quadratic time. In the case where the coevolving systems considered are composed of trees produced under a Uniform model we present the best space complexity solution for the DTR problem by almost a factor of $O(\sqrt{n})$, with a time complexity result which is comparable to current methods, although is shown to outperform them in practice.

2.1   The Number of Mapping Sites Required for Each Parasite Node

Node mapping algorithms construct a dynamic programming table by mapping each parasite node p into the host tree, from the leaves up to the root. This is contrasted by Bansal et al.'s algorithm which maps each parasite into the host phylogeny starting at the root, moving down to the tips, resulting in a significant reduction in the algorithm's time complexity. This is possible due to the application of a novel $O(n \log n)$ preprocessing step, executed for each parasite node which is processed. Prior research [5], however, has found that while asymptotically slower, bottom-up, taxa to root, approaches are capable of solving the DTR problem in sub-quadratic space. This is achieved by leveraging that not all elements stored within the dynamic programming table are populated when solving this problem optimally. This result cannot be applied to Bansal et al.'s algorithm, however, due to its quadratic space requirement for preprocessing [5].

Previous approaches which solve the DTR problem using sub-quadratic space have constructed their asymptotic space complexity bound by considering the number of mapping sites required at each level in the reconciliation process. In this context, the number of *mapping sites* is the number of unique positions where p may be mapped in the host phylogeny, that is the number of potential solutions which must be retained, and the *level* is the maximum distance that a node is from its leaves starting at zero with a maximum number of levels bound by $n - 1$.

Prior research has shown that the number of mapping sites required for a specific node p at level i can be defined as the function $f(i)$ as follows [5]:

$$f(i) = \begin{cases} 1 & \text{if } i = 0 \\ \min(3^{(2^i - 1)}, n) & \text{if } i \geq 1 \end{cases} \qquad (1)$$

We will show that while this exponential function does provide a bound for the number of mapping sites required for each node p at level i, it often significantly over–counts this number.

The function $f(i)$ considers two cases to infer the number of mapping sites for each parasite node, p, at a particular level. Either the node p may be mapped to all nodes in the host tree, that is there are n mapping sites required for node p, or there is only a subset of the host tree where p is mapped where the number of nodes in this subset is bound by $3^{(2^i - 1)}$ where $3^{(2^i - 1)} < n$. This subset can alternately be bound by the following recurrence as first defined in [5]:

$$a_0 = 1$$
$$a_i = 3 \times (a_{i-1})^2 \text{ for all } i \geq 1 \qquad (2)$$

Under this model all host nodes contained within the subtree bounded by a_i children's mapping sites are considered possible mapping site candidates. This model is in line with the original construction of the node mapping algorithm [7]. The function a_i was constrained, by noting that the number of mapping sites required does not exceed n, but did not consider any filters to reduce the rate of growth of the recurrence relation, a_i.

We will show that it is possible to bound the rate at which a_i grows by applying two filters. The first filter is derived from noting that only one optimal location for a codivergence or duplication needs to be retained for each parasite node p. This was not considered in the original construction of the node mapping algorithm [7] but has been adapted in subsequent methodologies [10,14]. By applying this filter only one additional mapping site is considered for codivergence and duplication events when computing a_i from its children, a_{i-1}.

The second filter that we apply is to leverage the previous proof [11] that while there are up to $O(n^2)$ optimal host switch locations for each parasite node p, that only one needs to be retained to guarantee that the reconciled map is optimal. Therefore, when selecting a host switch event only one additional mapping location needs to be retained for a_i. This is the case even though a host switch may be inferred in either direction during the construction of the dynamic programming table, as at least one of those two host switch events will be mapped to the same node as its child, a_{i-1} [11].

It is important to note that these two filters complement one another, and that by applying both filters we can guarantee that an optimal reconstruction will be recovered when applied to the node mapping algorithm [11]. That is that retaining only three mapping sites, a single codivergence or duplication event along with two host switch events for each parasite node p, ensures that the resultant map will be optimal [10,11,14].

By applying these two filters to the function a_i, we can infer an additive growth function as opposed to the initial multiplicative function as defined in Eq. (2). This drastically reduces the growth of a_i, which we have redefined in Eq. (3). We show this reduction translates to a significant reduction to the asymptotic space complexity bound for the node mapping algorithm.

$$a_0 = 1$$
$$a_1 = 3$$
$$a_i = a_{i-1} + a_{i-1} + 2 \text{ for all } i \geq 2 \tag{3}$$

From this recurrence relation (Eq. (3)) we can construct a closed form function as follows:

Theorem 1. *The maximum number of mapping sites required to solve the DTR problem optimally for each level, a_i, is bounded by the function $a_i = 5 \times 2^{(i-1)} - 2$ $\forall\ i \geq 1$*

Proof.

$$a_i = a_{i-1} + a_{i-1} + 2$$
$$= 2 \times 2 \times a_{i-2} + 4 + 2$$
$$= 2 \times 2 \times 2 \times a_{i-3} + 8 + 4 + 2$$
$$= \ldots$$
$$= 2^{(i-1)} a_1 + 2 \times (2^{(i-1)} - 1)$$
$$= 5 \times 2^{(i-1)} - 2 \tag{4}$$

Therefore this gives rise to a new function for $f(i)$ as:

$$f(i) = \begin{cases} 1 & \text{if } i = 0 \\ \min(5 \times 2^{(i-1)} - 2, n) & \text{if } i \geq 1 \end{cases} \tag{5}$$

This function can then be broken into three parts, where $i = 0$, and the values for i for which $5 \times 2^{(i-1)} - 2 < n$ and where $5 \times 2^{(i-1)} - 2 > n$ as follows:

Lemma 1. $5 \times 2^{(i-1)} - 2 < n\ \forall\ i < \lfloor \lg(n+2) \rfloor - 1$

Proof.

$$5 \times 2^{(i-1)} - 2 \leq n$$
$$2^{(i-1)} \leq \frac{n+2}{5}$$
$$i \leq \lg(n+2) - \lg 5 + 1$$
$$i \leq \lfloor \lg(n+2) \rfloor - \lfloor \lg 5 \rfloor + 1$$
$$i < \lfloor \lg(n+2) \rfloor - 1 \tag{6}$$

Using Lemma 1 we redefine $f(i)$ as:

$$f(i) = \begin{cases} 1 & \text{if } i = 0 \\ 5 \times 2^{(i-1)} - 2 & \text{if } 0 < i < \lfloor \lg(n+2) \rfloor - 1 \\ n & \text{if } i \geq \lfloor \lg(n+2) \rfloor - 1 \end{cases} \tag{7}$$

In line with [5] we are able to use our function for $f(i)$ to compute the space required to solve the DTR problem.

2.2 Space Complexity Reduction

To compute the total storage requirement to solve the DTR problem requires that the number of mapping sites stored for each subsolution, $f(i)$, be multipled by the number of nodes at each level. If we let the number of nodes at each level be defined by the function, $g(i)$, in line with prior analysis, then the total storage requirement for solving the DTR problem optimally is:

$$\sum_{i=0}^{h} (f(i) \times g(i)) \tag{8}$$

where h is the height of the tree. As we don't have an exact height for trees produced under either the Yule or Uniform models, we let h be equal to the maximum height of any possible tree $(n-1)$ for this complexity analysis. This approach will potentially over-count the total number of mapping sites, which is appropriate as our aim is to provide an asymptotic bound rather than the exact number of mapping sites required.

Previous analysis has derived explicit functions for $g(i)$ for both the Yule and Uniform models as $\frac{2^{(i-1)}n}{3^i}$ and $\frac{3^{(i-1)}n}{4^i}$ respectively [5], and therefore substituting our new $f(i)$ along with these known results for $g(i)$ gives us the following respective asymptotic space complexity bounds for solving the DTR problem:

$$O\left(n + n \sum_{i=1}^{\lfloor \lg(n+2) \rfloor - 1} \frac{2^{(i-1)}(5 \times 2^{(i-1)} - 2)}{3^i} + n^2 \sum_{i=\lfloor \lg(n+2) \rfloor}^{n-1} \frac{2^{(i-1)}}{3^i}\right) \tag{9}$$

$$O\left(n + n \sum_{i=1}^{\lfloor \lg(n+2) \rfloor - 1} \frac{3^{(i-1)}(5 \times 2^{(i-1)} - 2)}{4^i} + n^2 \sum_{i=\lfloor \lg(n+2) \rfloor}^{n-1} \frac{3^{(i-1)}}{4^i}\right) \tag{10}$$

By simplifying Eqs. (9) and (10) a new set of worst case bounds for the space required to solve the DTR problem may be established as follows:

Theorem 2. *The required space to solve the DTR problem for trees constructed under the expected Yule process is bounded by $O(n^{1.42})$.*

Theorem 3. *The required space to solve the DTR problem for trees constructed under the expected Uniform process is bounded by $O(n^{1.58})$.*

Theorems 2 and 3 (as proven in Appendices A and B respectively) are representative of the lowest worst case bounds for any approach capable of solving the DTR problem optimally, with the prior best space complexity result achieving a bound of only $O\left(\frac{n^2}{(\log n)^{0.58}}\right)$ and $O\left(\frac{n^2}{(\log n)^{0.42}}\right)$ for trees produced under the Yule and Uniform processes respectively [5]. What is even more significant is compared to Bansal et al.'s reconciliation algorithm [10], our approach offers almost an $O(\sqrt{n})$ improvement in the space required to solve the DTR problem optimally for all expected evolutionary data.

2.3   Time Complexity Reduction

Theorems 2 and 3 provide a new worst case bound for space required to solve the DTR problem. In this section we apply these asymptotic bounds to provide a reduction to the cubic time complexity bound faced by Node Mapping [11], and show that for specific tree topologies it is possible to solve the DTR problem in subquadratic time.

A subquadratic space requirement is achieved by storing a sublinear number of mapping sites for each parasite node. From Theorems 2 and 3 we can derive the average number of mapping sites stored for each parasite node for trees produced by both the expected Yule $(\psi(n))$ and Uniform $(\upsilon(n))$ models as:

$$\psi(n) \approx (n^{0.42}) \tag{11}$$

$$\upsilon(n) \approx (n^{0.58}) \tag{12}$$

The functions for the number of mapping sites stored for each parasite node, p, for trees produced under the Yule $(\psi(n))$ and Uniform $(\upsilon(n))$ models, are derived by dividing the total storage requirement for each tree topology, $O(n^{1.42})$ or $O(n^{1.58})$ as proven in Theorems 2 and 3, by the total number of subsolutions, $O(n)$, required to solve the DTR problem. As a result, $\psi(n)$ and $\upsilon(n)$ represent the average number of mapping sites stored for each subsolution. It is important to note that the worst case running time will occur when this average number of mapping sites is required to be stored for each subsolution [5].

Node Mapping [11] solves the DTR problem by mapping each parasite node, of which there are $O(n)$, into the host tree. The set of mapping sites for each parasite node, p, is inferred by comparing all possible permutations of both its left and right child mapping sites, that is either $(\psi(n))^2$ or $(\upsilon(n))^2$ for the Yule and Uniform models respectively. Therefore, the worst case time complexity bound for solving the DTR problem for trees produced under the Yule and Uniform models may be described by multiplying the number of mapping site comparisons and the number of parasite nodes, giving us a time complexity bound for solving the DTR problem for trees constructed under the expected Yule process of $O(n^{1.83})$, and a time complexity for solving the DTR problem for trees constructed under the expected Uniform process of $O(n^{2.17})$.

This result is of interest as for a select subset of tree topologies, that is those that conform to a Yule process, it is possible to solve the DTR problem in sub-quadratic time. In the case where trees conform to those produced by a Uniform process, the running time using our approach runs in $O(n^{2.17})$, which is slightly worse asymptotically than Bansal et al.'s algorithm which runs in $O(n^2 \log n)$. While our algorithm is asymptotically worse it performs better in practice. This is because $n^{2.17} < n^2 \log n \ \forall \ 2 < n < 353,830,149$. As there are only an estimated 8.7 million species [16], our algorithm for trees produced under both a Yule and Uniform process is the fastest algorithm for solving the dated tree reconciliation problem for all feasible data.

3   Results and Discussion

The analysis of how our theoretical complexity bounds translate to a time and space complexity reduction in practice is broken into three parts. The first presents the space complexity reduction offered by our proposed model over synthetic data sets constructed using both Yule and Uniform models. Following this we compare the running time of our proposed method against Bansal et al.'s reconciliation algorithm, over the same synthetic data sets.

In both cases the synthetic data applied was constructed using CoRe-PA's random nexus file generator [13] allowing a larger number of taxa to be considered. These data sets represent the largest coevolutionary data sets created for algorithmic analysis, with ten times more taxa than prior analyses [5]. Finally, we compare the time and space complexity of both our methods against Bansal et al.'s approach over 102 previously published biological data sets [15], ensuring our result translates to time and space reduction for biological data analysis.

In all three sections we compared a Java implementation of the Improved Node Mapping algorithm [11] against a Java implementation of Bansal et al.'s reconciliation algorithm [10]. RANGER-DTL was not used as the source code is not open source and our aim was to implement both algorithms using common code wherever possible, to provide an accurate comparison of each algorithm.

3.1   Analysis of Space Complexity Improvements (Synthetic Data)

The premise of the asymptotic reduction presented herein is that mapping a parasite phylogeny into a host phylogeny is achievable using asymptotically less than $O(n^2)$ mapping sites. Therefore it is important that this be validated before considering any time reductions that may be observed due to this result.

To validate that our theoretical result translates to a space complexity improvement in practice, we recorded the number of mapping sites stored when solving the DTR problem, for two sets of 250 synthetic coevolutionary systems, composed of trees ranging from 10 through to 2500 taxa that were constructed using either the Yule or Uniform model. The median space required for 100 replicates of each data set has been plotted in Fig. 2, where it can be seen that significantly less than $O(n^2)$ mapping sites are required. In fact our asymptotic bounds appear to actually grow at a rate slightly faster than the required space, meaning that an even lower asymptotic complexity bound may be achievable.

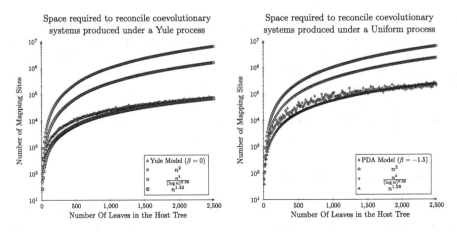

Fig. 2. The space required to solve the DTR problem for systems composed of trees produced under a Yule (left) or Uniform (right) model.

3.2   Analysis of Time Complexity Improvements (Synthetic Data)

The running time required to solve the DTR problem for each of the synthetic data sets applied in Sect. 3.1 was recorded for both Improved Node Mapping and Bansal et al.'s reconciliation algorithm. To ensure a robust comparison of each approach, 100 replicates were run for each system with the median running time plotted in Fig. 3. We only considered the time required to map the parasite into the host. This was to observe Bansal et al.'s reconciliation algorithm's running time variation for systems composed of specific tree topologies, without any potential noise that its quadratic preprocessing may introduce. This

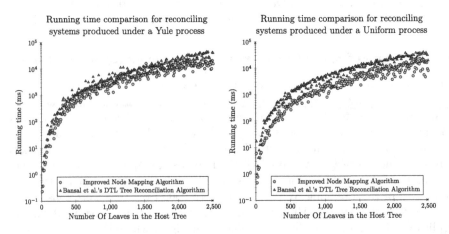

Fig. 3. A running time comparison using coevolutionary systems composed of trees produced under a Yule (left) or Uniform (right) model.

did not assist node mapping's observed improvement as it applies linear time preprocessing compared to Bansal et al.'s quadratic preprocessing requirement.

It can be seen in Fig. 3 that Improved Node Mapping out performs Bansal et al.'s reconciliation algorithm in both cases, with a median reduction of 43 % for systems composed of trees produced under the Yule model and 62 % for systems composed of trees produced under the Uniform model.

3.3   Time and Space Complexity Improvements for Biological Data

To ensure that our successful results translate to biological data, we repeated the same experiments from Sects. 3.1 and 3.2 over a set of previously published biological data [15]. These sets are smaller, however, with the largest only containing 53 taxa compared with 2500 in the synthetic data sets.

Even with smaller data sets our experiments show that the space required is generally less than $O(n^2)$. The exceptions seen in Fig. 4 are due to larger parasite phylogenies, often twice the size of the host, which therefore require more mapping sites as there are $2n$ subsolutions (parasite nodes). Overall however, as n grows it can be seen that asymptotically fewer mapping sites are required. In terms of running time, our proposed algorithm is observed to have a median reduction of 51 %, which is significant considering this is achieved while providing an asymptotic space reduction.

These results demonstrate that the asymptotic time and space reductions proven herein translate to an in-practice time and space complexity improvement over both biological and synthetic data sets. We have shown that the node mapping algorithm, while not achieving an asymptotic reduction from its bound of $O(n^3)$ for all data sets, is able to perform as well as Bansal et al.'s $O(n^2 \log n)$ algorithm for evolutionary data, with a worst case bound of $O(n^{2.17})$. This trans-

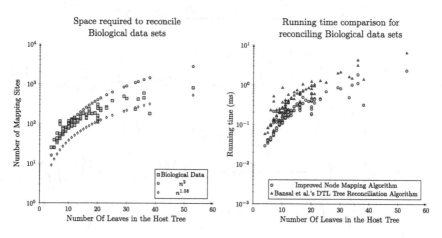

Fig. 4. The space complexity requirements for the Node Mapping algorithm (left) and a comparison of running times of Bansal et al.'s reconciliation algorithm and the Node Mapping algorithm (right), both over a set of previously published biological data

lates to a running time reduction of more than 50 % in practice, while providing almost an $O(\sqrt{n})$ reduction in the space required. If adopted, this approach will allow metaheuristic frameworks to execute a higher number of threads which are capable of finishing in less than half the time. This will result in a higher degree of confidence in the coevolutionary analysis performed due to a greater exploration of the exponential search space within the same fixed time period.

Acknowledgements. This work was supported by the Australian Postgraduate Award and William and Catherine McIlrath Scholarship awarded to BD. We would also like to thank Nicolas Wieseke for his assistance with generating the synthetic data used in this analysis, along with Anastasios Viglas for his guidance with the asymptotic complexity analysis performed herein.

Appendix A: Proof for Theorem 2

Proof. To infer the worst case space complexity required to solve the dated tree reconciliation problem for coevolutionary systems composed by phylogenetic trees produced under an expected Yule model requires that the set of geometric series defined in Eq. (9) be simplified. To do this we let the worst case asymptotic complexity defined in Eq. (9) be redefined as a function $\alpha(n)$:

$$\alpha(n) = n + n \sum_{i=1}^{\lfloor \lg(n+2) \rfloor - 1} \frac{2^{(i-1)}(5 \times 2^{(i-1)} - 2)}{3^i} + n^2 \sum_{i=\lfloor \lg(n+2) \rfloor}^{n-1} \frac{2^{(i-1)}}{3^i} \quad (13)$$

which may be simplified as follows:

$$\alpha(n) = n + n \sum_{i=1}^{\lfloor \lg(n+2) \rfloor - 1} \frac{2^{(i-1)}(5 \times 2^{(i-1)} - 2)}{3^i} + n^2 \sum_{i=\lfloor \lg(n+2) \rfloor}^{n-1} \frac{2^{(i-1)}}{3^i}$$

$$= n + n \left(\frac{5}{4} \sum_{i=1}^{\lfloor \lg(n+2) \rfloor - 1} \left(\frac{4}{3}\right)^i - \sum_{i=1}^{\lfloor \lg(n+2) \rfloor - 1} \left(\frac{2}{3}\right)^i + \frac{n}{2} \sum_{i=\lfloor \lg(n+2) \rfloor}^{n-1} \left(\frac{2}{3}\right)^i \right)$$

$$< n + n \left(\frac{5}{4} \sum_{i=1}^{\lfloor \lg(n+2) \rfloor - 1} \left(\frac{4}{3}\right)^i + \frac{n}{2} \sum_{i=\lfloor \lg(n+2) \rfloor}^{n-1} \left(\frac{2}{3}\right)^i \right)$$

$$= n + n \left(\frac{5}{4} \times \left(\frac{\frac{4}{3} - \frac{4}{3}^{\lfloor \lg(n+2) \rfloor}}{1 - \frac{4}{3}} \right) + \frac{n}{2} \times \left(\frac{\left(\frac{2}{3}\right)^{\lfloor \lg(n+2) \rfloor} - \left(\frac{2}{3}\right)^n}{1 - \frac{2}{3}} \right) \right)$$

$$= n + n \left(\frac{15}{4} \times \left(\frac{4}{3}^{\lfloor \lg(n+2) \rfloor} - \frac{4}{3} \right) + \frac{3n}{2} \times \left(\frac{2}{3}^{\lfloor \lg(n+2) \rfloor} - \left(\frac{2}{3}\right)^n \right) \right)$$

$$\leq n + n \left(\frac{15}{4} \times \left(\frac{4}{3}^{(\lg(n+2))} - \frac{4}{3} \right) + \frac{3n}{2} \times \left(\frac{2}{3}^{(\lg(n+2))} - \left(\frac{2}{3}\right)^n \right) \right)$$

$$= n + n \left(\frac{15}{4} \times \frac{4}{3}^{(\lg(n+2))} + \frac{3n}{2} \times \frac{2}{3}^{(\lg(n+2))} - n \times \left(\frac{2}{3}\right)^{n-1} - 5 \right)$$

$$< n + n \left(\frac{15}{4} \times \frac{4}{3}^{(\lg(n+2))} + \frac{3n}{2} \times \frac{2}{3}^{(\lg(n+2))} \right)$$

$$= n + n \left(\frac{15}{4} \times (n+2)^{(2-\lg 3)} + \frac{3n}{2} \times (n+2)^{(1-\lg 3)} \right)$$

$$\approx n \times n^{(2-\lg 3)}$$

$$\approx n^{1.42} \tag{14}$$

Appendix B: Proof for Theorem 3

Proof. To infer the worst case space complexity required to solve the dated tree reconciliation problem for coevolutionary systems composed by phylogenetic trees produced under an expected Uniform model requires that the set of geometric series defined in Eq. (10) be simplified. To do this we let the worst case asymptotic complexity defined in Eq. (10) be redefined as a function $\beta(n)$:

$$\beta(n) = n + n \sum_{i=1}^{\lfloor \lg(n+2) \rfloor - 1} \frac{3^{(i-1)} (5 \times 2^{(i-1)} - 2)}{4^i} + n^2 \sum_{i=\lfloor \lg(n+2) \rfloor}^{n-1} \frac{3^{(i-1)}}{4^i} \tag{15}$$

which may be simplified as follows:

$$\beta(n) = n + n \sum_{i=1}^{\lfloor \lg(n+2) \rfloor - 1} \frac{3^{(i-1)} (5 \times 2^{(i-1)} - 2)}{4^i} + n^2 \sum_{i=\lfloor \lg(n+2) \rfloor}^{n-1} \frac{3^{(i-1)}}{4^i}$$

$$= n + n \left(\frac{5}{6} \sum_{i=1}^{\lfloor \lg(n+2) \rfloor - 1} \left(\frac{3^i \times 2^i}{2^i \times 2^i} \right) - \frac{2}{3} \sum_{i=1}^{\lfloor \lg(n+2) \rfloor - 1} \left(\frac{3}{4} \right)^i + \frac{n}{3} \sum_{i=\lfloor \lg(n+2) \rfloor}^{n-1} \left(\frac{3}{4} \right)^i \right)$$

$$< n + n \left(\frac{5}{6} \sum_{i=1}^{\lfloor \lg(n+2) \rfloor - 1} \left(\frac{3}{2} \right)^i + \frac{n}{3} \sum_{i=\lfloor \lg(n+2) \rfloor}^{n-1} \left(\frac{3}{4} \right)^i \right)$$

$$= n + n \left(\frac{5}{6} \times \left(\frac{\frac{3}{2} - \frac{3}{2}^{\lfloor \lg(n+2) \rfloor}}{1 - \frac{3}{2}} \right) + \frac{n}{3} \times \left(\frac{\left(\frac{3}{4} \right)^{\lfloor \lg(n+2) \rfloor} - \left(\frac{3}{4} \right)^n}{1 - \frac{3}{4}} \right) \right)$$

$$= n + n \left(\frac{5}{3} \times \left(\frac{3}{2}^{\lfloor \lg(n+2) \rfloor} - \frac{3}{2} \right) + \frac{4n}{3} \times \left(\frac{3}{4}^{\lfloor \lg(n+2) \rfloor} - \left(\frac{3}{4} \right)^n \right) \right)$$

$$\leq n + n \left(\frac{5}{3} \times \left(\frac{3}{2}^{(\lg(n+2))} - \frac{3}{2} \right) + \frac{4n}{3} \times \left(\frac{3}{4}^{(\lg(n+2))} - \left(\frac{3}{4} \right)^n \right) \right)$$

$$= n + n\left(\frac{5}{3} \times \frac{3}{2}^{(\lg{(n+2))}} + \frac{4n}{3} \times \frac{3}{4}^{(\lg{(n+2))}} - n \times \left(\frac{3}{4}\right)^{(n-1)} - \frac{5}{2}\right)$$

$$< n + n\left(\frac{5}{3} \times \frac{3}{2}^{(\lg{(n+2))}} + \frac{4n}{3} \times \frac{3}{4}^{(\lg{(n+2))}}\right)$$

$$= n + n\left(\frac{5}{3} \times (n+2)^{(\lg 3 - 1)} + \frac{4n}{3} \times (n+2)^{(\lg 3 - 2)}\right)$$

$$\approx n \times n^{(\lg 3 - 1)}$$

$$\approx n^{1.58} \tag{16}$$

References

1. Steel, M., McKenzie, A.: Properties of phylogenetic trees generated by Yule-type speciation models. Math. Biosci. **170**(1), 91–112 (2001)
2. Yule, G.U.: A mathematical theory of evolution, based on the conclusions of Dr. JC Willis, FRS. Philos. Trans. R. Soc. Lond. Ser. B Contain. Pap. Biol. Charact. **213**, 21–87 (1924)
3. Harding, E.: The probabilities of rooted tree-shapes generated by random bifurcation. Adv. Appl. Probab. **3**(1), 44–77 (1971)
4. Rosen, D.E.: Vicariant patterns and historical explanation in biogeography. Syst. Biol. **27**(2), 159–188 (1978)
5. Drinkwater, B., Charleston, M.A.: A time and space complexity reduction for coevolutionary analysis of trees generated under both a Yule and Uniform model. Comput. Biol. Chem. **57**, 61–71 (2015)
6. Aldous, D.J.: Stochastic models and descriptive statistics for phylogenetic trees, from Yule to today. Statist. Sci. **16**, 23–34 (2001)
7. Libeskind-Hadas, R., Charleston, M.: On the computational complexity of the reticulate cophylogeny reconstruction problem. J. Comput. Biol. **16**(1), 105–117 (2009)
8. Ronquist, F.: Reconstructing the history of host-parasite associations using generalised parsimony. Cladistics **11**(1), 73–89 (1995)
9. Libeskind-Hadas, R., Wu, Y.C., Bansal, M.S., Kellis, M.: Pareto-optimal phylogenetic tree reconciliation. Bioinformatics **30**(12), i87–i95 (2014)
10. Bansal, M.S., Alm, E.J., Kellis, M.: Efficient algorithms for the reconciliation problem with gene duplication, horizontal transfer and loss. Bioinformatics **28**(12), i283–i291 (2012)
11. Drinkwater, B., Charleston, M.A.: An improved node mapping algorithm for the cophylogeny reconstruction problem. Coevolution **2**(1), 1–17 (2014)
12. Ovadia, Y., Fielder, D., Conow, C., Libeskind-Hadas, R.: The cophylogeny reconstruction problem is NP-complete. J. Comput. Biol. **18**(1), 59–65 (2011)
13. Merkle, D., Middendorf, M., Wieseke, N.: A parameter-adaptive dynamic programming approach for inferring cophylogenies. BMC Bioinform. **11**(Suppl 1), S60 (2010)
14. Doyon, J.P., Ranwez, V., Daubin, V., Berry, V.: Models, algorithms and programs for phylogeny reconciliation. Briefings Bioinform. **12**(5), 392–400 (2011)

15. Drinkwater, B., Charleston, M.A.: Introducing TreeCollapse: a novel greedy algorithm to solve the cophylogeny reconstruction problem. BMC Bioinform. **15**(Suppl 16), S14 (2014)
16. Mora, C., Tittensor, D.P., Adl, S., Simpson, A.G., Worm, B.: How many species are there on Earth and in the Ocean? PLoS Biol. **9**(8), e1001127 (2011)

Maximum Parsimony Analysis of Gene Copy Number Changes

Jun Zhou[1,2], Yu Lin[3], Vaibhav Rajan[4], William Hoskins[2], and Jijun Tang[1,2]($\boxtimes$)

[1] Tianjin Key Laboratory of Cognitive Computing and Application,
Tianjin University, Tianjin 300072, China
jitang@cse.sc.edu
[2] University of South Carolina, Columbia, SC 29205, USA
[3] University of California, San Diego, La Jolla, CA 92093, USA
[4] Xerox Research Centre India (XRCI), Bangalore, India

Abstract. Evolution of cancer cells are characterized by large scale and rapid changes in the chromosomal landscape. The fluorescence in situ hybridization (FISH) technique provides a way to measure the copy numbers of preselected genes in a group of cells and has been found to be a reliable source of data to model the evolution of tumor cells. Chowdhury et al. [1,2] recently develop a theoreticallly sound and scalable model for tumor progression driven by gains and losses in cell count patterns obtained by FISH probes. Their model aims to find the Rectilinear Steiner Minimum Tree (RSMT) that describes progression of FISH cell count patterns over its branches in a parsimonious manner. This model is found to effectively model tumor evolution and is also useful in tumor classification. However the RSMT problem is NP–complete and efficient heuristics are necessary to obtain useful solutions, especially for large datasets. In this paper we design a new algorithm for the RSMT problem, based on Maximum Parsimony phylogeny inference. Experimental results from both simulated and real tumor data show that our approach outperforms previous heuristics for the RSMT problem, thus obtaining better models for tumor evolution.

Keywords: FISH · Tumor Phylogenetics · Maximum Parsimony · Gene copy number · Rectilinear steiner minimum tree

1 Introduction

Cancer is recognized to be an evolutionary process driven by mutations in tumor cells [3]. These evolutionary processes include single-nucleotide variations, insertions and deletions, copy-number aberrations, structural variations and gene fusions [4]. Many experiments reveal considerable intra–tumor and inter–tumor heterogeneity [5], attributed to these evolutionary processes. Clinical implications of this heterogeneity, for example in drug resistance and disease diagnosis, has been well studied [5,6].

© Springer-Verlag Berlin Heidelberg 2015
M. Pop and H. Touzet (Eds.): WABI 2015, LNBI 9289, pp. 108–120, 2015.
DOI: 10.1007/978-3-662-48221-6_8

Rapid, simultaneous linear and branching evolution in multiple subclones of cancer cells can be modeled by a phylogenetic tree [7]. Inferring such phylogenies facilitates the study of cancer initiation, progression, treatment, and resistance [8]. They can help pinpoint important changes that lead to the recurrence of some genome aberrations [9]. Phylogenies also aid in identifying genes crucial for the evolution and hence may contribute to developing better cancer treatment [10–13].

Mutation patterns in cancer are characterized by frequent and widespread gains and losses of genomic material which is markedly different from what is observed in species or population level evolution [8]. In particular, gene copy number changes affect a larger fraction of the genome in cancers than do any other type of somatic genetic alteration [14,15]. During tumor development, the gene copy number can increase or decrease, due to failures in DNA repair mechanisms (e.g., translesion synthesis and non-homologous end joining) [16–23]. Another characteristic feature of tumor evolution is the high genetic heterogeneity found. Previous phylogenetic models for cancer, such as [11,24–28], either do not account for these unique characteristics of cancer evolution or are not scalable and hence not of practical use. Thus there is need for development of new phylogenetic models with scalable algorithms that can adequately model cancer evolution. A step towards a scalable model for inferring tumor phylogeny by copy number variation was taken by Chowdhury et al. [1,2] using FISH data.

FISH (Fluorescent In Situ Hybridization) was developed by bio-medical researchers in the early 1980 s and has been used to detect and localize the presence or absence of specific DNA sequences and to visualize the genomic diversity of chromosome aberrations [29]. While single cell sequencing (SCS) technique also has the potential to count the number of specific genes or specific regions for a group of cells, the highly non-uniform coverage, the admixture signal and relatively high cost make the current SCS technique unsuitable. By allowing us to count copies of gene probes across hundreds to thousands of cells, FISH provides a way to characterize tumor heterogeneity reliably.

Chowdhury et al. [1] model the progression of tumor cells from the FISH copy number data assuming that gene copy number changes occur by gain or loss of a single gene probe. They show that such a progression of FISH cell count patterns over a tree effectively models the evolution of tumor cells. They assume a parsimonious model describing evolution by single gene copy number changes and develop a theoretical foundation for handling large copy number datasets. They reduce the modeling problem to the NP-hard Rectilinear Steiner Minimum Tree (RSMT) problem, and develop heuristics to construct RSMT trees. Another heuristic for the same problem was proposed by Zhou et al. [30]. Apart from providing a good model for tumor evolution, RSMT topologies have been found to be useful in distinguishing primary and metastasis samples [1,2]. RSMT topologies and other tree-based statistics yield insights into selective pressure which simpler statistics (like cell counts) do not and provide independent support to findings such as in [31]. They also are useful as discriminatory features in downstream classification-based analyses. Experiments in [1,2] suggest that mpFISHtrees can potentially improve these analyses that rely on accurate RSMT inference.

Since FISH yields cell count patterns of gene copy numbers at single-cell resolution, parsimony-based phylogenetic approaches (designed previously for building phylogenies of species) can be applied to such data. Maximum parsimony approaches seek the tree and the cell count patterns (gene copy numbers) for the internal nodes that minimize the total number of events (in this case, single gene duplication or deletion) needed to produce the given input from a common ancestor. Although this also results in an NP hard formulation, several heuristics have been developed in the last decade to solve the Maximum Parsimony Phylogeny problem. Packages such as TNT [32] have largely overcome computational limitations and allow reconstructions of large trees, inferring accurate trees with hundreds of taxa within minutes, and the use of continuous characters [33].

In this paper we design a new approach based on Maximum Parsimony tree reconstruction. We use approximations to Maximum Parsimony trees for phylogenetic inference to obtain excellent approximations to Rectilinear Steiner Minimum Trees. Experimental results from both simulated and real tumor data show that our new approach outperforms previous heuristics by obtaining better solutions for the RSMT problem and thus provides a good model for cancer phylogenies using cell count patterns from FISH data.

2   Methods

In this section we describe the RSMT problem as proposed by Chowdhury et al. [1] for modeling the progression of FISH cell count patterns. We then describe our heuristic for obtaining an approximate solution to the problem.

2.1   The Rectilinear Steiner Minimum Tree (RSMT) Problem

The RSMT problem for gene copy number changes is defined as follows.
Definition: RSMT(n, d)
Input: FISH data of n cell count patterns on d gene probes for a given patient
Output: A minimum weight tree with the rectilinear metric (or L_1 distance) including all the observed n cell count patterns and, as needed, unobserved Steiner nodes along with their cell count patterns for d probes, Steiner nodes here is used to represent missing node during gene copy number change process

Each cell has some non-negative integer count of each gene probe. Given two cell count patterns $(x_1, x_2, \ldots, x_d)$ and $(y_1, y_2, \ldots, y_d)$, the pairwise distance under the rectilinear metric (or L_1 distance) is defined as $|x_1 - y_1| + |x_2 - y_2| + \ldots + |x_d - y_d|$, where $x_i, y_i \in \mathbb{N}$. The weight of a tree with nodes labeled by cell count patterns is defined as the sum of all branch lengths under the rectilinear metric. Since the distance between two cell count patterns under the rectilinear metric represents the number of single gene duplication and loss events between them, a minimum weight tree, including Steiner nodes if needed, explains the n observed cell count patterns of d probes with minimum total number of single gene duplication and loss events, from a single ancestor. The single ancestor could be, for example, cell count pattern with a copy number count of 2 for each gene probe (a healthy diploid cell) [1,2].

The RSMT problem is NP-complete [34]. Note that if all possible cell count patterns in cancer cells are present as the input, then the RSMT is simply the minimum spanning tree, since no additional Steiner nodes are needed. Since both the minimum spanning tree and the minimum spanning network (as the union of all minimum spanning trees) can be constructed efficiently, previous heuristics have approximated RSMT by adding additional Steiner nodes to the minimum spanning network [1,2] or to the minimum spanning tree [30]. For example, Fig. 1 shows an instance of 4 cell count patterns on 3 genes, and the RSMT can be obtained by adding a Steiner node to the minimum spanning tree. However, both the above heuristics are likely to be trapped in a local optimum if there are multiple possible Steiner nodes that can be introduced, since the order in which the Steiner nodes are added may affect the resulting tree weight [30].

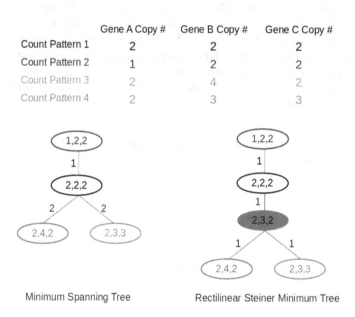

	Gene A Copy #	Gene B Copy #	Gene C Copy #
Count Pattern 1	2	2	2
Count Pattern 2	1	2	2
Count Pattern 3	2	4	2
Count Pattern 4	2	3	3

Minimum Spanning Tree   Rectilinear Steiner Minimum Tree

Fig. 1. (Top) the input data of 4 cell count patterns on 3 genes. (Bottom left) the minimum spanning tree has weight 5. (Bottom right) the RSMT has weight 4. The Steiner node in RSMT is colored in red (Color figure online).

2.2 The Maximum Parsimony Tree (MPT) Problem

The Maximum Parsimony Tree problem for phylogenetic inference of gene copy number changes is defined as follows.

Definition: MPT(n, d)

Input: FISH data of n cell count patterns on d gene probes for a given patient

Output: A minimum weight unrooted binary tree with the rectilinear metric (or L_1 distance) including all the observed n cell count patterns as leaves and n-2 unobserved internal nodes

The MPT problem is also NP complete [35] but heuristics like TNT [32], have largely overcome computational limitations and allow reconstructions of large trees and the use of continuous characters [33]. The copy number of each gene can be treated as continuous characters and TNT can be used to find the minimum weight phylogenetic tree. Note that, given n observed cell count patterns as the input (leaf nodes), MPT introduces $(n-2)$ unobserved internal nodes, while the minimum spanning tree does not introduce any unobserved nodes.

2.3   From MPT to RSMT

In general, there may be multiple optimal solutions for the MPT problem, e.g., the internal nodes labeled by different cell count patterns. In any MPT with all nodes labeled by cell count patterns, a branch is called *trivial* if its length is 0 under the rectilinear metric. For any MPT, an unobserved internal node is a Steiner node if and only if it is labeled by a distinct cell count pattern other than any input cell count patterns. If we contract all trivial branches in MPT, the remaining unobserved internal nodes will be the Steiner nodes in RSMT. See Fig. 2 for an example.

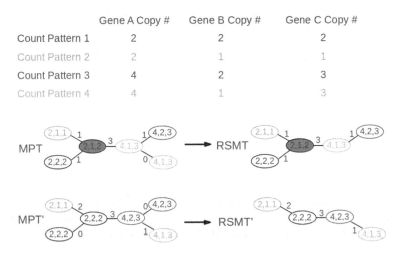

	Gene A Copy #	Gene B Copy #	Gene C Copy #
Count Pattern 1	2	2	2
Count Pattern 2	2	1	1
Count Pattern 3	4	2	3
Count Pattern 4	4	1	3

Fig. 2. (Top) the input data of 4 cell count patterns on 3 genes. (Bottom) two maximum parsimony trees MPT and MPT', both of weight 6, are shown on the left. Nodes with identical cell count patterns are shown in the same color in both MPT and MPT'. The corresponding RSMT and RSMT', both of weight 6, are shown on the right, and the Steiner node in RSMT is colored in red (Color figure online).

Minimizing Steiner Nodes. The MPT, as obtained above, may contain up to $(n-2)$ Steiner nodes. Following the philosophy of parsimony, we seek to minimize

Input: MPT with optimal weight W_{opt}
Output: RSMT with optimal weight W_{opt}
For each *Leaf* in MPT
  $Parent(Leaf)$: the parent node of *Leaf* in MPT
  $MPT \setminus Leaf$: the tree obtained by removing *Leaf*, rooted at $Parent(Leaf)$
        (Fig. 3(a))
  Compute the ranges of possible values in internal nodes in $MPT \setminus Leaf$
        (DP bottom-up phase; Fig. 3(b))
  Assign the cell count pattern of *Leaf* to $Parent(Leaf)$
  Determine all the values for all other internal nodes in MPT
        (DP top-down phase; Fig. 3(c))
  Contract all trivial branches in $MPT \setminus Leaf$ and derive $RSMT^*$
        (Fig. 3(d))
  If the weight of $RSMT^*$ is equal to W_{opt}
      Store $RSMT^*$ as a candidate RSMT
Return a candidate RSMT with the minimum number of Steiner nodes

Algorithm 1. Algorithm to derive RSMT from MPT

these artificially introduced nodes, although this step does not reduce the final tree weight and is not required by the formal definition of RSMT (which does not place any explicit constraints on the number of Steiner nodes). In fact, all the previous heuristics [1,2,30] also implicitly do not add unnecessary Steiner nodes and are biased towards a parsimonious solution due to their incremental way of adding Steiner nodes to an initial tree with no Steiner nodes.

Given any MPT, if the internal nodes are labeled by cell count patterns, the RSMT can be derived by contracting all its trivial edges; but the MPT obtained does not have labels assigned to the internal nodes. Hence the problem reduces to finding the best possible labels for internal nodes that does not increase the weight. The dynamic programming (DP) method of [36] can be adapted to find the internal labels, but modifications are needed to account for the rectilinear metric and its implications on the total tree weight. Our algorithm proceeds by finding whether a leaf label can be reused in its parent (or "lifted") for each leaf in the tree. The node with the lifted pattern is chosen to be the root node and the leaf is removed. In the bottom–up phase of the DP, labels from other leaves are propagated up the tree by using ranges of cell count patterns that can maintain the leaf cell counts without increasing the tree weight. In the top–down phase, cell count values are assigned to the internal nodes and a candidate tree is generated by contracting trivial edges. Several such candidate trees are generated by selecting different root nodes from lifted leaves. We choose a candidate tree with minimum number of Steiner nodes, with no increase in tree weight. The complete algorithm is presented in Algorithm 1 and a detailed example is shown in Fig. 3.

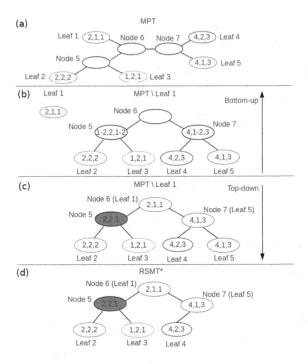

Fig. 3. An example to test whether $Leaf_1$ can be optimally "lifted"to its parent node $Node_6$ in MPT. (a) a MPT on 5 leaves and 3 internal nodes. (b) $Leaf_1$ and compute the ranges of possible values to internal nodes, except $Node_6$, in $MPT \backslash Leaf_1$ in a bottom-up phase. (c) Assign the cell count pattern of $Leaf_1$ to the root of $MPT \backslash Leaf_1$, and determine the values for other internal nodes in $MPT \backslash Leaf_1$ in a top-down phase. (d) Contract all trivial branches in $MPT \backslash Leaf_i$ and derive $RSMT^*$. Nodes with identical cell count patterns are shown in the same color and the Steiner node in RSMT* is colored in red (Color figure online).

3   Experimental Results

In the following, we refer to previous heuristics as FISHtree [1,2][1] and iFISHtree [30], and we refer to our Maximum-Parsimony based approach as mpFISHtree. We also refer to the exact method [1] as Exact.

3.1   Real Cancer Datasets

We use both the real cervical cancer and breast cancer data samples and simulation samples generated through the same process described by Chowdhury et al. [1,2]. The cervical cancer data contains four gene probes LAMP3, PROX1,

---

[1] We use the best result derived from the heuristic option in [1] and the option PLOIDY_LESS_HEURISTIC in [2] that also approximate RSMT under the case of gene copy number changes of single probes.

PRKAA1 and CCND1, and the breast cancer data contains eight gene probes COX-2, MYC, CCND1, HER-2, ZNF217, DBC2, CDH1 and p53. These genes are chosen because they are considered as important factors for cancer growth inhibition or promotion. The cervical cancer data is from 16 lymph positive patients (both primary and metastatic tumors) and 15 lymph negative patients, making 47 samples in total. The breast cancer data is from 12 patients with both IDC and DCIS and 1 patient with only DCIS, making 25 samples in total. More details of this FISH data set can be found in Chowdhury et al. [1,2].

Tables 1 and 2 summarize the comparison of FISHtree, iFISHtree and mpFISHtree for breast cancer samples and cervical cancer samples, respectively (and the best tree weights are shown in bold). Note that mpFISHtree of the three heuristic methods has the best performance in all the samples. Figure 4 shows three approximate RSMT trees for the cervical cancer sample of patient 29, constructed by FISHtree (Fig. 4(a), tree weight = 83), iFISHtree (Fig. 4(b), tree weight = 82) and mpFISHtree (Fig. 4(c), tree weight = 81), respectively.

Table 1. Comparison on the real datasets for breast cancer samples (**Exact** results are not available due to the time limitation). The best tree weights are shown in bold for each sample. The number of Steiner nodes is shown in parenthesis. 7 breast cancer samples have ties in tree weights and thus are not included due to the space limit.

Case #	Tree weight (# Steiner nodes)		
	FISHtree	iFISHtree	mpFISHtree
B1_IDC	213 (15)	212 (13)	**211** (19)
B1_DCIS	241 (14)	242 (15)	**239** (22)
B2_IDC	217 (15)	216 (20)	**211** (22)
B2_DCIS	56 (2)	56 (2)	**55** (3)
B3_DCIS	100 (7)	**98** (7)	**98** (10)
B4_IDC	214 (16)	**213** (17)	**213** (17)
B6_IDC	112 (4)	**111** (4)	**111** (6)
B7_IDC	116 (8)	**113** (12)	**113** (12)
B7_DCIS	186 (13)	184 (14)	**182** (22)
B9_IDC	222 (22)	217 (25)	**213** (30)
B9_DCIS	164 (12)	163 (13)	**161** (15)
B10_IDC	128 (4)	128 (4)	**127** (4)
B10_DCIS	146 (6)	**145** (8)	**145** (9)
B11_DCIS	136 (6)	135 (7)	**134** (7)
B12_IDC	201 (9)	200 (10)	**198** (15)
B12_DCIS	161 (9)	161 (10)	**158** (13)
B13_IDC	132 (7)	**131** (8)	**131** (8)
B13_DCIS	63 (3)	**62** (4)	**62** (4)

Table 2. Comparison on the real datasets for cervical cancer samples. The best tree weights are shown in bold for each sample. The number of Steiner nodes is shown in parenthesis. 24 cervical cancer samples have ties in tree weights and thus are not included due to the space limit.

Case #	Tree weight (# Steiner nodes)			
	FISHtree	iFISHtree	mpFISHtree	Exact
C5	195 (13)	196 (12)	**194 (13)**	**194 (13)**
C6	82 (2)	82 (2)	**81 (5)**	**81 (4)**
C8	103 (6)	103 (6)	**100 (9)**	**100 (8)**
C9	143 (1)	**142 (2)**	**142 (5)**	**142 (2)**
C10	87 (0)	**86 (1)**	**86 (1)**	**86 (1)**
C12	72 (1)	**71 (2)**	**71 (2)**	**71 (2)**
C13	150 (5)	150 (5)	**149 (7)**	**149 (7)**
C15	74 (1)	**73 (2)**	**73 (2)**	**73 (2)**
C18	127 (4)	127 (4)	**126 (6)**	**126 (6)**
C21	**73 (4)**	74 (3)	**73 (5)**	**73 (4)**
C27	59 (1)	**57 (3)**	**57 (2)**	**57 (3)**
C29	83 (2)	82 (3)	**81 (3)**	**81 (3)**
C30	118 (9)	118 (9)	**116 (9)**	**116 (10)**
C32	209 (7)	207 (9)	**205 (14)**	**205 (13)**
C34	83 (5)	**82 (6)**	**82 (6)**	**82 (6)**
C35	67 (1)	67 (1)	**66 (2)**	**66 (3)**
C42	199 (7)	198 (9)	**197 (12)**	**197 (11)**
C45	172 (10)	**169 (13)**	**169 (14)**	**169 (15)**
C46	110 (5)	109 (6)	**108 (8)**	**108 (7)**
C49	162 (4)	**161 (5)**	**161 (7)**	**161 (7)**
C53	80 (3)	**79 (4)**	**79 (4)**	**79 (4)**
C54	146 (6)	145 (7)	**144 (10)**	**144 (9)**

3.2 Simulated Cancer Data

We also test on simulated datasets generated for different number of gene probes (4, 6, 8) and for different tree growth factors (0.4 and 0.5) [1,2]. For each pair of parameters, we simulate 200 samples with the number of distinct cell count patterns varying from 120 to 150. Table 3 summarizes the number of times each of the methods, FISHtree, iFISHtree, mpFISHtree and Exact, obtains the best results on these simulation datasets. We observe that mpFISHtree outperforms both FISHtree and iFISHtree by a large margin. Thank to the very efficient implementation of TNT [37], the running time of mpFISHtree is comparable to that of FISHtree, iFISHtree, all of which is orders of magnitude faster than the exact method (i.e., we could not derive the optimal solutions within a reasonable amount of time when there are more than 4 gene probes, shown as N/A in Table 3)."

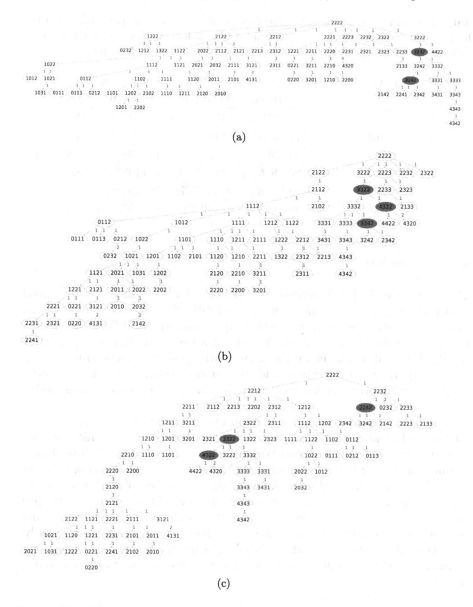

Fig. 4. Given the metastatic cervical cancer sample of patient 12, (a) approximate RSMT constructed by FISHtree with weight 83, (b) approximate RSMT constructed by iFISHtree with weight 82 and (c) approximate RSMT constructed by mpFISHtree with weight 81. Each node in the tree is labeled by a cell count pattern of four gene probes LAMP3, PROX1, PRKAA1 and CCND1. Each white node represents an input cell count pattern, and each red node represents an inferred Steiner node. Branch lengths are shown in blue (Color figure online).

Table 3. Comparison on simulated datasets: number of times and percentage that the best scoring tree (including ties) is obtained by the four methods. **Exact** results for datasets with over four gene probes are not available due to the time limitation.

Probe #	Growth factor	Best score count (Best score percentage)			
		FISHtree	iFISHtree	mpFISHtree	Exact
4	0.4	92 (46%)	137 (68.5%)	196 (98%)	200
6	0.4	70 (35%)	98 (49%)	194 (97%)	N/A
8	0.4	41 (20.5%)	69 (34.5%)	196 (98%)	N/A
4	0.5	93 (46.5%)	130 (65%)	194 (97%)	200
6	0.5	68 (34%)	99 (49.5%)	196 (98%)	N/A
8	0.5	40 (20%)	64 (32%)	195 (97.5%)	N/A

4   Discussion

The Rectilinear Steiner Minimum Tree (RSMT) has been shown to be a good model for progression of cancer cells using FISH cell count pattern data [1,2]. Efficient heuristics are necessary to obtain approximations to RSMT since finding the optimal solution is NP–hard. We present a new algorithm to approximate the RSMT based on Maximum Parsimony (MP) phylogeny reconstruction.

Our experiments on synthetic and real datasets demonstrate the superiority of our algorithm over previous methods in obtaining better parsimonious models of cancer evolution.

RSMT instances found by our method (as well as previous heuristics) have multiple solutions with the same tree weight and additional constraints are needed to choose one from them. We choose the parsimonious solution of minimizing the Steiner nodes introduced by MP reconstruction. Proving that our method produces the solution with the minimum number of Steiner nodes and exploring other strategies to choose from multiple RSMT solutions remain open problems. Chowdhury *et al.* recently extended the evolutionary model of tumor progression by gene copy number changes of single probes to all probes, jointly, on a gene, a chromosome and the whole genome [2]. Extensions of our new method to this more general model are possible, but remain to be thoroughly tested.

Acknowledgements. We thank Lingxi Zhou, Bin Feng,and Yan Zhang for helpful comments. JZ, WH and, JT were funded by NSF IIS 1161586 and an internal grant from Tianjin University, China. YL was supported by a fellowship of the Swiss National Science Foundation. The funders had no role in study design, data collection and analysis, decision to publish, or preparation of the manuscript.

References

1. Chowdhury, S.A., Shackney, S.E., Heselmeyer-Haddad, K., Ried, T., Schäffer, A.A., Schwartz, R.: Phylogenetic analysis of multiprobe fluorescence in situ hybridization data from tumor cell populations. Bioinformatics **29**(13), 189–198 (2013)
2. Chowdhury, S.A., Shackney, S.E., Heselmeyer-Haddad, K., Ried, T., Schäffer, A.A., Schwartz, R.: Algorithms to model single gene, single chromosome, and whole genome copy number changes jointly in tumor phylogenetics. PLoS Comput. Biol. **10**(7), 1003740 (2014)
3. Weinberg, R.: The Biology of Cancer. Garland Science, New York (2013)
4. Futreal, P.A., Coin, L., Marshall, M., Down, T., Hubbard, T., Wooster, R., Rahman, N., Stratton, M.R.: A census of human cancer genes. Nat. Rev. Cancer **4**(3), 177–183 (2004)
5. Swanton, C.: Intratumor heterogeneity: evolution through space and time. Cancer Res. **72**(19), 4875–4882 (2012)
6. Greaves, M., Maley, C.C.: Clonal evolution in cancer. Nature **481**(7381), 306–313 (2012)
7. Yates, L.R., Campbell, P.J.: Evolution of the cancer genome. Nat. Rev. Genet. **13**(11), 795–806 (2012)
8. Attolini, C.S.-O., Michor, F.: Evolutionary theory of cancer. Ann. N. Y. Acad. Sci. **1168**(1), 23–51 (2009)
9. Baudis, M.: Genomic imbalances in 5918 malignant epithelial tumors: an explorative meta-analysis of chromosomal cgh data. BMC Cancer **7**(1), 226 (2007)
10. Pleasance, E.D., Cheetham, R.K., Stephens, P.J., McBride, D.J., Humphray, S.J., Greenman, C.D., Varela, I., Lin, M.-L., Ordóñez, G.R., Bignell, G.R., et al.: A comprehensive catalogue of somatic mutations from a human cancer genome. Nature **463**(7278), 191–196 (2009)
11. Martins, F.C., De, S., Almendro, V., Gönen, M., Park, S.Y., Blum, J.L., Herlihy, W., Ethington, G., Schnitt, S.J., Tung, N., et al.: Evolutionary pathways in brca1-associated breast tumors. Cancer Discov. **2**(6), 503–511 (2012)
12. Navin, N., Krasnitz, A., Rodgers, L., Cook, K., Meth, J., Kendall, J., Riggs, M., Eberling, Y., Troge, J., Grubor, V., et al.: Inferring tumor progression from genomic heterogeneity. Genome Res. **20**(1), 68–80 (2010)
13. Cheng, Y.-K., Beroukhim, R., Levine, R.L., Mellinghoff, I.K., Holland, E.C., Michor, F.: A mathematical methodology for determining the temporal order of pathway alterations arising during gliomagenesis. PLoS Comput. Biol. **8**(1), 1002337 (2012)
14. Shlien, A., Malkin, D.: Copy number variations and cancer. Genome Med **1**(6), 62 (2009)
15. Zack, T.I., Schumacher, S.E., Carter, S.L., Cherniack, A.D., Saksena, G., Tabak, B., Lawrence, M.S., Zhang, C.-Z., Wala, J., Mermel, C.H., et al.: Pan-cancer patterns of somatic copy number alteration. Nat. Genet. **45**(10), 1134–1140 (2013)
16. De Bont, R., Van Larebeke, N.: Endogenous dna damage in humans: a review of quantitative data. Mutagenesis **19**(3), 169–185 (2004)
17. Schärer, O.D.: Dna interstrand crosslinks: natural and drug-induced dna adducts that induce unique cellular responses. ChemBioChem **6**(1), 27–32 (2005)
18. Sung, J.-S., Demple, B.: Roles of base excision repair subpathways in correcting oxidized abasic sites in dna. Febs J. **273**(8), 1620–1629 (2006)
19. Caldecott, K.W.: Single-strand break repair and genetic disease. Nat. Rev. Genet. **9**(8), 619–631 (2008)

20. Cleaver, J.E., Lam, E.T., Revet, I.: Disorders of nucleotide excision repair: the genetic and molecular basis of heterogeneity. Nat. Rev. Genet. **10**(11), 756–768 (2009)
21. Sale, J.E., Lehmann, A.R., Woodgate, R.: Y-family dna polymerases and their role in tolerance of cellular dna damage. Nat. Rev. Mo. Cell Biol. **13**(3), 141–152 (2012)
22. Chapman, J.R., Taylor, M.R., Boulton, S.J.: Playing the end game: Dna double-strand break repair pathway choice. Mo. Cell **47**(4), 497–510 (2012)
23. Wolters, S., Ermolaeva, M.A., Bickel, J.S., Fingerhut, J.M., Khanikar, J., Chan, R.C., Schumacher, B.: Loss of caenorhabditis elegans brca1 promotes genome stability during replication in smc-5 mutants. Genetics **196**(4), 985–999 (2014)
24. Pennington, G., Smith, C.A., Shackney, S., Schwartz, R.: Reconstructing tumor phylogenies from heterogeneous single-cell data. J. Bioinf. Comput. Biol. **5**(02a), 407–427 (2007)
25. Xu, X., Hou, Y., Yin, X., Bao, L., Tang, A., Song, L., Li, F., Tsang, S., Wu, K., Wu, H., et al.: Single-cell exome sequencing reveals single-nucleotide mutation characteristics of a kidney tumor. Cell **148**(5), 886–895 (2012)
26. Von Heydebreck, A., Gunawan, B., Füzesi, L.: Maximum likelihood estimation of oncogenetic tree models. Biostatistics **5**(4), 545–556 (2004)
27. Greenman, C.D., Pleasance, E.D., Newman, S., Yang, F., Fu, B., Nik-Zainal, S., Jones, D., Lau, K.W., Carter, N., Edwards, P.A., et al.: Estimation of rearrangement phylogeny for cancer genomes. Genome Res. **22**(2), 346–361 (2012)
28. Gerstung, M., Baudis, M., Moch, H., Beerenwinkel, N.: Quantifying cancer progression with conjunctive bayesian networks. Bioinformatics **25**(21), 2809–2815 (2009)
29. Langer-Safer, P.R., Levine, M., Ward, D.C.: Immunological method for mapping genes on drosophila polytene chromosomes. Proc. Natl. Acad. Sci. **79**(14), 4381–4385 (1982)
30. Zhou, J., Lin, Y., Hoskins, W., Tang, J.: An iterative approach for phylogenetic analysis of tumor progression using FISH copy number. In: Harrison, R., Li, Y., Mandoiu, I. (eds.) ISBRA 2015. LNCS, vol. 9096, pp. 402–412. Springer, Heidelberg (2015)
31. Wangsa, D., Heselmeyer-Haddad, K., Ried, P., Eriksson, E., Schäffer, A.A., Morrison,L.E., Luo, J., Auer, G., Munck-Wikland, E., Ried, T., et al.: Fluorescence insitu hybridization markers for prediction of cervical lymph node metastases. Am. J. Pathol. **175**(6), 2637–2645 (2009)
32. Goloboff, P.A., Farris, J.S., Nixon, K.C.: TNT, a free program for phylogenetic analysis. Cladistics **24**(5), 774–786 (2008)
33. Goloboff, P.A., Mattoni, C.I., Quinteros, A.S.: Continuous characters analyzed as such. Cladistics **22**(6), 589–601 (2006)
34. Garey, M.R., Johnson, D.S.: The rectilinear steiner tree problem is np-complete. SIAM J. Appl. Math. **32**(4), 826–834 (1977)
35. Day, W.H.: Computational complexity of inferring phylogenies from dissimilarity matrices. Bull. Math. Biol. **49**(4), 461–467 (1987)
36. Swofford, D.L., Maddison, W.P.: Reconstructing ancestral character states under wagner parsimony. Math. Biosci. **87**(2), 199–229 (1987)
37. Giribet, G.: Efficient tree searches with available algorithms. Evolutionary bioinformaticsonline 3, 341 (2007)

Multiple-Ancestor Localization for Recently Admixed Individuals

Yaron Margalit[1], Yael Baran[1], and Eran Halperin[1,2,3]([✉])

[1] The Blavatnik School of Computer Science, Tel Aviv University, Tel Aviv, Israel
[2] Department of Molecular Microbiology and Biotechnology, Tel-Aviv University,
Tel Aviv, Israel
[3] International Computer Science Institute, Berkeley, CA 94704, USA
heran@post.tau.ac.il

Abstract. Inference of ancestry from genetic data is a fundamental problem in computational genetics, with wide applications in human genetics and population genetics. The treatment of ancestry as a continuum instead of a categorical trait has been recently advocated in the literature. Particularly, it was shown that a European individual's geographic coordinates of origin can be determined up to a few hundred kilometers of error using spatial ancestry inference methods. Current methods for the inference of spatial ancestry focus on individuals for whom all ancestors originated from the same geographic location.

In this work we develop a spatial ancestry inference method that aims at inferring the geographic coordinates of ancestral origins of recently admixed individuals, *i.e.* individuals whose recent ancestors originated from multiple locations. Our model is based on multivariate normal distributions integrated into a two-layered Hidden Markov Model, designed to capture both long-range correlations between SNPs due to the recent mixing and short-range correlations due to linkage disequilibrium. We evaluate the method on both simulated and real European data, and demonstrate that it achieves accurate results for up to three generations of admixture. Finally, we discuss the challenges of spatial inference for older admixtures and suggest directions for future work.

Keywords: Admixture · Ancestry inference · Spatial model · Hidden Markov Model · Multivariate-normal distribution

1 Introduction

Determining the ancestry of individuals based on their DNA sequence is a common and useful task in the study of human genetics: In addition to its multiple applications in population genetics [5,7,9,10,24], it has a critical role in correcting for confounders in disease studies [19]: If different ancestral composition in cases and controls is not accounted for, ancestry-informative markers will appear

---

Y. Margalit and Y. Baran—Contributed equally to this work.

M. Pop and H. Touzet (Eds.): WABI 2015, LNBI 9289, pp. 121–135, 2015.
DOI: 10.1007/978-3-662-48221-6_9

to be disease-related. Admixed individuals, *i.e.* individuals whose ancestry is mixed, are of special importance in discovering disease-associated traits. The analysis of individuals from admixed populations, such as African-Americans and Latinos, have allowed for the detection of multiple disease association signals [8,12,28] using Mapping by Admixture Linkage Disequilibrium (MALD) [22] as well as other computational techniques [21]. An accurate characterization of the ancestral origins of study samples is therefore critical when population structure exists in the dataset; improved accuracy should lead to more accurate control of confounding, which in turn leads to increased statistical power to detect disease-related markers.

Until recently, the ancestry of admixed individuals was treated as a discrete trait, and individuals were classified into pre-defined classes (*e.g.* Tuscan, Sicilian, Catalan). For admixed individuals, different methods were developed to determine, under the assumption that their multiple ancestries are known, the fraction of genome originating from each one [1,20], and the ancestral origin of each genomic region [2,6,13,18]. In reality, however, ancestries are not discrete, because different populations go through constant mixing whose rate is determined to a large extent by the geographic distance between them. Ancestries are therefore better described as a continuum strongly correlated with geographic structure. One recent work attempts to deal with this challenge by modeling individuals as a mixture of a large panel of reference populations, in a procedure that does not require any prior knowledge about the number or identity of the mixture components; this approach moves closer to a continuous representation, but still suffers from the principal drawbacks of other discrete approaches. Other works tried to cope with this challenge by first classifying the parts of the genomes into continents of origins, and then obtaining within-continent continuous spatial separation using Principal Component Analysis (PCA) [11,15]. One serious problem with this approach is that when the mixing populations are from the same continent, *e.g.* when attempting to separate two different European origins [13], the preliminary classification stage is inaccurate. A more fundamental problem is that the use of PCA for localization is merely a heuristic, and although PCA maps sometimes fit well with the geographic map [17], this is not always the case [23].

Motivated by above difficulties, a few probabilistic spatio-genetic models have recently been developed [3,26,27]. These models describe the allele frequency of each SNP as a continuous function of the geographic coordinates. In addition to improved accuracy of localization compared with PCA, these models can be naturally extended to describe individuals of mixed origin. Indeed, one of these methods named SPA [26] included an immediate extension of its model for the inference of two different origins, paternal and maternal. As we show below, this trivial extension does not provide accurate localization. In addition, the inference becomes more challenging in the presence of multiple generations of mixture, *e.g.* when each of the four grandparents originates from a different ancestral location. Here too, a recent effort to address this problem has been made by the method SPAMIX [27], but as we show below this method does not provide accurate localization either. Multiple-ancestor geographic localization is therefore an open challenge.

In this paper we provide LIZARD (LocalIZAtion of Recently aDmixed individuals), a method for estimating the multiple geographic coordinates of origin for individuals of recent admixed ancestry. Our approach is based on modeling both the long- and short-range correlations that exist in the genetic sequence of admixed individuals. The long-range correlations, also termed *Admixture Linkage Disequilibrium*, result from the recent mixing, due to which the chromosomes become mosaics of long haplotypic segments originating from the same location; because origin affects sequence, the alleles of SNPs residing on the same ancestral haplotype are correlated. Next, given a haplotype's location of origin, short-range correlations exist between nearby SNPs due to Linkage Disequilibrium (LD). Our model captures the short-range correlations by modeling haplotypes within short genomic windows using the multivariate normal distribution (MVN). The long-range correlations are captured by combining these windows into Hidden Markov Models (HMMs) whose structure and between-window transition rates are determined by the mixing pattern.

We evaluate LIZARD on both simulated and real data of European individuals. LIZARD attains high accuracy in localizing the two parental locations of individuals whose father and mother originate from different geographic regions in Europe, with a median error of 374 km, considerably better than other existing approaches. We use real individuals of admixed ancestry from Europe whose multiple origins are known to validate the method on real data, and observe that LIZARD's localization accuracy remains high and similar to the simulation results. Next, we test our approach of individuals of 2 and 3-generation admixture. As expected, the method's performance deteriorates as g increases, though the results remain useful for downstream applications (median error of 478 and 571 km for 2 and 3 generations, respectively). We discuss the limitations of our approach for large values of g and suggest directions for future work. A software package implementing our method will be freely available upon publication of the manuscript.

2 Materials and Methods

We define an individual to be *of homogeneous ancestry* (or simply *homogenous*) if all of their ancestors originated from the same geographic location, *i.e.* from identical geographic coordinates. In addition, we say that an admixed individual resulted from a *g-generations admixture* if each of their 2^g ancestors from g generations ago is of homogeneous ancestry, and g is the smallest number for which this holds. In reality, locations are never identical and individuals are never homogenous, as we are all mixed to some extent, but we can use them as approximations when the scale of geographic variation in the study sample is large enough.

We assume that a reference panel of $2n$ haplotypes whose locations of origin are known is available through phasing the genotypes of n individuals of homogenous ancestry. We denote the haplotypes by $H = (h_1, ..., h_{2n})$ and the corresponding locations by $X = (x_1, ..., x_{2n})$. Each location is a vector which contains longitude and latitude coordinates, and since the individuals are homogenous $x_{2i-1} = x_{2i} \; \forall i \in \{1 \ldots n\}$. Given an individual of recent admixed ancestry, our goal is to predict their geographical origins by utilizing the information in the reference panel.

We begin by introducing our spatial model and a procedure for estimating its parameters from a group of homogenous individuals. We then present a model for individuals of 1-generation admixture, *i.e.* whose parents originate from different locations but are themselves of homogeneous ancestry, and show how it can be used to estimate the two parental origins. Finally, we extend this model to localize individuals of g-generations admixture (for instance, a case of two generations with four different ancestries).

2.1 Estimation of Spatial Parameters from Training Data

We split the haplotypes into L non-overlapping contiguous *windows* of l SNPs per window, and denote by h_{ij} the part of haplotype h_i confined to window j. Similarly to [3], our model assumes that given h_i's location of origin, x_i, h_{ij} is sampled from a multivariate normal distribution (MVN), with window-specific and location-dependent expectation $\beta_j x_i$ and window-specific covariance Σ_j. Here β_j is an $l \times d$ matrix and Σ_j is an $l \times l$ matrix, where d is the dimension of the spatial representation - 2 in our case, for latitude and longitude. The probability of observing haplotype h_i in window j given location x_i equals the multivariate normal likelihood:

$$L(h_{ij} \mid \beta_j, \Sigma_j, x_i) = MVN(h_{ij}; \beta_j x_i, \Sigma_j)$$

$$= \frac{1}{(2\pi)^{\frac{l}{2}} \mid \Sigma_j \mid^{\frac{1}{2}}} e^{-\frac{1}{2}(\beta_j x_i - h_{ij})^T \Sigma_j^{-1} (\beta_j x_i - h_{ij})} \tag{1}$$

Despite the fact that the multivariate normal distribution is continuous while genotypes are discrete, MVNs has been shown to model genotypic data well in multiple contexts [6,14,25], including localization [3]. The advantage of using MVNs is the ability to derive closed-form, efficiently-computable maximum likelihood solutions for the model parameters while accounting for pairwise correlations between SNPs.

The training stage in which we estimate the parameters of the multivariate normal distribution from the reference haplotypes has been previously derived [3], and we describe it here briefly. Denote by H_j the matrix whose ith column is haplotype h_{ij}, and by X the matrix whose ith column is the corresponding location vector x_i. Then the maximum likelihood estimator of β_j, Σ_j have the following closed form solution:

$$\hat{\beta}_j = H_j X^T (X X^T)^{-1} \tag{2}$$

$$\hat{\Sigma}_j = \frac{1}{2n} \sum_{i=1}^{2n} (\hat{\beta}_j x_i - h_{i,j})(\hat{\beta}_j x_i - h_{i,j})^T \tag{3}$$

We follow the standard procedure of adding a small positive constant ($\epsilon = 0.01$) to the diagonal of $\hat{\Sigma}_j$ in order to ensure it is full rank and eliminate potential overfitting of the covariance parameters due to the limited sample size.

2.2   Localization of Individuals of 1-Generation Admixture

We are given the two phased haplotypes h^1, h^2 of an individual whose parents are homogenous, and would like to determine the two corresponding locations of origin x_1, x_2. If phasing was error-free, the problem would be reduced to separately localizing each haplotype; to do that, we can use known methods that infer the location of haplotypes of homogeneous ancestry [3,26]. Unfortunately, perfect phasing is typically not feasible, and as a result, each haplotype is mosaic of parental and maternal segments.

Our model for 1-generation admixed individuals accounts for phasing errors by allowing for the two locations of origin to change along the haplotypes. In addition, the model assumes that there are no phasing errors within a window, and therefore h_j^1 and h_j^2 originate from a single location for every window j. As a result, we allow for one of two *phasing arrangements* per windows: Either h_j^1 originated from x_1 and h_j^2 from x_2, or the other way around (see Fig. 1 for illustration). Although the assumption of perfect phasing within windows may not hold for all windows, if the window size is set appropriately, it should hold for most of them.

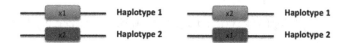

Fig. 1. The two possible phasing arrangements per window. Given the two phased haplotypes of a 1-generation admixed individual confined to a single genomic window and the individual's two ancestral locations x_1, x_2, for most windows it holds that either h^1 originated from x_1 and h^2 originated from x_2, or the other way around.

The exact structure of the model is as follows: We combine the window-specific parameters estimated from the reference panel as in Subsect. 2.1 into an HMM specified by the triplet (Q, ϵ, δ): Q is the set of states, δ are the transition probabilities, and ϵ are the emission probability functions. The set Q contains 2 x L states: For each window $j \in \{1 \ldots L\}$ there are two states, denoted $s_j = \{s_j^1, s_j^2\}$, for the two possible phasing arrangements. Given x_1 and x_2, the two states of window j emit the haplotypes in the window in probabilities that are determined by that MVN densities estimated in Eq. (2):

$$\epsilon_{s_j^1}(h_j^1, h_j^2; x_1, x_2) = MVN(h_j^1; \beta_j x_1, \Sigma_j) \cdot MVN(h_j^2; \beta_j x_2, \Sigma_j)$$

$$\epsilon_{s_j^2}(h_j^1, h_j^2; x_1, x_2) = MVN(h_j^2; \beta_j x_1, \Sigma_j) \cdot MVN(h_j^1; \beta_j x_2, \Sigma_j) \tag{4}$$

The between-states transition rate is constant across windows and is denoted p_s. This constant is determined by the phasing error rate and the window size, and its exact choice is discussed in Sect. 2.5. We therefore have

$$\delta(s_{(j-1)}^{k1} \rightarrow s_j^{k2}) = \begin{cases} 1 - p_s & \text{if } k1 = k2 \\ p_s & \text{otherwise.} \end{cases} \tag{5}$$

When performing localization of a 1-generation admixed individual, the HMM we have just described is nearly fully parameterized; the only missing parameters are the two location vectors (x_1, x_2). We use the Baum-Welch algorithm to estimate these parameters, thereby localizing the individual. Specifically, denote by (z_j^1, z_j^2) the indicator variables for the states (s_j^1, s_j^2):

$$z_j^1 = I_{s_j = s_j^1} = \begin{cases} 1 & \text{if } s_j = s_j^1 \\ 0 & \text{if } s_j = s_j^2 \end{cases} \tag{6}$$

and similarly for z_j^2, so that for each window $z_j^1 + z_j^2 = 1$. In iteration t of the algorithm we use the Forward-Backward algorithm to compute $(z_j^{1(t)}, z_j^{2(t)})$, the posterior probabilities of the indicator variables, for every window j. We then search for the location parameters that maximize the expected log likelihood:

$$(x_1^{(t)}, x_2^{(t)}) = \operatorname*{argmax}_{x_1, x_2} \sum_{j=1}^{L} \left[(z_j^{1(t)} \log(MVN(h_j^1; \beta_j x_1, \Sigma_j) \cdot MVN(h_j^2; \beta_j x_2, \Sigma_j)) \right.$$
$$\left. + z_j^{2(t)} \log(MVN(h_j^1; \beta_j x_2, \Sigma_j) \cdot MVN(h_j^2; \beta_j x_1, \Sigma_j)) \right] \tag{7}$$

The values of x_i, $i = 1, 2$ that maximize the above expression have the following closed form solutions:

$$x_i^{(t)^T} = \left(\sum_{j=1}^{L} \left(z_j^{i(t)} h_j^1 + (1 - z_j^{i(t)}) h_j^2 \right)^T \Sigma_j^{-1} \beta_j \right) \left(\sum_{j=1}^{L} (\beta_j^T \Sigma_j^{-1} \beta_j) \right)^{-1} \tag{8}$$

We repeat the expectation-maximization iterations until convergence of the log likelihood (change smaller than 10^{-8}), allowing for up to 100 iterations.

2.3   A Spatial Model for Individuals of g-generations Admixture

We now extend the model to describe individuals of g-generation admixture. For such individuals each of the chromosomes (paternal and maternal) has originated from a different (but potentially overlapping) set of up to 2^{g-1} locations. Because the admixture is recent, the pair of locations in window j are highly correlated with the pair of locations of window $j + 1$. In order to capture these correlations, we group each K consecutive windows into a *block*, and assume no recent recombinations within blocks. As a result, all windows within the block

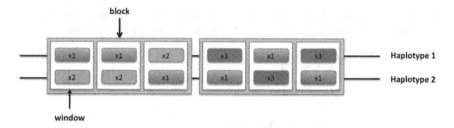

Fig. 2. Model Assumptions for g-generation admixed individuals. Two assumptions are made: (1) The two ancestral locations are constant within block (colored gray), (2) no phasing errors within windows (colored white). In the figure, all windows in the first block contain the locations x_1 and x_2, but different windows in the block have different phasing arrangements. In the second block x_3 replaces x_2 due to a recombination that occurred in the parental DNA (Color figure online).

contain the same two locations of origin, though perhaps with different phasing arrangements. These assumptions are illustrated in Fig. 2. We note that in our model windows and blocks have fixed sizes across the genome, but a more accurate approach would be to set their sizes according to the region-specific recombination rates.

We model these assumptions using a two-level HMM defined as follows (and illustrated in Fig. 3):

1. Per each block we construct $2^{2(g-1)}$ bottom-level HMMs, one for each unordered pair of parental locations - a paternal one and a maternal one. Each of these HMMs contains $K \times 2$ states, thereby capturing all possible phasing arrangements in the block's windows. These HMMs are identical to the HMM defined in Sect. 2.2, but are confined to a single block.
2. We combine the bottom-level HMMs into a single top-level HMM. Denote by B the number of blocks in the genome, then the top-level HMM has $B \times 2^{2(g-1)}$ states, corresponding to all possible assignments of paternal and maternal locations along the genome.

We estimate the 2^g location vectors using the Baum-Welch algorithm. The Expectation step in iteration t consists of the following two sub-steps:

1. For each block b and for each pair of locations r, we run the Forward-Backward algorithm on the bottom-level HMM. This computation yields the emission probability of block b and pair r in the top-level HMM. The computation also gives us $(z_{br1}^{1(t)}, z_{br1}^{2(t)}) \ldots (z_{brK}^{1}, z_{brK}^{2})$, the posterior probabilities of the per-window state variables in this block for this pair.
2. We run the Forward-Backward algorithm on the top level HMM. The transition probabilities between blocks are assumed to be fixed along the genome and are determined by the average genome-wide recombination rate and by g. For block b this computation gives us $(w_{b1}^{(t)} \ldots w_{b2^{2(g-1)}}^{(t)})$, the conditional probabilities of the per-block indicator variables corresponding to each of the $2^{2(g-1)}$ states.

In the maximization step we derive the locations that maximize the expected log likelihood. The optimal x_i has the following closed form solution:

$$x_i^T = \left(\sum_{b=1}^{B} \sum_{r \in R_i} \sum_{j=1}^{K} w_{br}^{(t)} (z_{brj}^{1(t)} h_j^1 + z_{brj}^{2(t)} h_j^2)^T \Sigma_j^{-1} \beta_j \right)$$
$$\times \left(\sum_{b=1}^{B} \sum_{r \in R_i} \sum_{j=1}^{K} w_{br}^{(t)} \beta_j^T \Sigma_j^{-1} \beta_j \right)^{-1} \tag{9}$$

where R_i is the set of location pairs that include x_i as one of the two locations.

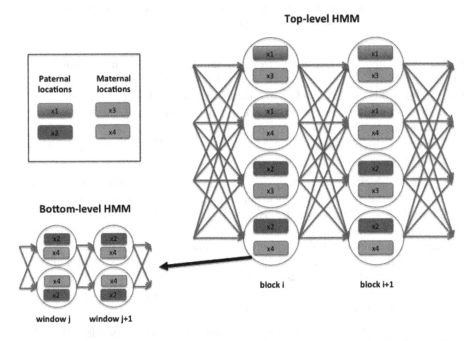

Fig. 3. Two-level HMM for modeling 2-generation admixed individuals. The top-level HMM includes four states per block - one for every unordered combination of paternal and maternal states. In addition, for every block we define a bottom-level HMM over the block's windows: Given the block's top-level state, each window has two states, one for every possible phasing arrangement.

2.4   Simulation Setup

We tested our methods on simulated data generated from the POPRES European samples [16]. Our dataset consists of 364,373 SNPs with minor allele frequency > 0.01 and no-call rate $< 10\%$. We used BEAGLE [4] for phasing and imputation. This data contains 1385 individuals whose four grandparents were reported to originate from the same country. Following phasing we obtained a set of 2×1385 *homogenous haplotypes*, one part of which was used for training the different methods, the other for simulating admixed individuals.

We simulated European admixed haplotypes as mosaics of homogenous haplotypes. Specifically, for a g-generation admixed individual we first drew the number of recombinations from a Poisson distribution with a parameter determined by g, and then uniformly sampled the paternal and maternal locations, separately for the paternal and maternal haplotypes. Finally, we used the homogenous haplotypes to fill out the genotype values according to the determined recombination events. For all simulated individuals we assumed all 2^g ancestral origins to be different from each over. Overall we simulated 1500, 500 and 300 Europeans of 1, 2 and 3-generation admixture, respectively. Finally, we introduced phasing errors into the phased haplotypes at a rate determined through simulations. Specifically, we joined pairs of phased homogenous haplotypes from different individuals to form artificial genotypes, used BEAGLE to phase them, and measured the error rate produced by the phasing procedure.

2.5   Method Comparison

We compared LIZARD to two recently published methods for the inference of ancestral locations in recently admixed individuals, SPA [26] and SPAMIX [27]. In a nutshell, SPA's model assumes that a SNP's frequency changes linearly across the geographic space, similarly to LIZARD. However, SPA does not model LD, and only works for 1-generation admixture. As for SPAMIX, it does not model LD either, but does model g-generation admixture as well as admixture LD. In accordance with these differences, SPA has only $\mathcal{O}(m)$ parameters (m is the number of SNPs), all of them spatial, while SPAMIX has $\mathcal{O}(2^g)$ additional parameters that determine the individual's admixture proportions. As for LIZARD, its model contains $\mathcal{O}(ml)$ spatial parameters (l is the window size) in order to capture local LD, and only a fixed number of parameters that determine the HMM transition rates. We note that both SPA and SPAMIX should perform approximately the same on 1-generation admixture due to the lack of admixture-LD in these individuals.

All methods were evaluated on simulated data (see Sect. 2.4) of 1-generation admixed Europeans; in addition, LIZARD and SPAMIX were evaluated on 2,3-generation admixed individuals, and LIZARD on real European samples. All methods were trained on the same set of POPRES homogenous individuals - SPA and SPAMIX on genotypes, LIZARD on haplotypes. LIZARD's window size (l), block size (K) and phasing switch rate (p_p) parameters were tuned on the training data and set to $l = 100$, $K = 20$ and $p_p = 0.1$. These parameters are likely to be optimized by other values in other datasets, but we observe that the method is robust to their setting within a wide range (results not shown). One possible strategy for adjusting these parameters is simulating admixed individuals using each study's specific SNP set and relevant recombination maps, and choosing the optimal values in a cross-validation scheme.

A first measure of performance is the distance in kilometers between the estimated geographical coordinates and the true coordinates; the latter was taken to be the center of the (known) country of origin. The calculation of the distance error involved applying a previously described transformation [3,17,26]. The per-individual error is computed as the average error over all locations. Since for each individual an unordered set of locations is estimated, we choose the permutation that produces the best match between the estimated and the true locations.

A second measure of performance is the fraction of accurate assignments to country of origin. Assignments were obtained by choosing the country whose center is closest to the estimated location. Since multiple countries are assigned per individual, the accuracy we report is the fraction of countries that were correctly detected, averaged over individuals. We emphasize, though, that our method aims at continuous assignment and not at classification, and we report classification here only as a proxy to the former, in the absence of exact location information for the POPRES individuals.

3   Results

3.1   Localization of Simulated 1-Generation Admixed Individuals

1-generation admixed individuals were simulated as described in Sect. 2.4. As expected, SPA and SPAMIX achieve approximately the same results, and LIZARD outperforms both of them, presumably due to its improved modeling approach. In terms of km error, LIZARD attains a median error of 374 km compared with SPA's 1141 km and SPAMIX's 1159 km (Table 1). These differences are reflected also in improved accuracy of assignment to country of origin, as shown in Fig. 4: LIZARD's average success rate is 57 %, while both SPA and SPAMIX attain an average success rate of 42 %. The two country pairs for which LIZARD did not attain the highest accuracy both involve Portugal, and we observe that when making these errors LIZARD localized the samples too far to the East, resulting in a mis-classification to Spain. More generally, LIZARD is more likely to make an error when the two countries involved are in geographic proximity; in some of these cases these supposedly wrong assignments may be artifacts of the assignment scheme, which assigns an individual to the country whose center is the closest.

Table 1. km error on simulated 1-generation admixed Europeans. For each method we given the 0.5 [0.25, 0.75] quantiles of km error over all simulated individuals.

Method	Error
LIZARD	374.95 [267.76, 519.05]
SPA	1141.4 [706.75, 1674.6]
SPAMIX	1159.6 [723.13, 1704.7]

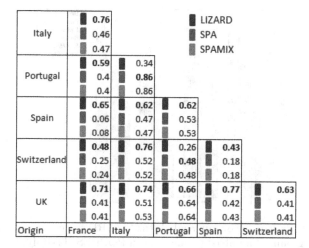

Italy	■ 0.76 ■ 0.46 ■ 0.47				■ LIZARD ■ SPA ■ SPAMIX
Portugal	■ 0.59 ■ 0.4 ■ 0.4	■ 0.34 ■ 0.86 ■ 0.86			
Spain	■ 0.65 ■ 0.06 ■ 0.08	■ 0.62 ■ 0.47 ■ 0.47	■ 0.62 ■ 0.53 ■ 0.53		
Switzerland	■ 0.48 ■ 0.25 ■ 0.24	■ 0.76 ■ 0.52 ■ 0.52	■ 0.26 ■ 0.48 ■ 0.48	■ 0.43 ■ 0.18 ■ 0.18	
UK	■ 0.71 ■ 0.41 ■ 0.41	■ 0.74 ■ 0.51 ■ 0.53	■ 0.66 ■ 0.64 ■ 0.64	■ 0.77 ■ 0.42 ■ 0.43	■ 0.63 ■ 0.41 ■ 0.41
Origin	France	Italy	Portugal	Spain	Switzerland

Fig. 4. Accuracy of assignment to country of origin for simulated 1-generation admixed Europeans. For each pair of locations we give the fraction of haplotypes correctly classified to their country of origin, per method. The panel includes only populations that are represented by at least 40 individuals in the training data. In bold is the method which achieved the highest accuracy.

3.2   Localization of Real Europeans from the POPRES Dataset

We used 254 real 1-generation admixed Europeans from the POPRES dataset to calibrate the performance estimates we obtained in the simulations. LIZARD's median km error on this data was 368 km, but this number cannot be directly compared to the simulations results due to the difference in ancestral composition between the two test panels. We therefore generated additional simulated datasets so that each real individual is matched with a simulated individual with identical locations of origin. We generated ten such simulated datasets so as to account for the variance resulting from the sampling of the haplotypes. Figure 5 shows that LIZARD's localization error on the real data is similar to the error observed in simulation: For example, LIZARD achieves median errors of 396 and 399 km on the real and simulated datasets, respectively, for individuals originating from Switzerland and France.

3.3   Localization of Simulated g-generation Admixed Individuals

We simulated individuals of g-generation admixture as described in Sect. 2.4. LIZARD localizes individuals of 2-generation admixture who originated from $2^g = 4$ different locations with a median error of 478 km, and individuals of 3-generation admixture and $2^g = 8$ different origins with a median error of 571 km (Fig. 6). SPAMIX's error is considerably higher - medians of 1170 and 1332 km for 2 and 3 generations, respectively. The deterioration in the performance of both methods is unsurprising, mostly because as g grows the number of locations increases exponentially with g, while the amount of data to estimate each

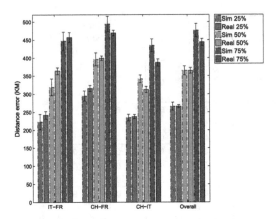

Fig. 5. LIZARD's km error on real data. LIZARD's accuracy was measured on the three main groups of admixed populations in POPRES (IT-Italy; FR-France; CH-Switzerland). The figure gives the 0.25, 0.50 and 0.75 quantiles of km error per group and per method. Error bars for simulated data give the standard error over ten draws of simulated datasets, and for real data the standard error over individuals. The resulting assignment accuracy are 0.68, 0.54 and 0.7 for IT-FR, CH-FR and CH-IT, respectively.

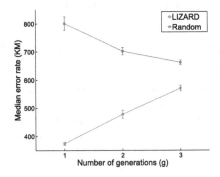

Fig. 6. LIZARD's km error for g-generation admixture. LIZARD is compared with a random assignment. Error bars give the standard error of the median as estimated from a 10-fold cross-validation experiment.

location decreases exponentially. The problem therefore becomes harder very fast if no additional assumptions about the locations are made, and we discuss possible solutions to this in the Discussion. We also note that because we choose the best match between true and estimated locations over all possible permutations (see Sect. 2.5), a method that produces random location estimates is expected to score better as g increases. When compared to a random method (Fig. 6), LIZARD can be seen to achieve significantly better results for up to 3 generations. Our results therefore suggest that LIZARD is suitable only for recently admixed individuals.

4   Discussion

In this paper we presented LIZARD, a new method for the inference of ancestral coordinates for individuals of recent admixed ancestry. LIZARD is capable of accurately inferring the origins of 1-generation admixtures, and with a lesser success of 2 and 3-generation mixtures. Its improved performance compared with existing approaches is achieved by modeling both long-range genomic correlations due to recent admixture, and short-range correlations due to linkage disequilibrium. As a result of using closed-form optimization formulae, LIZARD runs fast: Its training on a reference set of thousands of haplotypes take a few minutes, and localizing each 1-generation individual takes 45 seconds, on average. We note that LIZARD requires haplotype data, and phasing may take up to a few days, depending on the available computational resources; however, phasing is usually performed in any case as a routine part of the data analysis.

As the number of generations in admixture increase, LIZARD's performance deteriorates rapidly. The main reason is that the length of genomic sequence available for determining each origin decrease exponentially with g. Moreover, the average length of each single ancestral segment decreases, and hence the long-range correlations decay faster. Finally, in terms of efficiency, the complexity of our algorithm is exponential in g. We close this paper by suggesting a few enhancements that will enable better handling of larger g values.

First, it is often possible to utilize existing information about the ancestral coordinates. In some cases, priors distributions can be formulated, at least for some of the ancestors. In other cases, it is known that all ancestors from one side of the family originate from the same region, and this information can be easily integrated into our optimization. Second, our model assumes that within a parent's haplotype (paternal or maternal), a segment from any location is equally likely to be followed by a segment from any other location. In fact, the transition patterns between locations contain regularities due to the pedigree structure that induced the mixing, and future methods could model these regularities.

More generally, continuous localization as we have attempted here is a qualitatively more difficult task than classification to discrete ancestries. Spatiogenetic modeling of human data is a relatively new research direction, and much work remains to be done in refining the models that underlie current localization methods; we expect that such refinements will yield a significant improvement in localization accuracy of both homogenous and admixed individuals.

Acknowledgements. E.H. is a faculty fellow of the Edmond J. Safra Center at Tel Aviv University. Y.B. was supported in part by a fellowship from the Edmond J. Safra Center for Bioinformatics at Tel-Aviv University. E.H. and Y.B. were also supported in part by the United States Israel Binational Science Foundation (grant 2012304), and by the National Science Foundation (grant III-1217615), and by the Israeli Science Foundation (grant 989/08). E.H, Y.B, and Y.M were partially supported by the German-Israeli Foundation (grant 1094-33.2/ 2010). E.H was also supported by the Israel Science Foundation (grant 1425/13).

References

1. Alexander, D.H., Novembre, J., Lange, K.: Fast model-based estimation of ancestry in unrelated individuals. Genome Res. **19**(9), 1655–1664 (2009)
2. Baran, Y., Pasaniuc, B., Sankararaman, S., Torgerson, D.G., Gignoux, C., Eng, C., Rodriguez-Cintron, W., Chapela, R., Ford, J.G., Avila, P.C., et al.: Fast and accurate inference of local ancestry in latino populations. Bioinformatics **28**(10), 1359–1367 (2012)
3. Baran, Y., Quintela, I., Carracedo, Á., Pasaniuc, B., Halperin, E.: Enhanced localization of genetic samples through linkage-disequilibrium correction. Am. J. Hum. Genet. **92**(6), 882–894 (2013)
4. Browning, S.R., Browning, B.L.: Rapid and accurate haplotype phasing and missing-data inference for whole-genome association studies by use of localized haplotype clustering. Am. J. Hum. Genet. **81**(5), 1084–1097 (2007)
5. Bryc, K., Velez, C., Karafet, T., Moreno-Estrada, A., Reynolds, A., Auton, A., Hammer, M., Bustamante, C., Ostrer, H.: Genome-wide patterns of population structure and admixture among hispanic/latino populations. Proc. Nat. Acad. Sci. **107**(Supplement 2), 8954–8961 (2010)
6. Churchhouse, C., Marchini, J.: Multiway admixture deconvolution using phased or unphased ancestral panels. Genet. Epidemiol. **37**(1), 1–12 (2013)
7. Gravel, S., Henn, B., Gutenkunst, R., Indap, A., Marth, G., Clark, A., Yu, F., Gibbs, R., Bustamante, C., Altshuler, D., et al.: Demographic history and rare allele sharing among human populations. Proc. Nat. Acad. Sci. **108**(29), 11983–11988 (2011)
8. Haiman, C.A., Patterson, N., Freedman, M.L., Myers, S.R., Pike, M.C., Waliszewska, A., Neubauer, J., Tandon, A., Schirmer, C., McDonald, G.J., et al.: Multiple regions within 8q24 independently affect risk for prostate cancer. Nat. Genet. **39**(5), 638–644 (2007)
9. Hinch, A., Tandon, A., Patterson, N., Song, Y., Rohland, N., Palmer, C., Chen, G., Wang, K., Buxbaum, S., Akylbekova, E., et al.: The landscape of recombination in african americans. Nature **476**(7359), 170–175 (2011)
10. Jarvis, J., Scheinfeldt, L., Soi, S., Lambert, C., Omberg, L., Ferwerda, B., Froment, A., Bodo, J., Beggs, W., Hoffman, G., et al.: Patterns of ancestry, signatures of natural selection, and genetic association with stature in western african pygmies. PLoS Genet. **8**(4), e1002641 (2012)
11. Johnson, N.A., Coram, M.A., Shriver, M.D., Romieu, I., Barsh, G.S., London, S.J., Tang, H.: Ancestral components of admixed genomes in a mexican cohort. PLoS Genet. **7**(12), e1002410 (2011)
12. Kao, W.L., Klag, M.J., Meoni, L.A., Reich, D., Berthier-Schaad, Y., Li, M., Coresh, J., Patterson, N., Tandon, A., Powe, N.R., et al.: Myh9 is associated with nondiabetic end-stage renal disease in african americans. Nature Genet. **40**(10), 1185–1192 (2008)
13. Maples, B.K., Gravel, S., Kenny, E.E., Bustamante, C.D.: Rfmix: a discriminative modeling approach for rapid and robust local-ancestry inference. Am. J. Hum. Genet. **93**(2), 278–288 (2013)
14. Menelaou, A., Marchini, J.: Genotype calling and phasing using next-generation sequencing reads and a haplotype scaffold. Bioinformatics **29**(1), 84–91 (2013)
15. Moreno-Estrada, A., Gravel, S., Zakharia, F., McCauley, J.L., Byrnes, J.K., Gignoux, C.R., Ortiz-Tello, P.A., Martínez, R.J., Hedges, D.J., Morris, R.W., et al.: Reconstructing the population genetic history of the caribbean. PLoS Genet. **9**(11), e1003925 (2013)

16. Nelson, M.R., Bryc, K., King, K.S., Indap, A., Boyko, A.R., Novembre, J., Briley, L.P., Maruyama, Y., Waterworth, D.M., Waeber, G., et al.: The population reference sample, popres: a resource for population, disease, and pharmacological genetics research. Am. J. Hum. Genet. **83**(3), 347–358 (2008)

17. Novembre, J., Johnson, T., Bryc, K., Kutalik, Z., Boyko, A.R., Auton, A., Indap, A., King, K.S., Bergmann, S., Nelson, M.R., et al.: Genes mirror geography within europe. Nature **456**(7218), 98–101 (2008)

18. Price, A.L., Tandon, A., Patterson, N., Barnes, K.C., Rafaels, N., Ruczinski, I., Beaty, T.H., Mathias, R., Reich, D., Myers, S.: Sensitive detection of chromosomal segments of distinct ancestry in admixed populations. PLoS Genet. **5**(6), e1000519 (2009)

19. Price, A.L., Zaitlen, N.A., Reich, D., Patterson, N.: New approaches to population stratification in genome-wide association studies. Nat. Rev. Genet. **11**(7), 459–463 (2010)

20. Pritchard, J.K., Stephens, M., Donnelly, P.: Inference of population structure using multilocus genotype data. Genetics **155**(2), 945–959 (2000)

21. Seldin, M.F., Pasaniuc, B., Price, A.L.: New approaches to disease mapping in admixed populations. Nat. Rev. Genet. **12**(8), 523–528 (2011)

22. Smith, M.W., O'Brien, S.J.: Mapping by admixture linkage disequilibrium: advances, limitations and guidelines. Nat. Rev. Genet. **6**(8), 623–632 (2005)

23. Wang, C., Zöllner, S., Rosenberg, N.A.: A quantitative comparison of the similarity between genes and geography in worldwide human populations. PLoS Genet. **8**(8), e1002886 (2012)

24. Wegmann, D., Kessner, D., Veeramah, K., Mathias, R., Nicolae, D., Yanek, L., Sun, Y., Torgerson, D., Rafaels, N., Mosley, T., et al.: Recombination rates in admixed individuals identified by ancestry-based inference. Nat. Genet. **43**(9), 847–853 (2011)

25. Wen, X., Stephens, M.: Using linear predictors to impute allele frequencies from summary or pooled genotype data. Ann. Appl. Stat. **4**(3), 1158 (2010)

26. Yang, W.Y., Novembre, J., Eskin, E., Halperin, E.: A model-based approach for analysis of spatial structure in genetic data. Nat. Genet. **44**(6), 725–731 (2012)

27. Yang, W.Y., Platt, A., Chiang, C.W.K., Eskin, E., Novembre, J., Pasaniuc, B.: Spatial localization of recent ancestors for admixed individuals. G3: Genes, Genomes, Genet. **4**(12), 2505–2518 (2014)

28. Zhu, X., Luke, A., Cooper, R.S., Quertermous, T., Hanis, C., Mosley, T., Gu, C.C., Tang, H., Rao, D.C., Risch, N., et al.: Admixture mapping for hypertension loci with genome-scan markers. Nat. Genet. **37**(2), 177–181 (2005)

Association Mapping for Compound Heterozygous Traits Using Phenotypic Distance and Integer Programming

Dan Gusfield[1][(✉)] and Rasmus Nielsen[2]

[1] Computer Science Department, University of California, Davis, USA
gusfield@cs.ucdavis.edu
[2] Integrative Biology, University of California, Berkeley, USA

Abstract. For many important *complex* traits, Genome Wide Association Studies (GWAS) have only recovered a small proportion of the variance in disease prevalence known to be caused by genetics. The most common explanation for this is the presence of multiple rare mutations that cannot be identified in GWAS due to a lack of statistical power. Such rare mutations may be concentrated in relatively few genes, as is the case for many known Mendelian diseases, where the mutations are often *compound heterozygous (CH)*, defined below. Due to the multiple mutations, each of which contributes little by itself to the prevalence of the disease, GWAS also lacks power to identify genes contributing to a CH-trait. In this paper, we address the problem of finding genes that are causal for CH-traits, by introducing a discrete optimization problem, called the *Phenotypic Distance Problem*. We show that it can be efficiently solved on realistic-size simulated CH-data by using integer linear programming (ILP). The empirical results strongly validate this approach.

1 Biological Background and CH-Model

Identifying specific genetic variants that are associated with disease risk or other measurable phenotypes has been one of the major of objectives of modern human genetics. Today, the most commonly used technique is association mapping. Association mapping tries to detect correlations between genotypes and phenotypes in random population samples, or in case-control samples. Most commonly, association mapping is performed using so-called Genome Wide Association Studies (GWAS), in which each variable position in the genome, called a Single Nucleotide Polymorphism (SNP), is tested independently. There have been many successes using GWAS, but for many of the important *complex* traits, such as obesity, Type 2 Diabetes (T2D), cardio-vascular diseases, and many psychiatric disorders, GWAS have only recovered a minor proportion of the variance in disease prevalence known to be caused by genetics [12]. This problems is known as the 'missing heritability' problem [8].

Different explanation have been proposed for missing heritability, including epigentic factors, gene-environment interactions, and epistasis [12,15]. However,

© Springer-Verlag Berlin Heidelberg 2015
M. Pop and H. Touzet (Eds.): WABI 2015, LNBI 9289, pp. 136–147, 2015.
DOI: 10.1007/978-3-662-48221-6_10

the most common explanation is the presence of multiple rare mutations that could not be identified in GWAS due to a lack of statistical power [16]. Such rare mutations may be concentrated in relatively few genes affecting the trait in question. This is the case for many Mendelian diseases in which multiple mutations, sometimes hundreds or even thousands of rare mutations in the same gene, or genomic region, may contribute to the disease [2,9,11]. There may be a similar concentration of rare mutations in relatively few genes in complex diseases as well. If so, it might be possible to identify genes affecting the trait even though each individual mutation in the gene contributes very little to the population level variance. This insight has been one of the main motivations for the development of a number of different statistical methods for combining information from multiple mutations in the same gene, including SKAT [13], C-alpha [10], KBAC [6] and their derivatives. However, in many cases these tests have not been able to recover much more of the genetic variance than standard tests [1,5,7]. In this paper, we address this problem using a discrete optimization method rather than a purely statistical approach.

Compound Heterozygous Traits. Mendelian diseases caused by multiple muta-tions often have a mode of inheritance in which individuals are affected by the disease if they are homozygous or *compound heterozygous* for disease mutations. A compound heterozygous (CH) individual is an individual who carries disease-causing mutations in both copies (homologs) of their DNA, but *not* necessarily in the *same* exact position on their respective homologs. In fact, the two muta-tions rarely occur at the same position (hence each such site is heterozygous), although they typically will fall in the same gene. If both the copy of the gene received from the father and from the mother carry a disease mutation (although at different positions in the gene), the offspring will have a greater risk for the disease, relative to individuals without those mutations. Examples of compound heterozygous traits include *phenylketonuria* and *Tay-Sachs* disease.

Existing GWAS efforts have generally had difficulty identifying causal genes for CH-traits because the effect of each mutation is only observed when it occurs in combination with another mutation–by itself, each mutation may contribute very little to the disease. To address this problem, we propose modeling the phenomena of CH traits in terms of a discrete optimization model that we call *phenotypic distance (PD)* (defined in detail in below).

1.1 A Formal Model of a CH-trait at a Single Gene

Here we give a more formal definition of a *CH trait* at a single gene. The data for a single gene g consists of n pairs of binary vectors of length m each (the *SNP* haplotype pairs) from the two homologs of the gene. The two haplotypes in the i'th pair are denoted $H_{i,1}, H_{i,2}$ respectively; and jointly, the i'th haplotype pair is denoted H_i. The matrix of the n haplotype pairs is denoted H. For example, Table 1 shows data for $n = 2$ and $m = 7$.

Table 1. Vector X_g and two haplotype pairs. CH(1) is 1, and CH(2) is 0.

$$X_g : 0 \;\; 1\,1 \;\; 0 \;\; 0\,1\,1$$
$$H_1$$
$$H_{1,1} : 1\,0\,1 \;\; 1 \;\; 1\,0\,0$$
$$H_{1,2} : 1\,1\,0 \;\; 0 \;\; 1\,1\,0$$

$$H_2$$
$$H_{2,1} : 1\,0\,0 \;\; 1 \;\; 1\,0\,0$$
$$H_{2,2} : 0\,1\,1 \;\; 0 \;\; 1\,0\,1$$

We let binary vector X_g denote which of the m sites are *causal* (i.e., contribute to the CH-trait), and which are not. That is, $X_g(c) = 1$ if site c is causal, and $X_g(c) = 0$ otherwise. Then, given X_g and H, we define

$$CH(i) = [\bigvee_c (X_g(c) \wedge H_{i,1}(c))] \wedge [\bigvee_c (X_g(c) \wedge H_{i,2}(c))], \tag{1}$$

where $H_{i,1}(c)$ and $H_{i,2}(c)$ are the values of $H_{i,1}$ and $H_{i,2}$ at site c. So, given X_g and H, $CH(i)$ will have value 1 if and only if there is a site c with $X_g(c) = 1$, where site c in haplotype $H_{i,1}$ also has value 1; and there is also a site c' (possibly c) with $X_g(c') = 1$, where site c' in haplotype $H_{i,2}$ also has value 1 (see Table 1). We let CH denote the vector of length n, containing the values $CH(1), ..., CH(n)$. If $CH(i) = 1$, we say that individual i is CH, or is a CH-individual.

Hidden Phenotypic Distance. For a given CH-trait, we cannot observe which individuals are CH, although we can determine which individuals have a phenotype (say disease) that is hypothesized to be associated with the CH-trait. Those individuals are the *cases*, and the others are the *controls*. So, for each individual i, the input data contains a single bit, $T(i)$, (the *phenotype*), which determines whether the individual has been classified as a case (coded 1), or as a control (coded 0). We let T denote the vector of the n phenotypes.

Definition. Given vectors T and CH (which is a function of H and X_g) at a gene g, the *hidden phenotypic distance*, denoted $\text{HPD}(CH, T)$, is equal to the Hamming distance between the vectors CH and T. The Hamming distance is the number of positions where the values in the two vectors disagree. For example, with the data in Table 1, if $T(1)$ and $T(2)$ are both one, the Hamming distance between CH and T is one.

Thus, the hidden phenotypic distance at g reflects how well the data at gene g fits the CH-model. The word "hidden" is used because we generally don't know vector CH (or X_g), and so $\text{HPD}(CH, T)$ can't be determined from the known data, H and T. But, the hidden phenotypic distance can be determined in simulated data, where X_g is known.

2   The Phenotypic Distance Problem

The fact that vectors X_g and CH are unknown in real data, but a classification of the individuals into cases and controls is known (given as vector T), leads to the problem of *estimating X_g* (and CH). Informally, the phenotypic distance problem is to estimate vector X_g, given matrix H and vector T, so that the *implied CH* vector matches the phenotype vector T as *closely* as possible. More formally, for each SNP site c, we associate a variable $\widetilde{X}(c)$ that can be assigned either value 0 or 1; and let $\widetilde{X}$ denote the vector of those m values. Then, given H and $\widetilde{X}$, the CH model is reflected by the values of variables $\widetilde{CH}(i)$, for i from 1 to n, defined as:

$$\widetilde{CH}(i) = [\bigvee_c (\widetilde{X}(c) \wedge H_{i,1}(c))] \wedge [\bigvee_c (\widetilde{X}(c) \wedge H_{i,2}(c))]. \tag{2}$$

Vector $\widetilde{X}$ is an estimate of the unknown X_g, and indicates which of the m sites in the gene might contribute to (or be causal for) the CH-trait. Compare this to Eq. 1. We let $\widetilde{CH}$ denote the vector of all the $\widetilde{CH}(i)$ values.

Definition. Given the haplotype matrix H, and a phenotype vector T, the *Phenotypic Distance Problem* is the problem of setting the values of vector $\widetilde{X}$ to *minimize* the Hamming distance between the resulting vector $\widetilde{CH}$ and the phenotype vector T. We call that Hamming distance the *Phenotypic Distance* for H and T, and write it $\text{PD}(H, T)$.

Intuitively, small phenotypic distance at g (relative to the number of SNPs, and compared to other genes) suggests the *hypothesis* that gene g is causal for the CH-trait, and that the sites with value 1 in $\widetilde{X}$ are causal sites.

Computing Phenotypic Distance. When the number of sites, m, is small, it is feasible to explicitly enumerate all 2^m subsets of sites, and for each subset S, set the value of $\widetilde{X}(c)$ to 1 if and only if site c is in S. Finding the Hamming distance between each resulting vector $\widetilde{CH}$ and vector T solves the Phenotypic Distance Problem. However, this approach is infeasible for many values of m that are of realistic biological interest. For example, there are genes of interest with more than two hundred sites, and we cannot test 2^{200} possible values for $\widetilde{X}$. Further, Yufeng Wu has proved that the problem of computing the Phenotypic Distance is NP-hard [14]. For that reason, we have developed and explored an approach based on *integer linear programming (ILP)*.

In the next section, we discuss the formulation and solution of the Phenotypic Distance Problem through the use of ILP. Extensive testing of simulated data with up to $n = 4000$ haplotype pairs and $m > 200$ sites, shows that this approach is convincingly effective, i.e., both fast and accurate. Moreover, the Phenotypic distance can be used to effectively distinguish genes that are likely causal for the CH-trait, from genes that are not.

2.1   An ILP Formulation for the Phenotypic Distance Problem

Definition. For each haplotype pair H_i, the two entries $H_{i,1}(c)$ and $H_{i,2}(c)$ in a column c are called *type 0* if they are 0,0; *type 1* if they are 0,1; *type 2* if they are 1,0, and *type 3* if they are 1,1. In other words, the type of the two bits is determined by considering them as a binary number, reading top to bottom. Note that the type of a column is relative to H_i, so the same column can have a different type for two different values of i.

The ILP Variables. Overloading symbols a bit, for each column c, we will use the variable $\widetilde{X}(c)$ (from the Phenotypic Distance problem) as a binary ILP variable. Then, the value of $\widetilde{X}(c)$ in an optimal solution to the ILP formulation will be interpreted as the value of $\widetilde{X}(c)$ in the Phenotypic Distance Problem. Similarly, for each haplotype pair H_i, we will use the variable $\widetilde{CH}(i)$ (from the Phenotypic Distance problem) as a binary ILP variable; it's value in an optimal solution to the ILP formulation will be interpreted as its value in the Phenotypic Distance problem. There will also be two binary ILP variables $Z_{i,1}$ and $Z_{i,2}$ for each $H_i \in H^1$, where H^1 is the set of H_i pairs with $T(i) = 1$; similarly H^0 is the set of H_i pairs with $T(i) = 0$. Variables $Z_{i,1}$ and $Z_{i,2}$ have a technical use in the ILP, and will be discussed next. A binary ILP variable is restricted to have only value 0 or 1.

The ILP Inequalities. For each haplotype pair $H_i \in H^1$, the ILP formulation for the Phenotypic Distance will have the following inequalities:

$$\widetilde{CH}(i) - \sum_{c \ \text{is} \ \text{type} \ 2 \ \text{or} \ 3} \widetilde{X}(c) \leq 0$$

$$\widetilde{CH}(i) - \sum_{c \ \text{is} \ \text{type} \ 1 \ \text{or} \ 3} \widetilde{X}(c) \leq 0$$

The first inequality ensures that for any $H_i \in H^1$, $\widetilde{CH}(i)$ can be set to 1 *only* if some $\widetilde{X}(c)$ is set to 1 for a column c where $H_{i,1}(c) = 1$. The second inequality says the similar thing for $\widetilde{CH}(i)$ and $H_{i,2}(c)$. So, for any $H_i \in H^1$, $\widetilde{CH}(i)$ will be set to 1 *only* when the values of $\widetilde{X}$ and H_i satisfy Eq. 2.

The converse, that for $H_i \in H^1$, $\widetilde{CH}(i)$ will be set to 1 *if* Eq. 2 is satisfied, will be enforced through the objective function that will be defined below. That is, the objective is to minimize the sum of several terms, one of which is $-\sum_{H_i \in H^1} \widetilde{CH}(i)$, so in any *optimal* solution to the ILP formation for the Phenotypic Distance Problem, any $\widetilde{CH}(i) \in H^1$ *will* be set to 1 *unless* doing so violates one of the two inequalities above. The result is that in an optimal ILP solution, $(|H^1| - \sum_{H_i \in H^1} \widetilde{CH}(i))$ is the number of haplotype pairs H_i where $T(i) = 1$ but $\widetilde{CH}(i)$ is set to 0.

Now we consider the inequalities for a haplotype pair $H_i \in H^0$. Let A_i be the number of type 2 or type 3 columns in H_i, and let B_i be the number of type 1 or type 3 columns in H_i. For each haplotype pair $H_i \in H^0$, the ILP formulation will have the three inequalities:

$$\sum_{c \ \text{is} \ \text{type} \ 2 \ \text{or} \ 3} \widetilde{X}(c) - |A_i| Z_{i,1} \leq 0$$

$$\sum_{c \ \text{is} \ \text{type} \ 1 \ \text{or} \ 3} \widetilde{X}(c) - |B_i| Z_{i,2} \leq 0$$

$$Z_{i,1} + Z_{i,2} - \widetilde{CH}(i) \leq 1$$

The first inequality ensures that $Z_{i,1}$ will be set to 1 *if* there is a column c where $\widetilde{X}(c)$ is set to 1 and $H_{i,1}(c) = 1$. The second inequality ensures that $Z_{i,2}$ will be set to 1 *if* there is a column c where $\widetilde{X}(c)$ is set to 1 and $H_{i,2}(c) = 1$. The third inequality ensures that $\widetilde{CH}(i)$ will be set to 1 *if both* $Z_{i,1}$ and $Z_{i,2}$ are set to 1.

The converse, that for $H_i \in H^0$, $\widetilde{CH}(i)$ will be set to 1 *only if* those inequalities are satisfied, will be enforced through the objective function. That is, the objective function has the term $+\sum_{H_i \in H^0} \widetilde{CH}(i)$, and since the objective is a *minimization*, any $\widetilde{CH}(i) \in H^0$ *will* be set to 0 *unless* doing so violates one of the three inequalities above. The result is that in an optimal ILP solution, $\sum_{H_i \in H^0} \widetilde{CH}(i)$ is the number of H_i pairs where $T(i) = 0$, but $\widetilde{CH}(i)$ is set to 1.

It follows that in an optimal ILP solution, the Hamming Distance between $\widetilde{CH}$ and T is $(|H^1| - \sum_{H_i \in H^1} \widetilde{CH}(i)) + \sum_{H_i \in H^0} \widetilde{CH}(i)$. So, the ILP formulation optimizes the objective function

$$\texttt{Minimize}[(|H^1| - \sum_{H_i \in H^1} \widetilde{CH}(i)) + \sum_{H_i \in H^0} \widetilde{CH}(i)],$$

and hence the optimal solution has value exactly $\texttt{PD}(H, T)$. The formulation has at most $3n + m$ variables and at most $3n$ inequalities, and so has modest size.

3   Simulated Data

The ILP formulation was extensively tested on simulated data under a range of biological assumptions and choices of parameters. Here we describe how data was generated to model DNA with CH-traits.

Realistic simulations of genetic data from case-control studies are complicated by the fact that the patterns of allele frequencies in different SNPs are correlated, with a complex structure that depends on the specifics of the population history (see e.g. [4]). To simulate realistic data for a single gene, we use the program MS [3], which uses an explicit population genetic model to simulate data from multiple individuals sampled from a population. The parameters specified to MS are: s (*segsites*), the number of SNP sites; r, the population recombination rate (a parameter that determines the degree of correlation among SNPs); and N (*nsam*), the MS sample size.

Population samples created by MS are then processed to produce data mimicking case-control samples from a typical association mapping study using a disease model of CH-traits. The parameters specified are pp, the population prevalence of the phenotype (disease) of interest; a, the proportion of cases desired in the case-control sample (often 0.5); α, the disease prevalence among individuals who are CH; β, the disease prevalence among individuals who are not CH; and $n \leq N$, the number of individuals in the case-control sample ($n \leq N$).

A case-control sample for a single gene g is created from the MS output in four steps: (1) First each of the SNP sites is given a value of 0. Then (2) an iterative algorithm determines which SNPs to declare as causative (and given value 1) until the proportion of individuals with the phenotype is equal to or larger than the desired population prevalence (pp). In more detail, at each iteration, a SNP site with value 0 is chosen uniformly at random to be switched from 0 to 1; then Eq. 1 is applied to determine the current vector CH and N_{CH}, the number of individuals that are now CH. The process stops when $N \times pp \leq \alpha \times N_{CH} + \beta \times (1 - N_{CH})$. This yields the vector X_g, and the final vector CH. (3) Each individual is then assigned to be a case with probability α if the individual is CH, and with probability β otherwise. This yields the vector T. Notice that unless $\alpha = 1$ and $\beta = 0$, T will likely not be equal to CH, and so the data will contain false positives and false negatives. (4) A sample of n individuals is randomly chosen from the N individuals. For case-control data, na and $n(1 - a)$ cases and controls, respectively, are randomly chosen. If these specifications cannot be satisfied with this sample, it is rejected. The advantage of this method is that it can simulate realistic case-control data, while controlling the relative risk ($\frac{\alpha}{\beta}$) and the phenotype prevalence in the population. The phenotype (disease) prevalence is often known for specific phenotype. However, the proportion of causative mutations is typically not known, but is here controlled by α, β, and pp.

Note that since the simulation creates the vectors X_g and CH, the *hidden* Hamming distance between CH and T, *HPD(CH,T)*, can be computed in the simulation. However, neither X_g nor CH is part of the input to the Phenotypic Distance Problem.

Genomic Data. To simulate genomic data, we generate one dataset with a causal gene, g, as discussed above. Let T_g be the phenotype vector created for gene g. T_g represents the observed cases and controls. Then, we generate additional datasets with the same number of haplotype pairs, but possibly differing numbers of sites. These are the *non-causal* genes. For each non-causal gene g', we replace its phenotype vector T with T_g (from the chosen causal gene g). This models what would be encountered in a true genomic context, i.e., the observed phenotypes would be produced by the causal gene.

Significance Tests and Biological Fidelity. After computing PD(H,T) for some gene, we want to evaluate the statistical significance of that distance. There are several natural approaches. In one approach, we repeatedly, and randomly, *permute* the mapping of the phenotype values in T to the haplotype pairs in H. We use T^p to denote a permuted vector T. For each permutation, we compute $\text{PD}(H, T^p)$. Then the p-value of $\text{PD}(H, T)$ is simply the number of permuted mappings where $\text{PD}(H, T^p) \leq \text{PD}(H, T)$, divided by the total number of permuted mappings examined. The p-value can be computed both for simulated and real data.

When using simulated data, another reflection of the biological fidelity of an ILP result is the Hamming Distance between the computed $\tilde{X}$ vector, and the original vector X_g. This Hamming Distance is called the *SNP-distance* between $\tilde{X}$ and X_g.

Tests in a Genomic Context. As described above, data for one causal gene g is generated, and we let T_g denote the phenotype vector at that gene. Many non-causal genes are also generated, and we solve the Phenotypic Distance Problem at each of those genes, using T_g in place of their generated phenotype vector. For each gene, causal and non-causal, we permute T_g, creating T^p, and solve the Phenotypic Distance Problem at the gene, using T^p. What we expect is that the values $\text{PD}(H, T_g^p)$ and $\text{PD}(H, T_g)$ will be very similar at the non-causal genes, but $\text{PD}(H, T_g^p)$ will be significantly larger than $\text{PD}(H, T_g)$ at the causal gene. Hence the p-value at a causal gene will be significantly smaller than at a non-causal gene. Also, we expect that $\text{PD}(H, T_g)/(\text{number of SNPs in gene } g)$ will be significantly lower when g is the causal gene than when g is a non-causal gene. These difference allow us to distinguish the causal gene from the rest of the set.

4   The Most Striking and Positive Empirical Results

Empirical testing has shown that modeling CH-traits using the concept of phenotypic distance is very effective, and that the phenotypic distance problem can be solved convincingly fast in practice, by integer linear programming.

The most striking *computational* result is how quickly phenotypic distance can be computed via integer linear programming, particularly at causal genes, compared to the time needed for explicit enumeration and testing of all possible values for the vector $\tilde{X}$. For example, Table 2 shows that for every simulated causal gene with 4000 haplotype pairs and more than 200 sites, the ILP always finds the phenotypic distance in under *three* seconds (running GUROBI 6.0 on a 2.3 GH Macbook Pro laptop with 4 processors).

A related significant empirical result is that the time used to compute $\text{PD}(H, T_g)$ (via the ILP), is consistently less, and often overwhelmingly less, than the time used to compute $\text{PD}(H, T_g^p)$, i.e., when the phenotype vector is permuted. In those cases, the time needed is typically more than ten times that needed for the non-permuted vector. In the context of computing p-values at non-causal genes, the time can be reduced as detailed in Sect. 4.1.

Table 2. The first ten of 50 datasets generated to be causal genes, as explained in Sect. 3. In these simulations, the parameters of MS were N = 40,000 individuals, s = 400 sites, and recombination parameter of 20 (specifically, the call was: ms 40000 50 -s 400 - r 20 1000). Then, simulated CH data was created with parameters $pp = .2$, $\alpha = .9$, $\beta = .1$, and $n = 4000$. Each resulting dataset has 4000 haplotype pairs (hp), with an equal number of cases and controls, and more than 200 sites in each dataset. The column labeled HPD shows hidden phenotypic distance between CH and T_g, and the column labeled PD shows $PD(H, T_g)$ for that dataset. The time to compute the phenotypic distance was less than three seconds in each dataset. The forty datasets not shown are statistically similar to these ten.

hp	no. sites	HPD	PD	case, con	secs
4000	241	933	919	2000, 2000	1.27
4000	223	776	771	2000, 2000	1.71
4000	264	890	859	2000, 2000	1.72
4000	218	874	868	2000, 2000	0.25
4000	244	877	870	2000, 2000	1.58
4000	253	871	859	2000, 2000	2.25
4000	229	841	826	2000, 2000	2.49
4000	250	871	864	2000, 2000	0.40
4000	255	807	794	2000, 2000	1.54
4000	237	885	870	2000, 2000	1.60

Biological Fidelity. With respect to the fidelity of the phenotypic distance computations, the most striking empirical results are that at a *causal* gene g, $PD(H, T_g)$ is typically very close, and *often equal*, to $HPD(CH, T_g)$ (which we know in simulated data); and that there is typically a very large difference between $PD(H, T_g)$ and $PD(H, T_g^p)$. See Table 3. At a non-causal gene g', vector T_g acts like a random phenotype vector, so that the values of $PD(H, T_g^p)$ at g' are typically close to $n/2$ (when there is an equal number of cases and controls in T), which is a value obtainable by setting $\widetilde{X}$ to the all-zero vector (or in some cases the all-1 vector). Such settings of $\widetilde{X}$ have no biological meaning, illustrating that essentially no structural relationship between T_g^p and CH remains at a non-causal gene. In the genomic context, this means that we can easily distinguish a causal gene from non-causal genes, and it means that p-values computed at non-causal genes are much larger than at causal genes (where the p-value is essentially zero).

An additional striking empirical result is that the observed *SNP-distance* is typically (but not always) lower when the input T is used, compared to when T^p is used, and is lower at causal genes than at non-causal genes. These empirical results (the large differences between $PD(H, T)$ and $PD(H, T^p)$, the differences in computation times, and differences in SNP-distances) are very strong validations that the *Phenotypic Distance Problem* does reflect the CH-model used to generate the data.

Table 3. Typical results using simulated genomic data, as explained in Sect. 3, using parameters specified in the caption of Table 2. The first dataset is the causal gene, with associated phenotype vector T_g. The following datasets are non-causal genes, also using the phenotype vector T_g from the causal gene. At non-causal genes, the computation was terminated early; the computed ub is shown on the first line for each non-causal gene, and the computed lb, and percentage difference between the ub and lb are shown on the second line for each non-causal gene. As expected, phenotypic distance at the causal gene is substantially lower than the ub and lb values at each non-causal gene; the computation times are greater at the non-causal genes. Also (not shown), at every non-causal gene, the phenotypic distance when the phenotypes in T_g are permuted is essentially the same as for T_g, while at the causal gene, the phenotypic distance is substantially higher when T_g is permuted. These differences allow the causal gene to be identified in a genomic setting. In these simulated data, we can also compute the SNP-distances, and as expected, we see that the SNP-distance at the causal gene is substantially lower than at any non-causal gene.

	hp	sites	HPD	PD/ub	case, con	secs	SNP-dist
causal gene	4000	219	953	948	2000, 2000	0.34	65
non-causal	4000	218	2020	1864	2000, 2000	149.37	110
lb/gap				1785, 4.23 %			
non-causal	4000	226	2017	1864	2000, 2000	180.02	110
lb/gap				1728, 7.29 %			
non-causal	4000	237	1989	1853	2000, 2000	180.01	113
lb/gap				1693, 8.63 %			
non-causal	4000	210	2009	1915	2000, 2000	181.67	94
lb/gap				1649, 13.89 %			
non-causal	4000	231	1958	1868	2000, 2000	180.13	102
lb/gap				1648, 11.77 %			
non-causal	4000	240	1985	1871	2000, 2000	181.66	105
lb/gap				1718, 8.17 %			
non-causal	4000	217	1987	1925	2000, 2000	180.00	108
lb/gap				1713, 11.01 %			
non-causal	4000	228	1985	1848	2000, 2000	170.40	120

4.1 Speeding Up the Computations for Non-causal Genes and Permuted Data

ILP solvers solve a *minimization* problem by alternately focusing on finding better *solutions* (i.e., reducing the value, ub, of the best feasible solution at hand), and by finding better *lower bounds* on the value of an optimal ILP solution, i.e., by producing a number lb, where it is *guaranteed* that the optimal ILP solution has value at least lb. Therefore, when computing p-values, at any point during the computation of $\mathrm{PD}(H, T^p)$, it is guaranteed that $lb \leq \mathrm{PD}(H, T^p) \leq ub$, for the current values of lb and ub. In fact, the ILP solver only determines that

$PD(H, T^p)$ has been found when it has computed values of lb and ub that are equal.

The common, empirically observed phenomena of ILP solvers, is that they fairly quickly compute a ub that is equal or very close to the optimal solution, in this case $PD(H, T^p)$. The majority of the computation time is taken by computing a matching lb. In our simulations, the phenotypic distance at causal genes is significantly lower than the phenotypic distance at non-causal genes, so that even the computed lb at a non-causal gene quickly exceeds the phenotypic distance at the causal gene. Since the phenotypic distance at a causal gene is computed very rapidly, if the computation of a phenotypic distance at a gene (which we do not know is causal or non-causal) takes significant time, we can conclude that it is non-causal, or we can terminate the computation and use the computed ub in place of the actual phenotypic distance. In our genomic simulations, we use several conditions to terminate early. Table 3 shows that this strategy works exceedingly well; the computed lb values at non-causal genes are significantly larger than the phenotypic distance at the causal gene, and the computed ub is close to the optimal for that problem instance. Hence, in the context of a GWAS, the computation at any gene will take a bounded amount of time (limited to three minutes in our simulations).

In the context of computing p-values at a causal locus g, where $PD(H, T_g)$ has been computed, any computation of $PD(H, T_g^p)$ can be terminated when lb for the permuted data is larger than $PD(H, T_g^p)$. Moreover, experimentation shows that at that point, the computed ub value is almost always equal to $PD(H, T^p)$.

Acknowledgements. We thank Yufeng Wu and Charles Langley for helpful conversations and suggestions. Research partially supported by grants IIS-0803564, CCF-1017580, IIS-1219278 from the National Science Foundation.

References

1. Bang, S.Y., Na, Y.J., Bae, S.C., et al.: Targeted exon sequencing fails to identify rare coding variants with large effect in rheumatoid arthritis. Arthritis Res. Ther. **16**(5), 447 (2014)
2. Bobadilla, J.L., Macek, M., Fine, J.P., Farrell, P.M.: Cystic fibrosis: a worldwide analysis of CFTR mutations-correlation with incidence data and application to screening. Hum. Mutat. **19**(6), 575–606 (2002)
3. Hudson, R.: Generating samples under the Wright-Fisher neutral model of genetic variation. Bioinformatics **18**(2), 337–338 (2002)
4. Hudson, R., Kaplan, N.: Statistical properties of the number of recombination events in the history of a sample of DNA sequences. Genetics **111**, 147–164 (1985)
5. Hunt, K.A., Mistry, V., van Heel, D.A., et al.: Negligible impact of rare autoimmune-locus coding-region variants on missing heritability. Nature **498** (7453), 232–235 (2013)
6. Liu, D.J., Leal, S.M.: A novel adaptive method for the analysis of next-generation sequencing data to detect complex trait associations with rare variants due to gene main effects and interactions. PLoS Genet. **6**(10), e1001156 (2010)

7. Lohmuellers, K.E., Sparso, T., Pedersen, O., et al.: Whole-exome sequencing of 2,000 Danish individuals and the role of rare coding variants in type 2 diabetes. Am. J. Hum. Genet. **93**(6), 1072–1086 (2013)

8. Manolio, T.A., Collins, F.S.: The HapMap and genome-wide association studies in diagnosis and therapy. Ann. Rev. Med. **60**, 443–456 (2009)

9. Myerowitz, R.: Tay-Sachs disease-causing mutations and neutral polymorphisms in the Hex A gene. Hum. Mutat. **9**(3), 195–208 (1997)

10. Neal, B.M., Rivas, M.A., Daly, M.J., et al.: Testing for an unusual distribution of rare variants. PLoS Genet. **7**(3), e1001322 (2011)

11. Saenko, E.L., Ananyeva, N.N., Pipe, S., et al.: Molecular defects in coagulation Factor VIII and their impact on Factor VIII function. Vox Sang. **83**(2), 89–96 (2002)

12. Visscher, P.M., Brown, M.A., McCarthy, M.I., Yang, J.: Five years of GWAS discovery. Am. J. Hum. Genet. **90**, 7–24 (2012)

13. Wu, M.C., Lee, S., Cai, T., Li, Y., Boehnke, M., Lin, X.: Rare-variant association testing for sequencing data with the sequence kernel association test. Am. J. Hum. Genet. **89**(1), 82–93 (2011)

14. Yufeng, Wu.: Personal communication (2014)

15. Zuk, O., Hechter, E., Sunyaev, S.R., Lander, E.S.: The mystery of missing heritability: Genetic interactions create phantom heritability. Proc. Nat. Acad. Sci. (USA) **109**, 1193–1198 (2012)

16. Zuk, O., Schaffner, S.F., Lander, E.S., et al.: Searching for missing heritability: designing rare variant association studies. Proc. Nat. Acad. Sci. (USA) **111**(4), E455–E464 (2014)

Semi-nonparametric Modeling of Topological Domain Formation from Epigenetic Data

Emre Sefer$^{(\boxtimes)}$ and Carl Kingsford

School of Computer Science, Carnegie Mellon University, Pittsburgh, USA
esefer@andrew.cmu.edu, carlk@cs.cmu.edu

Abstract. Hi-C experiments capturing the 3D genome architecture have led to the discovery of topologically-associated domains (TADs) that form an important part of the 3D genome organization and appear to play a role in gene regulation and other functions. Several histone modifications have been independently suggested as the possible explanations of TAD formation, but their combinatorial effects on domain formation remain poorly understood at a global scale. Here, we propose a convex semi-nonparametric approach called $nTDP$ based on Bernstein polynomials to explore the joint effects of histone markers on TAD formation as well as predict TADs solely from the histone data. We find a small subset of modifications to be predictive of TADs across species. By inferring TADs using our trained model, we are able to predict TADs across different species and cell types, without the use of Hi-C data, suggesting their effect is conserved. This work provides the first comprehensive joint model of the effect histone markers on domain formation.

1 Introduction

The emerging evidence suggests that 3D nuclear architecture is important for the regulation of gene expression and it is tightly linked to the function of the genome. For instance, expression in the beta-globin locus is mediated by folding to bring an enhancer and associated transcription factors within close proximity of a gene [2,28]. Similarly, loci of mutations that affect expression of genomically far-away genes (eQTLs) are significantly closer in 3D to their regulated genes [26], indicating that 3D genome structure plays a wide-spread role in gene regulation. Lastly, spatial regions that interact with nuclear lamina are generally inactive [11]. Measuring and modeling the 3D shape of a genome is thus essential to obtain a more complete understanding of how cells function.

Chromatin interactions obtained from a variety of recent chromosome conformation capture experimental techniques such as Hi-C [17] have resulted in significant advances in our understanding of the geometry of chromatin structure [10,24]. These experiments yield matrices of counts that represent the frequency of cross-linking between restriction fragments of DNA at a certain resolution. Analysis of the resulting matrix by Dixon et al. [6] led to the discovery of topologically-associated domains (TADs) which correspond to consecutive, highly-interacting matrix regions typically a few megabases in size that are

© Springer-Verlag Berlin Heidelberg 2015
M. Pop and H. Touzet (Eds.): WABI 2015, LNBI 9289, pp. 148–161, 2015.
DOI: 10.1007/978-3-662-48221-6_11

closely embedded in 3D. TADs have been identified across different cell cycle phases and in prokaryotes [15]. Several lines of evidence suggest that TADs are a building block of genomic regulatory architecture [14,27]. Segmental packaging of genome via TADs likely have critical roles in cell dynamics such as long-range transcriptional regulation and cell differentiation [22,23].

The mechanism by which these TADs form and are demarcated is still largely unknown. A plethora of epigenetic modifications have been identified in metazoan genomes that are associated with 3D genome shape [5], and thus TADs. Several modifications have been found to be specifically correlated with TAD boundaries [6]. For instance, histone modifications with insulator roles such as H3K4me3 and H3K27ac are enriched within TAD boundaries [25], although the causal direction of these associations is still unknown [22]. Despite these analyses, the complete picture of how histone modifications are related to TAD formation is missing. This is partially because previous analyses relating histone marks to domain boundaries have often considered each histone mark independently, without accounting for their combined affects. It is unknown to what extent relationships between the histone markers are important or whether there is a small set of markers that are of primary importance.

Here, we develop and train a joint model, which we call $nTDP$, of how histone modifications are associated with domain boundaries and interiors. We show that we are able to train this model optimally in polynomial time because its likelihood function is convex. The model does not make any assumptions about the effect of each histone mark on domain formation, and instead fits the histone-domain relationship nonparametrically. Using this model, we systematically identify a small set of histone markers that in combination appear to explain TAD boundaries. We find a small number of epigenetic elements account for a large proportion of the accuracy of TAD prediction. All of these identified marks fail to predict domain boundaries when considered independently. We show that these markers are conserved across species and cell types in a very strong way: models trained on mouse continue to work well on human, and models trained on IMR90 cells continue to work on embryonic stem cells.

Our approach, $nTDP$, can form the basis of a unified, explanatory model of the relationship between epigenetic marks and topological domain structures. It can be used to predict domain boundaries for cell types, species, and conditions for which no Hi-C data is available. The model may also be of use for improving Hi-C-based domain finders.

1.1 Additional Related Work

Previous work mainly focused on analyzing epigenetic data in an unsupervised way. Segway [13] and ChromHMM [8] take as input a collection of genomics datasets and learn chromatin states that exhibit similar epigenetic activity patterns which then have different interpretations such as transcriptionally active, Polycomb-repressed. Libbrecht et al. [16] improve Segway predictions by integrating Hi-C data which is not as abundant as histone data, whereas [12] jointly infers chromatin state maps in multiple genomes by a hierarchical model.

However, none of these methods deal directly with TADs. Even though a subset of their chromatin states overlap with TADs, predicting TADs from them heuristically does not perform well. Additionally, they either ignore the histone densities, or make parametric distribution assumptions such as geometric or normal which are not always reflected in the true data. When modified to run in a supervised setting, they cannot capture the most informative subset of epigenetic elements.

The recent approach [3] proposes a supervised learning method based on random forests to predict TAD boundaries from histone modifications and chromatin proteins. In general, this approach is reported to perform quite accurately in predicting boundaries. However, it does not model interior TAD segments and it treats each segment independently ignoring the fact that TADs form as a result of the joint effects of multiple segments. Lastly, it also uses an error function based on gini index ignoring that the marker distributions may not be gaussian.

2   The $nTDP$ Model

2.1   The Likelihood Function

Let V be the ordered set of genome restriction fragments (bins), where each bin v represents the interval $[vr - r + 1 \ , vr]$, where r is the Hi-C resolution. Let M be the set of histone modifications (markers) over V. The marker data $H = (h_{vm})$ is a $|V| \times |M|$-matrix where its (v, m)'th entry h_{vm} is the count of the occurrences of marker m inside segment v. Let $d = [s, e]$ be a domain (interval) where s and e are its start and end boundaries respectively, $\{s+1, \ldots, e-1\}$ are the segments inside d, and let $D = \{[s_1, e_1], [s_2, e_2], \ldots, [s_i, e_i]\}$ be a partition of V where none of the domains overlap.

We propose a supervised, semi-nonparametric, high-dimensional model $nTDP$ that uses H to model and predict D. Our model can be seen as a generalization of Conditional Random Field [21,31] where we have continuous weights instead of binary features and where we model the marker effects semi-nonparametrically.

Specifically, we assume there are 3 types of segments in V that are relevant for modeling: those that are at the domain boundaries ($V_{\mathbf{b}}$), those that are in the interior of domains ($V_{\mathbf{i}}$), and those that are not part of a domain ($V_{\mathbf{e}}$), and we have $V = V_{\mathbf{b}} \cup V_{\mathbf{i}} \cup V_{\mathbf{e}}$. For each marker type m, we have 3 types of *effect functions*, $f_m^b(c, \mathbf{w_m^b})$, $f_m^i(c, \mathbf{w_m^i})$, $f_m^e(c, \mathbf{w_m^e})$, that will describe the relationship between marker count c and the fragment type (b, i, e) for marker type m. Here, $\mathbf{w_m^b}, \mathbf{w_m^i}, \mathbf{w_m^e}$ are parameters that we will fit to determine the shape of the effect function. Thus, for example, $f_m^i(c, \mathbf{w_m^i})$ will describe how a count of c for marker m influences whether the fragment is in the interior (i) of a domain.

We assume that these effect functions combine linearly. Therefore, let

$$E_{vq}^b = \sum_{m \in M} f_m^b(c_{vm}^q, \mathbf{w_m^b}) \tag{1}$$

be the total effect of all the markers on fragment v for boundary formation (b). Summations E_{vq}^i and E_{vq}^e are defined analogously for interior (i) and inter-domain fragments (e).

Let W be the union of model parameters $\mathbf{w_m^b}, \mathbf{w_m^i}, \mathbf{w_m^e}$, and let $D^{\text{train}} = \{D^q : q = 1, \ldots, Q\}$ be several domain decompositions (in different sequences or conditions) and let $H^{\text{train}} = \{H^q : q = 1, \ldots, Q\}$ be a set of corresponding histone markers. Under the assumption that the training pairs are independent, the log-likelihood of parameters W given D^{train} is

$$\log\left(P(D^{\text{train}}|W, H^{\text{train}})\right) = \sum_q \log\left(P(D^q|W, H^q)\right). \tag{2}$$

We define the probability $P(D^q, W, H^q) = \frac{\exp^{F(D^q, W, H^q)}}{\sum_{F'} \exp^{F'}}$ where $F(D^q, W, H^q)$ is the total quality of partition D^q and marker data H^q under model parameters W. Let V^q be the set of segments in pair q. Due to the independence of segments:

$$\log\left(P(D^q|W, H^q)\right) = \overbrace{\sum_{d=[s,e] \in D^q}\left(\sum_{v \in \{s,e\}} \bar{c}_b E_{vq}^b + \sum_{v=s+1}^{e-1} \bar{c}_i E_{vq}^i\right) + \sum_{v \in V_e^q} \bar{c}_e E_{vq}^e}^{\log\left(P(D^q, W, H^q)\right) = F(D^q, W, H^q)}$$
$$- \log(Z_{|V^q|}^q) \tag{3}$$

where $Z_{|V^q|}^q = \sum_{D'} P(D', W, H^q)$ is the partition function defined over all possible nonoverlapping partitions D', $\bar{c}_b, \bar{c}_i, \bar{c}_e$ are relative weights of different types of fragments to account for unbalanced training set, and V_e^q is the set of fragments that do not belong to any domain in D^q.

2.2 Nonparametric Form of the Effect Functions

Because the shape of the marker effect function is unknown, we choose the f functions from the nonparametric family of Bernstein basis polynomials. Bernstein polynomials can approximate any effect function and additionally can handle imposed shape constraints such as monotonicity and concavity.

Let A be the chosen dimension of these polynomials; larger A results in a more expressive family, but more parameters to fit. Let $m_{\max}$ be the maximum possible density of marker m. This is is used to transform the input c_{vm}^q to the range $[0, 1]$; therefore define $p_{vm}^q = c_{vm}^q/m_{\max}$. We model $f_m^b(c_{vm}^q, \mathbf{w_m^b})$ for segment v by a Bernstein polynomial $B_A(p_{vm}^q, \mathbf{w_m^b})$ as in:

$$f_m^b(c_{vm}^q, \mathbf{w_m^b}) = B_A\left(p_{vm}^q, \mathbf{w_m^b}\right) = \sum_{i=0}^A w_m^b[i] \overbrace{\binom{A}{i}(p_{vm}^q)^i(1-p_{vm}^q)^{A-i}}^{b_{i,A}(p_{vm}^q)} \tag{4}$$

where $b_{i,A}(p_{vm}^q)$ are the base Bernstein kernels.

3   Optimal Algorithms for Training and Inference

We must train the parameters W for the above model using data of the form $D^{\mathrm{train}}, H^{\mathrm{train}}$. We will examine these trained parameters (and several good solutions for them) for insights into which markers are most informative for describing D^{train} and thus topological domains.

Problem 1. Training: Given a set of marker data H^{train}, likely from several chromosomes and cell conditions, and corresponding set of TAD decompositions D^{train}, we estimate the most likely parameters W according to Eq. 2.

Problem 2. Inference: Given marker data H model parameters W, we estimate the best domain partition D of the track.

3.1   Training

A nice feature of the objective (3) is that it is convex in its arguments, $\{\mathbf{w}_{\mathbf{m}}^{\mathbf{b}}, \mathbf{w}_{\mathbf{m}}^{\mathbf{i}}, \mathbf{w}_{\mathbf{m}}^{\mathbf{e}}\}_{m \in M}$, which follows from linearity, composition rules for convexity, and convexity of the negative logarithm. However, training involves several challenges: (a) computing the partition function $Z_{|V^q|}^q$ in (3), and (b) estimating W so that the weights are sparse. We solve each of these challenges next.

Estimating the partition function. We estimate $Z_{|V^q|}^q$ in (3) recursively in polynomial time since each segment can belong to one of 4 states: domain start (sb), inside a domain (i), domain end (eb), non-domain (e), and state of each segment depends only on the previous segment's state. Let $Y = \{sb, i, eb, e\}$, and $Z_{|V^q|}^q = Z_{|V^q|,eb}^q + Z_{|V^q|,e}^q$ which components can be estimated by:

$$Z_{v,x}^q = \sum_{y \in Y} Z_{v-1,y}^q T_{y,x} \exp^{E_{vq}^x} \tag{5}$$

where $Z_{v,sb}^q$, $Z_{v,i}^q$, $Z_{v,eb}^q$, $Z_{v,e}^q$ represent the partition function up to segment v ending with sb, i, eb and non-domain respectively. T is a 4×4 binary state transition matrix where $T_{y,x} = 1$ if a segment can be assigned to x given previous segment is assigned to state y such as $(y, x) \in \{(sb, i), (sb, eb), (i, i), (i, eb), (eb, sb), (eb, e), (e, sb), (e, e)\}$, otherwise 0. Initial conditions are $Z_{0,sb}^q = Z_{0,i}^q = Z_{0,eb}^q = 0$, $Z_{0,e}^q = 1$. To avoid overflow in estimating $Z_{|V^q|}^q$ and speed it up, we estimate $\log(Z_{|V^q|}^q)$ by expressing it in terms of log of the sum of exponentials and forward and backward variables (α, β) similar to Hidden Markov Model [21].

Estimating a sparse set of good histone effect parameters. We would like to augment objective function (2) so that we select a sparse subset of markers from the data and avoid overfitting. If the coefficients $\mathbf{w}_{\mathbf{m}}^{\mathbf{b}} = 0$, then there is no influence of marker m. For this purpose, we will impose grouped lasso type of regularization on the coefficients w_{mk}. Grouped lasso regularization has the tendency to select a small number of groups of non-zero coefficients but push other groups of coefficients to be zero.

We introduce two types of regularization. First, we require that many of the weights be 0 using an L_2-norm regularization term. Second, we want the effect functions $\{f\}$ to be smooth. Let $P = \{b, i, e\}$. We modify our objective to trade off between these goals:

$$\underset{W}{\text{argmin}} \ -\sum_q \log\left(P(D^q|W, H^q)\right) + \overbrace{\lambda_1 \sum_{p \in P}\left(\sum_{m \in M}\|\mathbf{w_m^P}\|\right)^2 + \lambda_2 \sum_{p \in P}\sum_{m \in M} R(f_m^p)}^{\text{Regularization}} \quad (6)$$

where λ_1, λ_2 are the regularization parameters, and $R(f_m^p)$ is the smoothing function for effect of marker m at $p \in P$:

$$R(f_m^p) = \int_x \left(\frac{\partial^2 f_m^p(x, \mathbf{w_m^P})}{\partial x^2}\right)^2 dx \quad (7)$$

Group lasso in (6) uses the square of block l_1-norm instead of l_2-norm group lasso which does not change the regularization properties [1]. Second-order derivative in (7) can be expressed more explicitly as a convex quadratic function of $\mathbf{w_m^P}$. Its derivation can be found in Appendix.

We note that (6) is convex, but it is a nonsmooth optimization problem because of the regularizer. We solve it efficiently by using an iterative algorithm from multiple kernel learning [1]. By Cauchy-Schwarz inequality:

$$\sum_{p \in P}\left(\sum_{m \in M}\|\mathbf{w_m^P}\|\right)^2 \leq \sum_{p \in P}\sum_{m \in M}\frac{\|\mathbf{w_m^P}\|^2}{\gamma_{mp}} \quad (8)$$

where $\gamma_{mp} \geq 0$, $\sum_{m \in M}\gamma_{mp} = 1$, $p \in P$, and the equality in (8) holds when

$$\gamma_{mp} = \frac{\|\mathbf{w_m^P}\|}{\sum_{m \in M}\|\mathbf{w_m^P}\|}, \quad p \in P \quad (9)$$

This modification turns the objective into the following which is jointly convex in both $\mathbf{w_m^P}$ and γ_{mp}:

$$\underset{W}{\text{argmin}} \ -\sum_q \log\left(P(D^q|W, H^q)\right) + \sum_{p \in P}\sum_{m \in M}\left(\lambda_1 \frac{\|\mathbf{w_m^P}\|^2}{\gamma_{mp}} + \lambda_2 R(f_m^p)\right) \quad (10)$$

$$\text{s.t.} \ \sum_{m \in M}\gamma_{mp} = 1.0, \quad p \in P \quad (11)$$

$$\gamma_{mp} \geq 0, m \in M, p \in P \quad (12)$$

We solve this by alternating between the optimization of $\mathbf{w_m^P}$ and γ_{mp}. When we fix γ_{mp}, we can find the optimal $\mathbf{w_m^P}$ by any quasi-newton solver such as L-BFGS [18] which runs faster than the other solvers such as iterative scaling or conjugate gradient. When we fixed $\mathbf{w_m^P}$, we can obtain the best γ_{mp} by the closed form equation (9). Both steps iterate until convergence.

3.2   Training Extensions

We can model a variety of shape-restricted effect functions by Bernstein polynomials that cannot be easily achieved by other nonparametric approaches such as smoothing splines [19]. We add the following constraints to ensure monotonicity:

$$w_m^b[i] \leq w_m^b[i+1], \quad i = 0, \ldots, A-1 \tag{13}$$

which is a realistic assumption since increasing the marker density should not decrease its effect. We can also ensure concavity of the effect function by:

$$w_m^b[i-1] - 2w_m^b[i] + w_m^b[i+1] \leq 0, \quad i = 1, \ldots, A-1 \tag{14}$$

which has a natural diminishing returns property: the increase in the value of the effect function generated by an increase in the marker density is smaller when output is large than when it is small. Our problem is different than smoothing splines since our loss function is more complicated than traditional spline loss functions due to partition function estimation in (5) which makes it hard to directly apply the smoothing spline methods [29]. In addition, these nonnegativity and other shape constraints can be naturally enforced in our method.

3.3   Inferring Domains Using the Trained Model

Given marker data H over a single track and W, the inference log-likelihood is:

$$\underset{D}{\text{argmax}} \, \log\left(P(D|W,H)\right) = \sum_{d=[s,e]\in\overline{D}} r_{se} x_{se} + \sum_{v\in V} E_v^e y_v \tag{15}$$

where $\overline{D} = \{[s,e] \,|\, s,e \in V, e-s \geq 1\}$ is the set of all potential domains of length at least 2 and $r_{se} = E_s^b + E_e^b + \sum_{v=s+1}^{e-1} E_v^i$. The intuition is that variable $x_{se} = 1$ when the solution contains interval $[s,e]$, and variable $y_v = 1$ if v is not assigned to any domain. The $\log(Z_{|V|})$ term is removed during inference since it is same for all D. We solve (16)–(17) to find the best partition D:

$$\underset{D}{\text{argmax}} \sum_{d=[s,e]\in\overline{D}} r_{se} x_{se} + \sum_{v\in V} E_v^e \left(1 - \sum_{[s,e]\in M[v]} x_{se}\right) \tag{16}$$

$$\text{s.t. } x_{se} + x_{s'e'} \leq 1 \quad \forall \text{ domains } [s,e], [s',e'] \text{ that overlap} \tag{17}$$

where $M[v]$ is the set of intervals that span fragment v. We replace y_v in (15) with $1 - \sum_{[s,e]\in M[v]} x_{se}$ since each segment can be assigned to at most a single domain. (17) ensures that inferred domains do not overlap. Problem (16)–(17) is *Maximum Weight Independent Set* in interval graph defined over domains which can be solved optimally by dynamic programming in $O(|V|^2)$ time.

4   Results

4.1   Experimental Setup

We binned ChIP-Seq histone modification and DNase-seq data at 40 kb resolution, estimate RPKM (Reads Per Kilobase per Million) measure for each bin, and transform values x in each bin by $\log(x + 1)$, which reduces the distorting effects of high values. In the case of 2 or more replicates, the RPKM-level for each bin is averaged to get a single histone modification file, in order to minimize batch-related differences. We convert any data mapped to hg19 (mm8) to hg18 (mm9) using UCSC liftOver tool. We define TADs over human IMR90, human embryonic stem (ES), and mouse ES cells Hi-C data [6] at 40 kb resolution after normalization by [30]. We use consensus domains from Armatus [9] as the true TAD partition by selecting threshold γ where maximum Armatus domain size is closest to the maximum Dixon et al. [6] domain size ($\gamma = 0.5$ for IMR90, $\gamma = 0.6$ for human ES, and $\gamma = 0.2$ for mouse ES cells).

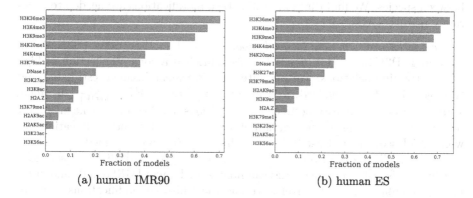

(a) human IMR90   (b) human ES

Fig. 1. Fraction of histone modifications appearing in a best scoring four-modification model in (a) human IMR90, (b) human ES. Best scoring is defined as reaching at least 95 % of NVI score of the model with all modifications.

We solved the training optimization problem by L-BFGS [18]. We use the public implementation of *Armatus* [9], and obtain histone modifications from NIH Roadmap Epigenomics [4] and UCSC Encode [7]. Code and datasets can be found at http://www.cs.cmu.edu/~ckingsf/research/ntdp. *nTDP* is reasonably fast: we train on all human IMR90 chromosomes in less than 3 h on a MacBook Pro with 2.5 Ghz processor and 8 Gb Ram. The iterative procedure in general converges in fewer than 10 iterations.

We prevent overfitting by following a two-step nested cross-validation which has inner and outer steps. The outer K-fold cross-validation, for example, trains on all autosomal human chromosomes except the one to be predicted. Within each loop of outer cross-validation, we perform $(K-1)$-fold inner cross-validation to estimate the regularization parameters.

4.2 *nTDP* Finds a Small Subset of Modifications Predictive of TADs

We identified a minimal set of histone marks that can model TADs as follows: we run *nTDP* independently on each chromosome of human IMR90 to obtain 21 sets of marks. These sets overlap significantly across all chromosome pairs (hypergeometric $p < 0.05$ for all pair-wise comparisons), and a total of 16 modifications cover all chromosomes. Despite the regularization, the weights of several of these marks are still very close to 0, so we identify a non-redundant subset of the modifications by Bayesian information criterion (BIC) [21] which penalizes model complexity more strongly.

As we increase the number of included modifications from 1 to 16, the BIC decrease nearly stops after 4 modifications, with some additional small reduction up to 6 modifications. The sets of 4 and 6 modifications that were most informative are: {H3K36me3, H3K4me1, H3K4me3, H3K9me3} and {H3K4me3, H3K79me2, H3K27ac, H3K9me3, H3K36me3, H4K20me1}. These non-redundant set of elements are preserved when we repeat this procedure between species. We find that only these $4-6$ modifications are needed to accurately predict TADs.

These marks are common in good models. The 4 modifications {H3K36me3, H3K4me1, H3K4me3, H3K9me3} are also enriched among a collection of high quality training solutions. We measure the agreement between estimated and true partitions by normalized variation of information $NVI = \frac{VI}{\log |V|}$ [20] where VI measures the similarity between two partitions and lower score means better performance. We analyze the fraction of models with 4 histone modifications for which NVI score is at least 95 % of optimum NVI score obtained by running *nTDP* over all modifications as in Fig. 1a and b. We find 161, 139 solutions satisfying this criteria among 1820 candidates for human IMR90 and human ES histone modifications respectively. We find the 4 histone modifications above to be significantly overrepresented in the set of models for both human IM90 and ES cells (hypergeometric $p < 0.0001$). These significance values combined with the results above suggest the importance of the identified modifications in TADs.

These marks have nearly optimal coherence score. We assess the performance of various subset of modifications by the coherence score which is the exponential of the negative mean log-likelihood of each chromosome on the test set, and it is normalized by the best model coherence score as in Table 1. As such it is a relative measure of the quality of various models. The coherence score using only the set {H3K36me3, H3K4me1, H3K4me3, H3K9me3} is almost as high as the score for all 28 histone modifications in human IMR90. Restricting the effect function shape to be nonnegative and concave slightly improves the score. Our analysis indicates that the remaining modifications carry either redundant information or are less important for TADs.

Table 1. Normalized coherence scores of various marker subsets

Allowed modifications (human IMR90 to IMR90)	Coherence score (Normalized)
28 histone modifications + Concave + Nonnegative *	1.00
28 histone modifications + Concave	0.99
28 histone modifications	0.97
H3K4me3, H3K79me2, H3K27ac, H3K9me3, H3K36me3, H4K20me1	0.94
H3K36me3, H3K4me1, H3K4me3, H3K9me3 + Concave + Nonnegative	0.94
H3K36me3, H3K4me1, H3K4me3, H3K9me3 + Concave	0.93
H3K36me3, H3K4me1, H3K4me3, H3K9me3	0.92

4.3 Predicting TADs from Histone Marks in Human

$nTDP$ is able to predict domain boundaries accurately using 4 histone marks alone in both human IMR90 and human ES cells. We compare TAD prediction performance of $nTDP$ with the chromatin state partition predicted by Segway [13] in terms of NVI. Even though Segway does not predict TADs directly, its chromatin state partition can still be used as a baseline. Training with all 28 histone modifications instead of with the identified 4 modifications does not lead to a major performance increase as shown in Fig. 2a even though it increases the training time approximately 4 times for human IMR90 cells. Restricting the effect function to be monotonic and concave only slightly increases the performance. Chromatin states inferred by Segway do not directly correspond to TADs which leads to a lower TAD prediction performance even though they have other meaningful interpretations.

We find combinatorial effects of histone modifications to be important for accurate domain prediction since none of the modifications can achieve NVI score better than 0.2 when considered independently. To verify that there are not inherent structures in the data that can lead to an easy prediction, we randomly shuffle domains in the training set by preserving their lengths without shuffling modifications, which NVI score is never better than 0.3 in all chromosomes showing the importance of histone modification distributions in TADs.

$nTDP$ also predicts TADs accurately across different species as well as across different cell types as in Figs. 2(b)–(d). For example, if we train on human IMR90 data, the model we obtain is still able to recover domains in human ES cells (Fig. 2a). This holds true across species as well: training on human ES data, for example, produces a model that can work well on mouse ES data.

4.4 Multiscale Analysis of the Predicted TADs

We find that our predicted TADs match true TADs more accurately at different scales defined by different *Armatus* γ's as in Fig. 3a. We observe a slight performance improvement if we define true TAD partition at lower *Armatus* γ values

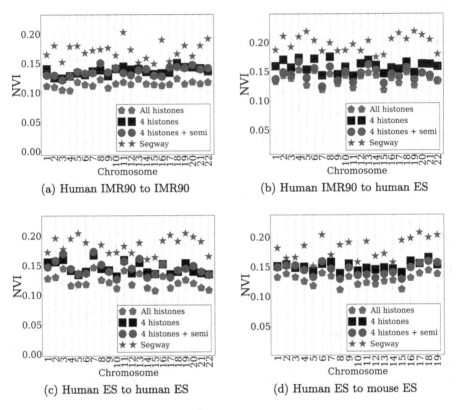

Fig. 2. TAD prediction performance on different chromomes (a) human IMR90 to human IMR90: infer each human IMR90 chromosome by training with all IMR90 chromosomes except the one to be inferred, (b) human IMR90 to human ES, (c) human ES to human ES, (d) human ES to mouse ES are defined similarly.

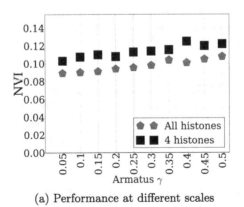

(a) Performance at different scales

Fig. 3. Multiscale analysis of the predicted TADs (a) Performance over true TAD partitions at different scales obtained via different *Armatus* γ in human IMR90.

in human IMR90 which correspond to longer TADs. This figure suggests that some of our wrong TAD predictions may actually correspond to longer TAD blocks which we erroneously interpret as incorrect due to a scale mismatch.

5   Conclusion

We formulate semi-nonparametric modeling of TADs in terms of histone modifications, and propose an efficient provably optimal solution $nTDP$ for training and inference. Experimental results on human and mouse cells show that a common subset of histone modifications can accurately predict TADs across cell types and species. Via our trained model, we also accurately predict TADs without using any Hi-C data which is especially useful for understanding the 3D genome conformation on species with limited Hi-C data.

Funding. This research is funded in part by the Gordon and Betty Moore Foundations Data-Driven Discovery Initiative through Grant GBMF4554 to Carl Kingsford, by the US NSF (1256087, 1319998), and by the US NIH (HG006913, HG007104). C.K. received support as an Alfred P. Sloan Research Fellow.

Appendix

$R(f_m^p)$ can be written more explicitly as in (18) according to [19]:

$$\frac{\partial^2 f_m^p(x, \mathbf{w_m^p})}{\partial x^2} = A(A-1) \sum_{i=0}^{A-2} (w_m^p[i+2] - 2w_m^p[i+1] + w_m^p[i]) \binom{A-2}{i} x^i(1-x)^{A-2-i}$$

(18)

which turns $R(f_m^p)$ into (19):

$$\int_0^1 \left(\frac{\partial^2 f_m^p(x)}{\partial x^2}\right)^2 dx = A^2(A-1)^2 \sum_{i=0}^{A} \sum_{j=i}^{A} (w_m^p[i]w_m^p[j])$$

$$\left(\sum_{q=\bar{e}_i}^{\min(i,2)} \sum_{r=\bar{e}_j}^{\min(j,2)} (-1)^{q+r} \binom{2}{q}\binom{2}{r} T_{j-r}^{i-q}(x)\right)$$

(19)

where $\bar{e}_p = \max(0, 2 - A + p)$, $T_{j-r}^{i-q}(x)$ is defined below and $\beta(i+j-q-r+1, 2A-3-i-j+q+r)$ is the beta function:

$$T_{j-r}^{i-q}(x) = \binom{A-2}{i-q}\binom{A-2}{j-r} \underbrace{\int_0^1 x^{i-q}(1-x)^{A-2-i+q} x^{j-r}(1-x)^{A-2-j+r} dx}_{\beta(i+j-q-r+1, 2A-3-i-j+q+r)}$$

(20)

$R(f_m^p)$ is convex which follows from semidefiniteness of the resulting polynomial.

References

1. Bach, F.R.: Exploring large feature spaces with hierarchical multiple kernel learning. In: Advances in Neural Information Processing Systems, pp. 105–112 (2009)
2. Baù, D., Marti-Renom, M.A.: Structure determination of genomic domains by satisfaction of spatial restraints. Chromosome Res. **19**(1), 25–35 (2011)
3. Bednarz, P., Wilczyński, B.: Supervised learning method for predicting chromatin boundary associated insulator elements. J. Bioinform. Computat. Biol. **12**(06), 1442006 (2014)
4. Bernstein, B.E., et al.: The NIH roadmap epigenomics mapping consortium. Nat. Biotechnol. **28**(10), 1045–1048 (2010)
5. Bickmore, W.A., van Steensel, B.: Genome architecture: Domain organization of interphase chromosomes. Cell **152**(6), 1270–1284 (2013)
6. Dixon, J.R., Selvaraj, S., Yue, F., Kim, A., Li, Y., Shen, Y., Hu, M., Liu, J.S., Ren, B.: Topological domains in mammalian genomes identified by analysis of chromatin interactions. Nature **485**(7398), 376–380 (2012)
7. ENCODE Project Consortium, et al.: An integrated encyclopedia of DNA elements in the human genome. Nature **489**, 57–74 (2012)
8. Ernst, J., Kellis, M.: ChromHMM: automating chromatin-state discovery and characterization. Nat. Methods **9**(3), 215–216 (2012)
9. Filippova, D., Patro, R., Duggal, G., Kingsford, C.: Identification of alternative topological domains in chromatin. Alg. Mol. Biol. **9**(1), 14 (2014)
10. Gibcus, J.H., Dekker, J.: The hierarchy of the 3D genome. Mol. Cell **49**(5), 773–782 (2013)
11. Guelen, L., et al.: Domain organization of human chromosomes revealed by mapping of nuclear lamina interactions. Nature **453**(7197), 948–951 (2008)
12. Ho, J.W., et al.: Comparative analysis of metazoan chromatin organization. Nature **512**(7515), 449–452 (2014)
13. Hoffman, M.M., Buske, O.J., Wang, J., Weng, Z., Bilmes, J.A., Noble, W.S.: Unsupervised pattern discovery in human chromatin structure through genomic segmentation. Nat. Methods **9**, 473–476 (2012)
14. Hou, C., Li, L., Qin, Z.S., Corces, V.G.: Gene density, transcription, and insulators contribute to the partition of the drosophila genome into physical domains. Mol. Cell **48**(3), 471–484 (2012)
15. Le, T.B.K., et al.: High-resolution mapping of the spatial organization of a bacterial chromosome. Science **342**(6159), 731–734 (2013)
16. Libbrecht, M.W., Ay, F., Hoffman, M.M., Gilbert, D.M., Bilmes, J.A., Noble, W.S.: Joint annotation of chromatin state and chromatin conformation reveals relationships among domain types and identifies domains of cell type-specific expression. Genome Res. **25**, 544–557 (2015)
17. Lieberman-Aiden, E., et al.: Comprehensive mapping of long-range interactions reveals folding principles of the human genome. Science **326**(5950), 289–293 (2009)
18. Liu, D.C., Nocedal, J.: On the limited memory BFGS method for large scale optimization. Math. Program. **45**(1–3), 503–528 (1989)
19. McKay Curtis, S., Ghosh, S.K., et al.: A variable selection approach to monotonic regression with Bernstein polynomials. J. Appl. Stat. **38**(5), 961–976 (2011)
20. Meilă, M.: Comparing clusterings–an information based distance. J. Multivar. Anal. **98**(5), 873–895 (2007)
21. Murphy, K.P.: Machine Learning: A Probabilistic Perspective. MIT Press, Cambridge (2012)

22. Nora, E.P., et al.: Segmental folding of chromosomes: a basis for structural and regulatory chromosomal neighborhoods? BioEssays **35**(9), 818–828 (2013)
23. Phillips-Cremins, J.E., Sauria, M.E., Sanyal, A., Gerasimova, T.I., Lajoie, B.R., Bell, J.S., Ong, C.T., Hookway, T.A., Guo, C., Sun, Y., Bland, M.J., Wagstaff, W., Dalton, S., McDevitt, T.C., Sen, R., Dekker, J., Taylor, J., Corces, V.G.: Architectural protein subclasses shape 3D organization of genomes during lineage commitment. Cell **153**(6), 1281–1295 (2013)
24. Rao, S.S., et al.: A 3D map of the human genome at kilobase resolution reveals principles of chromatin looping. Cell **159**(7), 1665–1680 (2014)
25. Sefer, E., Duggal, G., Kingsford, C.: Deconvolution of ensemble chromatin interaction data reveals the latent mixing structures in cell subpopulations. In: Przytycka, T.M. (ed.) RECOMB 2015. LNCS, vol. 9029, pp. 293–308. Springer, Heidelberg (2015)
26. Sefer, E., Kingsford, C.: Metric labeling and semi-metric embedding for protein annotation prediction. In: Bafna, V., Sahinalp, S.C. (eds.) RECOMB 2011. LNCS, vol. 6577, pp. 392–407. Springer, Heidelberg (2011)
27. Sexton, T., Yaffe, E., Kenigsberg, E., Bantignies, F., Leblanc, B., Hoichman, M., Parrinello, H., Tanay, A., Cavalli, G.: Three-dimensional folding and functional organization principles of the Drosophila genome. Cell **148**(3), 458–472 (2012)
28. Tolhuis, B., Palstra, R.J., Splinter, E., Grosveld, F., de Laat, W.: Looping and interaction between hypersensitive sites in the active β-globin locus. Mol. Cell **10**(6), 1453–1465 (2002)
29. Wahba, G.: Spline models for observational data, vol. 59. SIAM (1990)
30. Yaffe, E., Tanay, A.: Probabilistic modeling of Hi-C contact maps eliminates systematic biases to characterize global chromosomal architecture. Nat. Genet. **43**(11), 1059–1065 (2011)
31. Zhou, J., Troyanskaya, O.G.: Global quantitative modeling of chromatin factor interactions. PLoS Comput. Biol. **10**(3), e1003525 (2014)

Scrible: Ultra-Accurate Error-Correction of Pooled Sequenced Reads

Denise Duma[1]([⊠]), Francesca Cordero[4], Marco Beccuti[4],
Gianfranco Ciardo[5], Timothy J. Close[3], and Stefano Lonardi[2]

[1] Baylor College of Medicine, Houston, TX 77030, USA
dum@bcm.edu
[2] Department of Computer Science and Engineering,
University of California, Riverside, CA 92521, USA
[3] Department of Botany and Plant Sciences, University of California,
Riverside, CA 92521, USA
[4] Department of Computer Science, Università di Torino, 10149 Torino, Italy
[5] Department of Computer Science, Iowa State University, Ames, IA 50011, USA

Abstract. We recently proposed a novel clone-by-clone protocol for *de novo* genome sequencing that leverages combinatorial pooling design to overcome the limitations of DNA barcoding when multiplexing a large number of samples on second-generation sequencing instruments. Here we address the problem of correcting the short reads obtained from our sequencing protocol. We introduce a novel algorithm called SCRIBLE that exploits properties of the pooling design to accurately identify/correct sequencing errors and minimize the chance of "over-correcting". Experimental results on synthetic data on the rice genome demonstrate that our method has much higher accuracy in correcting short reads compared to state-of-the-art error-correcting methods. On real data on the barley genome we show that SCRIBLE significantly improves the decoding accuracy of short reads to individual BACs.

1 Introduction

We have recently demonstrated how to take advantage of combinatorial pooling (also known as *group testing*) for clone-by-clone *de-novo* genome sequencing [1,8,9]. In our sequencing protocol, subsets of non-redundant genome-tiling BACs are chosen to form intersecting pools, then groups of pools are sequenced on an Illumina sequencing instrument via standard multiplexing (DNA barcoding). Sequenced reads can be assigned to specific BACs by relying on the structure of the pooling design: since the identity of each BAC is encoded within the pooling pattern, the identity of each read is similarly encoded within the pattern of pools in which it occurs. Finally, BACs are assembled individually, simplifying the problem of resolving genome-wide repetitive sequences.

An unforeseen advantage of our sequencing protocol is the potential to correct sequencing errors more effectively than if DNA samples were not pooled. This paper investigates to what extent our protocol enables such error correction.

© Springer-Verlag Berlin Heidelberg 2015
M. Pop and H. Touzet (Eds.): WABI 2015, LNBI 9289, pp. 162–174, 2015.
DOI: 10.1007/978-3-662-48221-6_12

Due to obvious needs in applications of high-throughput sequencing technology, including *de-novo* assembly, the problem of correcting sequencing errors in short reads has been the object of intense research. Below, we briefly review some of these efforts, noting that our approach is substantially different, due to its original use of the pooling design.

Most error correction methods take advantage of the high sequencing depth provided by second-generation sequencing technology to detect erroneous base calls. For instance, SHREC [13] carries out error correction by building a generalized weighted suffix tree on the input reads, where the weight of each tree node depends on its coverage depth. If the weight of a node deviates significantly from the expectation, the substring corresponding to that node is corrected to one of its siblings. HITEC [4] builds a suffix array of the set of reads and uses the longest common prefix information to count how many times short substrings are present in the input. These counts are used to decide the correct nucleotide following each substring. HITEC was recently superseded by RACER [5], from the same research group, which improves the time- and space-efficiency by using a hash table instead of a suffix array.

Several other error-correction methods are based on k-mer decomposition of the reads, e.g., SGA [14], REPTILE [17] and QUAKE [6]. SGA uses a simple frequency threshold to separate "trusted" k-mers from "untrusted" ones, then performs base changes until untrusted k-mers can become trusted. REPTILE builds a k-mer tiling across reads, then corrects erroneous k-mers based on contextual information provided by their trusted neighbor in the tiling. QUAKE uses a coverage-based cutoff to determine erroneous k-mers, then corrects the errors by applying the set of corrections that maximizes a likelihood function. The likelihood of a set of corrections is defined by taking into account the error model of the sequencing instrument and the specific genome under study. Other methods are based on multiple sequence alignments. For example, CORAL [12] builds a multiple alignment for clusters of short reads, then corrects errors by majority voting.

2   Preliminaries

Our algorithm corrects *reads*, short strings over the alphabet $\Sigma = \{A, C, G, T\}$. With $r[i]$ we denote the i-th symbol in read r. A k-mer α is any substring of a read r such that $|\alpha| = k$. A *BAC (clone)* is a 100–150 kb fragment of the target genome replicated in a *E. coli* cell.

2.1   Pooling Design

After the selection of the BACs to be sequenced (see [8,9] for more details), we *pool* them according to a scheme that allows us to *decode* (assign) sequenced reads back to their corresponding BACs.

The design of a pooling scheme reduces to the problem of building a *disjunctive* matrix Φ where columns correspond to BACs to be pooled and rows correspond to pools. Let w be a subset of the columns (BACs) of the design matrix Φ and $p(w)$ be the set of rows (pools) that contain at least one BAC in w: the matrix Φ is said to be d-*disjunct* (or d-*decodable*) if, for any choice of w_1 and w_2 with $|w_1| = 1$, $|w_2| = d$, and $w_1 \not\subset w_2$, we have that $p(w_1) \not\subseteq p(w_2)$. Intuitively, d represents the maximum number of positives that are guaranteed to be identified by the pooling design.

We pool BACs using a combinatorial pooling scheme called *Shifted Transversal Design* (STD) [16]. STD is a *layered* design: the rows of the design matrix Φ are organized into multiple redundant layers such that each pooled variable (BAC) appears exactly once in each layer, that is, a layer is a partition of the set of variables. STD is defined by parameters (q, L, Γ) where L is the number of layers, q is a prime number equal to the number of rows (pools) in each layer and Γ is the *compression level* of the design. To pool n variables, STD uses $m = q \times L$ pools. The set of L pools defines a unique pooling pattern for each variable, and can be used to retrieve its identity. We call this set the *signature* of the variable. The compression level Γ is defined to be the smallest integer such that $q^{\Gamma+1} \geq n$. STD has the desirable property that any two variables co-occur in at most Γ pools, therefore by choosing a small value for Γ one can make STD pooling very robust to decoding errors. The parameter Γ is also related to the decodability of the design through the equation $d = \lfloor (L-1)/\Gamma \rfloor$. Therefore, Γ can be seen as a trade-off parameter: the larger it is, the more items can be tested, up to $q^{\Gamma+1}$, but fewer positives can be reliably identified, up to $\lfloor (L-1)/\Gamma \rfloor$. For more details on this pooling scheme and its properties, see [16].

2.2   Read Decoding

As the read decoding problem is presented elsewhere [1,8,9], here we only provide a brief overview to motivate the necessity of correcting reads before decoding them. Given a set of pools $\mathcal{P}$ and a set of BACs $\mathcal{B}$, the *signature* for a BAC $b \in \mathcal{B}$ is the subset $\mathcal{A} \subset \mathcal{P}$ of pools ($|\mathcal{A}| = L$) to which BAC b is assigned. Given a set $\mathcal{R}_p$ of reads for each pool $p \in \mathcal{P}$ and the set of all BAC signatures, the *read decoding* problem is to determine, for each read $r \in \mathcal{R}_p$, the BAC(s) from which r originated. In [8] we solved the read decoding problem with a combinatorial algorithm, while in [1] we proposed a compressed sensing approach. In [9] we further improved the decoding by a "data slicing" approach. A similar slicing strategy was used in [10] to improve the assembly quality for ultra-deep sequencing data. In all cases, we first decompose reads into their constituent k-mers and compute, for each k-mer, the number of times it occurs in each pool (the k-mer *frequency vector*).

The problems of decoding and error-correction are mutually dependent: correcting sequencing errors will improve the accuracy of decoding; a more accurate decoding can help correcting the reads more effectively (as it will become clear later).

3   Methods

3.1   Indexing k-mers

We first preprocess all reads r ($|r| \geq k$) in each pool $p \in \mathcal{P}$ by sliding a window of size k over each read $r \in \mathcal{R}_p$ to produce $|r| - k + 1$ k-mers. The result of this pre-processing is encoded in a function $poolcount: \Sigma^k \times \mathcal{P} \to \mathbf{N}$, where $poolcount(\alpha, p)$ is the number of times k-mer α (or its reverse complement) appears in pool p. We also define three additional functions, namely (i) $pools: \Sigma^k \to \mathbf{N}$ where $pools(\alpha)$ is the number of distinct pools where k-mer α appears at least once, (ii) $count:$ $\Sigma^k \to \mathbf{N}$ where $count(\alpha)$ is the total number of times α appears in any of the pools, and (iii) $bacs: \Sigma^k \to 2^{\mathcal{B}}$ where $bacs(\alpha)$ is the set of BACs corresponding to $pools(\alpha)$, i.e., the BACs whose signature is included in $pools(\alpha)$. Observe that $pools(\alpha) = |\{p \in \mathcal{P} : poolcount(\alpha, p) > 0\}|$, $count(\alpha) = \sum_{p \in \mathcal{P}} poolcount(\alpha, p)$, and that $bacs(\alpha)$ can be determined by matching $pools(\alpha)$ against the set of all BAC signatures. As a consequence, we only need to explicitly store $poolcount$ into a hash table.

3.2   Identification and Correction of Sequencing Errors

Taking advantage of the pooling design, we can assume that any k-mer α such that $|pools(\alpha)| < L$ (i.e., α occurs in a few pools, less than the expected L) is erroneous. This assumption holds when the sequencing depth is sufficiently high, so that each genomic location is covered by several correct k-mers possibly mixed with a few corrupted k-mers. In practice, the depth of sequencing can vary significantly along the genome. When it is particularly low, it is possible (although unlikely) for a correct k-mer to appear in fewer than L pools.

We observed that these low-frequency k-mers are responsible for the large majority of the entries in the hash table for $poolcount$. To save memory, we do not store a k-mer in the hash table during the pre-processing phase if it appears in fewer than l pools, where $1 \leq l < L$ is a user-defined parameter. At the other end, a k-mer α is deemed *repetitive* if $|pools(\alpha)| > h$, where h is another user-defined parameter such that $dL < h \leq qL$. Discarding low-frequency k-mers requires two passes over the data. In the first we only build the hash table for the function $|pools|$, and in the second we determine $poolcount$ by discarding any k-mer such that $|pools(\alpha)| < l$.

After building the hash table for $poolcount$, we process the reads one by one. If a k-mer α in read r is absent from the hash table, it is assumed to be incorrect. Our algorithm attempts to correct α by changing either its first or its last nucleotide into the other three possible nucleotides (we discuss below how to determine which one). The three variants are searched in the hash table: if only one is present, then it is the correct version of α, assuming that α contains only one error. If multiple variants of α are found in the hash table, the algorithm analyzes the read r to which α belongs. For any correct k-mer β in r, we expect $pools(\beta)$ to match a single BAC signature or the union of up to d BAC signatures. Furthermore, any other correct k-mer γ in r either satisfies $pools(\gamma) = pools(\beta)$, or $pools(\gamma) \cap pools(\beta)$ is equal to a BAC signature. The second condition takes into account the case when a portion of r originates from the overlapping region between BACs.

Given a read r, let $\mathcal{C} = \alpha_1, \alpha_2, \ldots, \alpha_{|r|-k+1}$ be the set of its k-mers in left to right order, i.e., $\alpha_1 = r[1, k]$, $\alpha_2 = r[2, k + 1]$, etc. We define a *correct set* (or *c-set*) $\mathcal{C}'$ as a maximal contiguous subset of $\mathcal{C}$ such that all k-mers in $\mathcal{C}'$ are either repetitive or all share the same BAC signature or the union of up to d BAC signatures.

If r has only one non-empty c-set $\mathcal{C}'$, we can use $\mathcal{C}'$ as a starting point to correct the remaining k-mers. Any c-set $\mathcal{C}'$ contains at least one *border* k-mer α_i such that $\alpha_{i-1} \notin \mathcal{C}'$ when $i > 1$ (or $\alpha_{i+1} \notin \mathcal{C}'$ when $i < |r| - k + 1$). Without loss of generality, assume $\alpha_{i-1} \notin \mathcal{C}'$. The c-set is "interrupted" at position i because $r[i - 1]$ is a sequencing error. Thus, we can attempt to change $r[i - 1]$ to any of the other three nucleotides, and search the variant k-mer α'_{i-1} in the hash table. If $bacs(\alpha'_{i-1})$ match the shared signature in the c-set $\mathcal{C}'$, we have found the right correction for nucleotide $r[i - 1]$ so we add α'_{i-1} to $\mathcal{C}'$ and let it become the new border k-mer. We then repeat the process of extending $\mathcal{C}'$ by correcting the k-mer preceding its new border k-mer (of course, the one correction we just made might actually suffice to correct multiple k-mers, up to k of them, in fact). This iterative process continues until $\mathcal{C}'$ has been extended to encompass the whole read r. Note that when correcting a read from pool p, we also update the hash table: for each k-mer α corrected into α', we increase $poolcount(\alpha', p)$ by one and decrease $poolcount(\alpha, p)$ by one. This process is expected to lead to erroneous k-mers having all their pool counts drop below l in which case they are removed from the hash table.

If the read contains multiple c-sets with conflicting signatures, we first assume that the first c-set is correct and try to correct the entire read accordingly. If we succeed, we have identified the correct c-set. Otherwise we assume that the second c-set is correct, and so on. If none of the c-sets leads to a successful correction of the entire read, we do not correct the read. Figure 1 illustrates an example with two c-sets.

To deal with an arbitrary number of c-sets, we employ an *iterative deepening depth first search* (IDDFS) heuristic strategy [11]. For each read, IDDFS searches for the correction path with the smallest number of nucleotide changes by starting with a small search depth (maximum number of base changes allowed at the current iteration) and by increasing the depth until either a solution is found or we reach the maximum number of base changes allowed.

Our proposed k-mer based error correction is sketched as CORRECTION (Algorithm 1). For each read, CORRECTION starts by determining the set of all repetitive and non-repetitive *csets* which will be extended one by one from left to right until all the k-mers of the read are considered. If conflicting c-sets are detected, they are removed from *csets* one at a time when attempting correction. For a given *cset*, we denote by *begin* and *end* its left and right border k-mers respectively. Also, we denote by *bacs* the BACs whose signature is shared by all the k-mers in *cset*. Starting at line 7, we iteratively call the recursive func-

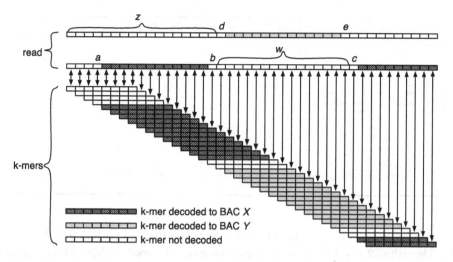

Fig. 1. An illustration of our error-correction strategy. C-sets are colored with dark and light gray. First, we assume that the read belongs to BAC X; in this case, the read positions corresponding to the first nucleotide of the k-mer starting at a, the last nucleotide of the k-mer starting at b, the first nucleotide of the k-mer starting at c and at least one more position in the portion w of the read (because the length of w is between k and $2k - 1$) must be corrected. If we assume that the read belongs to BAC Y, the read positions corresponding to the first nucleotide of the k-mer starting at d, the last nucleotide of the k-mer starting at e and at least two more positions in the portion z of the read (because the length of z is between $2k$ an $3k - 1$) must be corrected.

tion IDDFSEARCH with the maximum number of corrections *corr* allowed at the current iteration. If we can correct the entire read with exactly *corr* base changes, we output the corrected read and stop. Otherwise we increment *corr* and repeat the search.

Algorithm 2 sketches the recursive function IDDFSEARCH. When the entire read is covered by a single c-set or by several non-conflicting repetitive and non-repetitive c-sets, the corrected read is produced and the algorithm stops (lines 2-5). Otherwise, the algorithm tries to extend the current *cset* either to the left (line 9) or to the right (line 13). If *cset* is extended to the left, the algorithm needs to correct the read position *errPos* corresponding to the first nucleotide of the k-mer starting at $cset.begin - 1$ (lines 10–11). If *cset* is extended to the right, the algorithm needs to correct the read position *errPos* corresponding to the last nucleotide of the k-mer starting at $cset.end + 1$ (lines 14–15). Line 17 calls the function KMERCORRECT (Algorithm 3) with the *kmer* which is currently being corrected. KMERCORRECT searches for *kmer* in the hash table (line 2) and if found, verifies that the BAC(s) associated with it, *bacs(kmer)*, agree with the shared BAC signature(s) in *cset* (line 11). If this is the case, *kmer* is assumed to be correct and *read* is updated with the current base change (line 16 or line 21, depending on the direction). The variable *cset* is also extended

---

ALGORITHM 1. CORRECTION (*read-set* R, *hashtable* H)

```
 1   for each read in R do
 2        Determine csets, the set of all repetitive and non-repetitive c-sets;
 3        if conflicts detected then
 4            Remove conflicting c-sets from csets one at a time when attempting correction ;

 5        corr ← 0;
 6        cset ← leftmost element of csets ;

 7        while corr ≤ MaxCorrections do
 8            numCorrections ← 0 ;
 9            IDDFSEARCH (corr, numCorrections, read, cset) ;
10            if numCorrections = corr then
11                break;

12            corr ← corr + 1;

13        output read ;
14        update hashtable H by decreasing the counts of erroneous k-mers and increasing the counts of corrected
          k-mers;
```

---

by one position either to the left or to the right (line 18 or line 23). Line 17 in Algorithm 2 checks the same k-mer as found in *read* without changing it. This is necessary after every successful correction, so that, when extending the current c-set, we know that the k-mers we add contain no further errors. When all k-mers not initially covered by c-sets are checked and all c-sets are extended without detecting further errors, we have corrected the entire read, and we return from the recursive call and produce the solution. If instead we detect additional errors by checking the k-mers not initially covered by c-sets, the algorithm tries all three alternative nucleotides at the erroneous position detected, *errPos* (line 23 in Algorithm 2) and calls KMERCORRECT with the modified k-mer. Upon a successful return from KMERCORRECT, we recursively call IDDFSEARCH (line 31). The nucleotide changes at *errPos* and the calls to the two functions are only made if the total number of corrections so far does not exceed *corr*, the maximum number of corrections allowed at the current iteration (line 21).

4   Experimental Results

We present an experimental evaluation of our method on short reads derived from BACs belonging to a *Minimum Tiling Path* (MTP) of the rice and barley genomes. As its name suggests, an MTP is a set of BACs which cover the genome with minimum redundancy. The construction of an MTP for a given genome requires a physical map but we do not discuss either procedure here (see, e.g., [3,8] for details).

The use of an MTP allows us to assume that at most two (or three, to account for imperfections) BACs overlap each other. This assumption leads to a 3-decodable pooling design. To achieve $d = 3$ for STD [16], we choose parameters $L = 7$, $\Gamma = 2$ and $q = 13$, so that $d = \lfloor (L-1)/\Gamma \rfloor = 3$, $m = qL = 91$, and $n = q^{\Gamma+1} = 2,197$. With this parameter choices, we can handle up to 2,197 BACs using 91 pools organized in 7 redundant layers of 13 pools each. Since each layer is a partition of the set of pooled BACs, each BAC is pooled in exactly 7 pools (which is its *signature*). In addition, pools are well-balanced, as each pool contains exactly $q^\Gamma = 169$ BACs. By the properties of this pooling

---

ALGORITHM 2. IDDFSEARCH(corr, numCorrections, read, cset)

```
1   while true do
2       while cset joins its neighbors do
3           cset ← next c-set
4       if exhausted all c-sets then
5           return

6       prevEnd ← end of previous c-set or 0 if none
7       nextBegin ← begin of next c-set or (|r| − |k| + 1) if none
8       if cset.begin − 1 ≠ prevEnd then
9           direction ← left
10          errKmer ← cset.begin − 1
11          errPos ← errKmer

12      else
13          direction ← right
14          errKmer ← cset.end + 1
15          errPos ← errKmer + kmerSize − 1

16      kmer ← k-mer starting at position errKmer in read
17      corrected ← KMERCORRECT (read, cset, direction, kmer, errPos)
18      if not corrected then
19          break

20  numCorrections ← numCorrections + 1
21  if numCorrections > corr then
22      return

23  toTry ← the three alternative nucleotides at errPos
24  for nt ∈ toTry do
25      if direction = left then
26          kmer[1] ← nt

27      else
28          kmer[kmerSize] ← nt

29      corrected ← KMERCORRECT (read, cset, direction, kmer, errPos)
30      if corrected then
31          IDDFSEARCH (corr, numCorrections, read, cset)
32          if numCorrections = corr then
33              break
```

---

design, any two BAC signatures share at most $\Gamma = 2$ pools and any three BAC signatures share at most $3\Gamma = 6$ pools [16].

Once the set of MTP BACs has been pooled, we sequenced the resulting pools and used the read decoding algorithm HASHFILTER [8,9] to assign the reads back to their source BACs, and finally assemble each BAC individually. Error correction is applied prior to read decoding. All experiments were carried out on an Intel Xeon X5660 2.8 GHz server with 12 CPU cores and 192 GB of RAM.

For our correction algorithm SCRIBLE, we used parameters $k = 31$, $l = 3$, $h = 45$, and a maximum of 4 corrections per read, unless otherwise noted. An analysis of other choices of k is carried out later in Fig. 2. The other methods corrected all the reads in the 91 pools together (not pool-by-pool).

4.1 Results on Synthetic Reads for the Rice Genome

We tested our error correction method on short reads from the rice genome (*Oryza sativa*) which is a fully sequenced 390 Mb genome. We started from an MTP of 3,827 BACs selected from a real physical map library of 22,474 BACs. The average BAC length in the MTP was ≈ 150 kB. Overall, the BACs in the MTP spanned 91 % of the rice genome. We pooled a subset of 2,197 of these BACs into 91 pools according to the parameters defined above.

ALGORITHM 3. KMERCORRECT(read, cset, direction, kmer, errPos)
___

```
1  correct ← false

2  if kmer ∉ hashtable then
3  |   return false

4  pools ← pools(kmer) // from hashtable

5  if SIZE (pools) > h then
6  |   correct ← true

7  if SIZE (pools) ≥ l and SIZE (pools) ≤ h then
8  |   bacs ← bacs(kmer)
9  |   if SIZE (bacs) = 0 then
10 |   |   return false

11 |   if bacs agree with cset.bacs then
12 |   |   correct ← true

13 if correct then
14 |   if direction = left then
15 |   |   // update read with base change
16 |   |   read[errPos] ← kmer[1]
17 |   |   // extend cset left
18 |   |   cset.begin ← cset.begin − 1;

19 |   else
20 |   |   // update read with base change
21 |   |   read[errPos] ← kmer[kmerSize]
22 |   |   // extend cset right
23 |   |   cset.end ← cset.end + 1;

24 |   return true

25 return false
```

Table 1. Percentage of error-corrected reads that map to the rice genome for increasing number of allowed mismatches; execution times (preprocessing + correction) are *per* 1M reads; boldface values highlight the best result in each column.

	0 mm	1 mm	2 mm	3 mm	execution time (min)	space (GB)
Original reads	14.56 %	55.98 %	85.43 %	96.36 %	N/A	N/A
SGA	62.68 %	69.58 %	72.31 %	73.23 %	228.73 + 0.87	**6**
RACER	87.10 %	93.25 %	96.00 %	97.10 %	**19.76 + 0.11**	120
SCRIBLE	**95.00 %**	**97.77 %**	**98.70 %**	99.11 %	600 + 2.76	50

The resulting BAC pools were "sequenced" *in silico* using SIMSEQ, which is a high-quality short read simulator used to generate the synthetic data for Assemblathon [2]. SIMSEQ uses error profiles derived from real Illumina data to inject realistic substitution errors. We used SIMSEQ to generate ≈1M paired-end reads per pool with a read length of 100 bases and an average insert size of 300 bases. A total of ≈ 200M bases gave an expected ≈ 8× sequencing depth for a BAC in a pool. Since each BAC is present in 7 pools, this is an expected ≈ 56× combined coverage before decoding. The average error distribution for the first read in a paired-end read is: 48.42 % reads with no error, 34.82 % with 1 error, 12.96 % with 2 errors, 3.14 % with 3 errors, 0.57 % with 4 errors, 0.08 % with 5 errors and 0.01 % with 6 errors. For the second read, the error distribution is: 32.85 % no errors, 35.71 % with 1 error, 20.75 with 2 errors, 7.91 % with 3 errors, 2.20 % with 4 errors, 0.48 % with 5 errors, 0.09 % with 6 errors and 0.01 % with 7 errors.

We compared the performance of SCRIBLE against the state-of-the-art error-correction method RACER [5]. The authors of RACER performed extensive experimental evaluations and determined that RACER is superior in all aspects (performance, space, and execution time) to HiTEC, SHREC, REPTILE, QUAKE, and CORAL. We also compare SCRIBLE against SGA [14], as it was not evaluated in [5] but we had evidence it performs well. For both tools we used their default parameter setting.

Table 1 reports correction accuracy as well as time and space used by each method. As it was done in [5], correction accuracy is determined by mapping the corrected reads to the reference using BOWTIE [7] in stringent mode (paired-end, end-to-end alignment). Columns 2, 3, 4, and 5 report the fraction of reads mapped when 0, 1, 2, and 3 mismatches were allowed, respectively. The second row of Table 1 reports the mapping results for the original set of uncorrected reads. The last two columns report time (pre-processing + correction) and memory requirements for each method. The pre-processing time is method dependent: in our method it is the time to build the hash table (which currently is not multi-threaded), for SGA it is the time to build the FM-index, and, for RACER it is the time to compute witnesses and counters. For SCRIBLE and SGA we chose a $k = 31$ because (1) our method performs better with larger k-mer sizes and (2) this is the default choice for SGA. RACER determines the best k-mer size from the data.

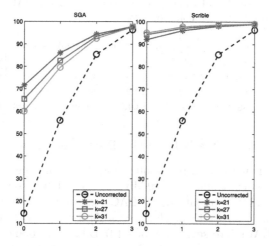

Fig. 2. Correction accuracy for different k-mer sizes for SGA and SCRIBLE (x-axis: number allowed mismatches for mapping, y-axis: percentage of reads mapped)

Table 1 shows that SCRIBLE is by far the most accurate. The difference between SCRIBLE and state-of-the-art RACER on a small number of allowed mismatches (which is what matters here) is very significant. In this application domain, accuracy is much more important than time- and space-complexity (as long as time and space are reasonable). Also observe in Table 1 that SGA's corrected reads do not map as well as uncorrected reads at two and three mismatches. SGA's results are puzzling, and currently do not have a good explanation for its behavior. In terms of execution time, the bottleneck in SCRIBLE is the preprocessing stage (building the hash table), which is currently single-threaded.

Figure 2 shows correction accuracy (percentage of mapped reads) for SGA and our method for other choices of k-mer size. As the k-mer size increases, our method corrects more reads, whereas the opposite is true for SGA.

An alternative way to assess the performance of error correction is to assemble the corrected reads, after decoding the reads with HASHFILTER. For this purpose, we used VELVET [18] with hash parameter $k = 49$ (chosen based on our extensive experience on this dataset). Table 2 reports the averaged results of 2,197 BACs in each of the four datasets. The table shows average coverage, the percentage of reads used by VELVET in the assembly, the number of contigs with length of at least 100 bases, the N50 value, the ratio of the sum of all contigs sizes over their BAC length, and the coverage of the BAC sequence by the assembly. The slightly different coverages depend on the ability of HASHFILTER to decode reads for each dataset. Despite the fact that VELVET employs its own error-correction algorithm, assemblies corrected by SCRIBLE have better statistics (except for the number of contigs) than those obtained using the outputs of the other tools.

Table 2. Assembly statistics for rice BACs on datasets of reads uncorrected and corrected via various methods; all values are an average of 2,197 BACs; boldface values highlight the best result in each column.

	coverage	reads used	#contigs	N50 (bp)	sum/size	BAC coverage
VELVET (uncorrected)	88.66x	92.40 %	52.27	35,846	89.05 %	82.22 %
SGA+VELVET	88.51x	97.46 %	51.90	36,603	90.33 %	82.88 %
RACER+VELVET	85.48x	98.22 %	**46.29**	29,770	88.59 %	81.82 %
SCRIBLE+VELVET	**88.76x**	98.43 %	53.92	**39,422**	**91.50 %**	**82.97 %**

4.2   Results on Real Reads from the Barley Genome

We also tested our method on real sequencing data from the barley (*Hordeum vulgare*) genome, which is about 5,300 Mb and not yet fully sequenced [15]. We started from an MTP of about 15,000 BACs selected from a subset of nearly 84,000 gene-enriched BACs for barley (see [8,9] for more details). We divided the set of MTP BACs into seven sets of 2,197 BACs and pooled each set using the STD parameters previously defined. We assessed the performance of error correction on three of these data sets, namely Hv4, Hv5 and Hv6 (with an average BAC length of about 116 kb). Each of the 91 pools in Hv4, Hv5 and Hv6 were sequenced on a flowcell of the Illumina HiSeq2000 by multiplexing 13 pools on each lane. After each sample was demultiplexed, we quality-trimmed and cleaned the reads of spurious sequencing adapters and vectors. We ended up with high quality reads of about 87–89 bases on average. The number of reads in a pool ranged from 3.37 M to 16.79 M (total of 706 M) in Hv4, from 2.32 M to 15.65 M (total of 500 M) in Hv5 and from 1.3 M to 6.33 M (total of 280 M) in Hv6.

We then compared SCRIBLE's ability to decode reads against HASHFILTER's. In other words, while HASHFILTER decodes erroneous reads using the improved

Table 3. Percentage of real barley reads (decoded by HASHFILTER and error-corrected/decoded by SCRIBLE) that map to 454-based high-quality BAC assemblies for increasing number of allowed mismatches; boldface values highlight the best result in each data set.

	dataset	0 mm	1 mm	2 mm	3 mm
HASHFILTER	Hv4	88.41 %	91.66 %	92.54 %	93.25 %
SCRIBLE	Hv4	**92.04 %**	**93.67 %**	**94.62 %**	**95.23 %**
HASHFILTER	Hv5	87.37 %	90.45 %	91.61 %	92.63 %
SCRIBLE	Hv5	**93.26 %**	**94.66 %**	**95.56 %**	**96.14 %**
HASHFILTER	Hv6	84.72 %	88.46 %	90.05 %	91.38 %
SCRIBLE	Hv6	**92.13 %**	**93.48 %**	**94.23 %**	**94.73 %**

method described in [9], SCRIBLE corrects and decodes the reads. Once the reads were decoded to individual BACs by HASHFILTER and SCRIBLE, we mapped them using BOWTIE [7] in stringent mode (paired-end, end-to-end alignment) to the subset of BACs for which a high-quality 454-based assembly was available (151 high-quality assemblies for HV4, 141 for HV5 and 121 for HV6). Table 3 presents the results of this comparison (using $k = 32$). Note that SCRIBLE achieves significantly higher mapping percentages than HASHFILTER. Higher mapping percentages indicate more error-free reads, and higher decoding accuracy.

To assess the impact of error correction on the final assemblies, we corrected/decoded reads using SCRIBLE then assembled them BAC-by-BAC using VELVET [18]. We compared the resulting assembly against the BAC assemblies on uncorrected reads decoded by HASHFILTER. Table 4 shows the average assembly statistics over 2,197 BACs in each barley set. Observe that assemblies obtained from SCRIBLE have consistently lower number of contigs, however the N50 is not always the highest.

Table 4. Average assembly statistics for barley BACs. All values are averages over 2,197 BACs. Boldface values highlight the best result for each data set.

	dataset	coverage	reads used	#contigs	N50 (bp)	sum/size
HASHFILTER+VELVET	Hv4	205.9x	93.9 %	56	**28,341**	114.5 %
SCRIBLE+VELVET	Hv4	205.9x	**96.7 %**	**44**	21,001	**110.9 %**
HASHFILTER+VELVET	Hv5	155.5x	94.9 %	72	**20,863**	101.3 %
SCRIBLE+VELVET	Hv5	155.5x	**96.1 %**	**35**	19,708	93.4 %
HASHFILTER+VELVET	Hv6	81.1x	**92.9 %**	44	25,194	**89.4 %**
SCRIBLE+VELVET	Hv6	81.1x	89.1 %	**34**	**27,631**	87.2 %

References

1. Duma, D., et al.: Accurate decoding of pooled sequenced data using compressed sensing. In: Darling, A., Stoye, J. (eds.) WABI 2013. LNCS, vol. 8126, pp. 70–84. Springer, Heidelberg (2013)

2. Earl, D., et al.: Assemblathon 1: a competitive assessment of de novo short read assembly methods. Genome Res. **21**(12), 2224–2241 (2011)

3. Engler, F., et al.: Locating sequence on FPC maps and selecting a minimal tiling path. Genome Res. **13**(9), 2152–2163 (2003)

4. Ilie, L., et al.: HiTEC: accurate error correction in high-throughput sequencing data. Bioinform. **27**(3), 295–302 (2011)

5. Ilie, L., Molnar, M.: RACER: rapid and accurate correction of errors in reads. Bioinform. **29**(19), 2490–2493 (2013)

6. Kelley, D.R., et al.: Quake: quality-aware detection and correction of sequencing errors. Genome Biol. **11**(11), R116 (2010)

7. Langmead, B., et al.: Ultrafast and memory-efficient alignment of short dna sequences to the human genome. Genome Biol. **10**(3), R25 (2009)

8. Lonardi, S., et al.: Combinatorial pooling enables selective sequencing of the barley gene space. PLoS Comput. Biol. **9**(4), e1003010 (2013)

9. Lonardi, S., et al.: When less is more: "slicing" sequencing data improves read decoding accuracy and *De Novo* assembly quality. Bioinform. **31**, 12 (2015). doi:10. 1093/bioinformatics/btv311

10. Mirebrahim, H., et al.: *De Novo* meta-assembly of ultra-deep sequencing data. Bioinform. **31**(12), i9–i16 (2015)

11. Russell, S., Norvig, P., et al.: Artificial Intelligence: A Modern Approach. Prentice-Hall Inc, Upper Saddle River, NJ, USA (1996). ch. 3

12. Salmela, L., Schroder, J.: Correcting errors in short reads by multiple alignments. Bioinform. **27**(11), 1455–1461 (2011)

13. Schroder, J., et al.: SHREC: a short-read error correction method. Bioinform. **25**, 2157–2163 (2009)

14. Simpson, J.T., Durbin, R.: Efficient de novo assembly of large genomes using compressed data structures. Genome Res. **22**(3), 549–556 (2012)

15. The International Barley Genome Sequencing Consortium: Nature. A physical, genetic and functional sequence assembly of the barley genome. **491**(7426), 711–716 (2012)

16. Thierry-Mieg, N.: A new pooling strategy for high-throughput screening: the shifted transversal design. BMC Bioinform. **7**, 28 (2006)

17. Yang, X., et al.: Reptile: representative tiling for short read error correction. Bioinform. **26**(20), 2526–2533 (2010)

18. Zerbino, D., Birney, E.: Velvet: Algorithms for de novo short read assembly using de Bruijn graphs. Genome Res. **8**(5), 821–829 (2008)

Jabba: Hybrid Error Correction for Long Sequencing Reads Using Maximal Exact Matches

Giles Miclotte, Mahdi Heydari, Piet Demeester, Pieter Audenaert, and Jan Fostier[✉]

Department of Information Technology, Internet Based Communication Networks and Services (IBCN), Ghent University - IMinds, Gaston Crommenlaan 8 (bus 201), 9050 Gent, Belgium
{giles.miclotte,mahdi.heydari,piet.demeester,pieter.audenaert, jan.fostier}@intec.ugent.be
http://www.ibcn.intec.ugent.be

Abstract. Third generation sequencing platforms produce longer reads with higher error rates than second generation sequencing technologies. While the improved read length can provide useful information for downstream analysis, underlying algorithms are challenged by the high error rate. Error correction methods in which accurate short reads are used to correct noisy long reads appear to be attractive to generate high-quality long reads. Methods that align short reads to long reads do not optimally use the information contained in the second generation data, and suffer from large runtimes. Recently, a new hybrid error correcting method has been proposed, where the second generation data is first assembled into a de Bruijn graph, on which the long reads are then aligned. In this context we present Jabba, a hybrid method to correct long third generation reads by mapping them on a corrected de Bruijn graph that was constructed from second generation data. Unique to our method is that this mapping is constructed with a seed and extend methodology, using maximal exact matches as seeds. In addition to benchmark results, certain theoretical results concerning the possibilities and limitations of the use of maximal exact matches in the context of third generation reads are presented.

Keywords: Sequence analysis · Error correction · de Bruijn graph · Maximal exact matches

1 Introduction

The accurate determination of the DNA sequence of an organism, i.e., establishing the precise order of the nucleotides A, C, G and T in a DNA molecule, is a fundamental and challenging problem in biology. Essentially this process consists of two steps: (i) sequencing the DNA by means of a chemical process, resulting in a large number of reads and (ii) genome assembly, where the reads are processed to

© Springer-Verlag Berlin Heidelberg 2015
M. Pop and H. Touzet (Eds.): WABI 2015, LNBI 9289, pp. 175–188, 2015.
DOI: 10.1007/978-3-662-48221-6_13

reconstruct the complete DNA sequence. Every sequencing technology results in reads that contain errors, with error profiles varying greatly between platforms. There is a clear distinction between *second generation* reads and *third generation* reads, where the latter are characterized by vastly improved read lengths albeit with much higher error rates. Processing such reads usually involves mapping them to other sequences, either by aligning the reads to each other to establish potential overlap, or by mapping them to a reference genome. Errors in the reads introduce noise to these alignments, leading to weaker alignments than the corresponding error-free reads would have. Lower rated alignments may then be discarded for further analysis, potentially discarding crucial information. This can be especially problematic when dealing with low quality reads in a region with low coverage. To deal with this sequencing noise, error correction methods can be applied. By correcting the errors in the reads, the optimal alignments can be more accurately identified and more appropriately rated, leading to better downstream analysis, as shown in e.g. [1] for *de novo* assembly.

In Jabba third generation reads are aligned to a de Bruijn graph built from second generation reads, using a seed-and-extend approach. The resulting paths in the graph dictate the read correction. The seeds are maximal exact matches (MEM) between an individual read and a node of the graph. The usage of MEMs as seeds has several advantages over k-mers. Firstly, the seeds can be longer. Even though long seed occur only rarely, few longer seed can be sufficient to have a rough estimate of how the read should be aligned to the graph. Shorter seeds can then be used to further refine this. Secondly, given an enhanced suffix array, seeds of arbitrary lengths can be sought without the need to rebuild this index. This is not the case for a k-mer index (e.g. a hash table). In case different values for k are to be used during the alignment process, different k-mer indexes need to be build. Finally the MEMs are not required to have the same size as the nodes. Since the high error rates of the third generation reads are the limiting factor on the minimal seed size, this allows the use of a larger value of k to build the de Bruijn graph, resulting in a less complex de Bruijn graph.

1.1   Related Work

For second generation sequencing we mainly consider Illumina. The different Illumina technologies produce many short (100–250 nucleotides) reads with a high accuracy ($<2\,\%$ errors, mainly substitutions) with high throughput and at a low financial cost. New algorithms, based on de Bruijn graphs, were specifically developed to efficiently deal with the assembly of huge amounts of second generation sequencing data. Overlap between short reads is then established in linear time between reads that share a k-mer, i.e., a substring of length k.

Algorithms to correct second generation reads have been classified [2] into three types. The k-mer spectrum-based methods rely on coverage thresholds to determine whether a k-mer represents part of the actual DNA sequence, these methods have been used for second generation error correction [3,4], but also in *de novo* genome assembly algorithms [5] and hybrid error correction methods [6]. The suffix tree-based methods [7,8] generalize the k-spectrum methods by handling multiple k values at once, while the multiple sequence alignment-based methods [9] correct the reads after aligning several similar reads.

Recently, third generation sequencing technologies (Pacific Biosciences 2013; Oxford Nano Technologies 2014) began to emerge. Pacific Biosciences SMRT sequencing results in much longer reads (avg. >5000 nucleotides), albeit with significantly higher error rates (15 %, mostly insertions and deletions and to a lesser extent substitutions). Despite this high error rate, the errors are uniformly distributed over the read, leading to a very high consensus accuracy if the coverage is sufficiently high and overlap between the reads is correctly established. Computing such overlap can not be efficiently achieved by means of a de Bruijn graph, because the high error rate leads to an overabundance of incorrect k-mers. Other efficient methods have been developed to compute pairwise alignments between third generation reads [10,11]. However, comparing each pair of reads results in a quadratic computational complexity, which is impractical for larger genomes and/or higher coverages. However, the coverage required for high accuracy consensus-based correction can still lead to a prohibitively high financial cost for many sequencing projects.

From this perspective, the use of hybrid error correction methods appears to be an attractive alternative. The goal is to correct long third generation reads using the more accurate sequence information contained in second generation reads. The idea is that a sufficient coverage in (cheap) second generation data might be sufficient to correct the long reads, regardless of the coverage of third generation data. This may result in a reduced financial cost for sequencing as low coverage third generation data might suffice. It should be noted that lower third generation coverage can directly result in lower assembly quality, no matter the quality of the reads, because of the uneven length distribution of the reads [12]. However, hybrid error correction methods also appear attractive from a computational point of view as they avoid pairwise comparisons between long reads, thus circumventing the quadratic computational complexity. The first type of hybrid error correction methods [13–15] rely on mapping short reads to long reads, and then calling the consensus sequence from this multiple alignment. However, such methods map short reads individually and do not exploit the context in which the short read occurs. A newer hybrid error correction method, LoRDEC [6], first constructs a de Bruijn graph from the short reads and then maps the long reads on this graph. The sequence implied by the path in the graph to which the long read aligns then represents the corrected read. The use of a de Bruijn graph has the advantage that overlap between short reads is established prior to mapping them to long reads. In [6], it was shown that LoRDEC achieves similar accuracy as other error correction methods, but with significantly improved runtimes.

2 Methods

2.1 Overview

In this work, we further build upon the idea of using a de Bruijn graph for hybrid error correction of long reads. Specifically, our main goal is the use of

Illumina data to correct Pacific Biosciences SMRT reads. A de Bruijn graph is then constructed from Illumina data and corrected using standard procedures (see further). Subsequently, long reads are aligned along a certain path in the graph in order to correct them. Whereas LoRDEC relies on shared k-mers to align long reads to a de Bruijn graph, we explore the idea of using maximal exact matches (MEMs). MEMs are exact matches between two sequences that can not be extended in either direction, this as opposed to common k-mers, which are exact matches of a fixed length k, which may or may not be extendable. Alignment methods based on maximal exact matches have been developed for read mapping [16–18]. It is shown in [16] that these methods can be more efficient than alignment techniques based on k-mers and Burrows-Wheeler transforms [19,20]. From the definition of a MEM, it is clear that every MEM of size $l \geq k$ can be represented as a consecutive sequence of k-mers, and vice versa. As such, finding and storing MEMs can be achieved in a more efficient manner, since MEMs can compactly represent multiple k-mers. The remainder of this section is dedicated to a more in-depth description of all steps involved (Fig. 1).

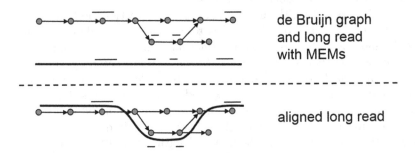

Fig. 1. To align a read to the de Bruijn graph, a seed-and-extend algorithm is used. First MEMs are found between the read and the graph, then a path in the graph is found between these seeds, creating the final alignment.

2.2   Correction of the de Bruijn Graph

Errors in short reads lead to erroneous paths in the de Bruijn graph. The errors in the graph can be corrected as described in [5]. Two types of errors can be discerned based on their position in the read. An error that is located at least $k - 1$ nucleotides away from both ends of the read will lead to k erroneous k-mers. In turn, this leads to the formation of a 'bubble', i.e. a path of length k that runs parallel to the real path. Assuming a sufficiently low error rate and a high coverage the correct path will typically have a higher coverage than the parallel erroneous paths, and the graph can be corrected by removing the erroneous path. On the other hand, errors positioned close to the ends of the read lead to the creation of less than k erroneous k-mers, thus forming 'dead ends' (tips) in the de Bruijn graph. These can be identified and removed based on topology and coverage considerations. Errors in the reads may also result in erroneous connections between unrelated parts of the graph, and because of coverage biases certain paths could be absent or underrepresented in the graph. This vastly complicates the graph correction procedure and erroneous paths may remain present in the final corrected graph.

2.3 Aligning Reads to a de Bruijn Graph

To align the reads to the graph a seed-and-extend approach is applied. By properly indexing the graph the seeds can be found in $O(m)$ time, where m is the size of the read that is being mapped.

Finding Maximal Exact Matches. To rapidly find MEMs between the nodes of the graph and the long reads, essaMEM [21] is used. These MEMs will be used as seeds for our alignment. By concatenating the sequences of every node and their reverse complement, a sequence is constructed. From this sequence an enhanced sparse suffix array is built by essaMEM. The sparseness factor of the index sharply reduces the space requirement for the index, compared to traditional suffix trees or enhanced suffix arrays, but this comes at the cost of a small increase in runtime. The required space could be even further reduced by only indexing the nodes and not their reverse complements. This would however double the search time, since the reverse complements of the reads then also have to be matched.

Chaining Seeds. To chain the seeds several passes over the read are performed. In each iteration the algorithm considers every region of the read that has not yet been corrected. For every such region separately, the largest seeds are considered. From these seeds it is determined to which nodes the current region of the read could map. For each such node the list of all seeds between this node and the current region of the read is considered, and an optimal placement of these seeds is decided, removing the ones that do not fit. Seeds are compatible if the distance between the two seeds on the read is contained in an interval determined by the estimated error rates and the distance of the seeds in the node.

Generally larger MEMs are less likely to be noise than shorter seeds, since the number of all k-mers increases exponentially if k increases and the number of k-mers contained in a sequence is similar to the size of the sequence, independent of k. There can still be noisy long seeds, especially when the genome contains imperfect repeats. In this case, the correct seeds can usually be recognized amidst the noisy seeds by considering the context. Firstly, the local context is considered, by comparing the seeds in the same node. This way seeds that occur in the same order in a node and in the read can be chained together to form inexact matches. Secondly, if the situation is still ambiguous, the global context is considered, by comparing the alignments in the neighborhood of the ambiguous region. If this neighborhood has not yet been chained in previous passes, the chaining of the current region is delayed to the next pass.

After obtaining the presumed layout of the seeds, the quality of the alignment is checked. Large gaps or a relatively large amount of mismatches may indicate incorrect alignments. The following cases are filtered:

1. Local mappings that are not super maximal, i.e., local mappings that are on the read contained in a larger local mapping.
2. Local mappings that cover less than 20 % of a node. The absence of any seeds in the rest of the node makes it less likely that this is actually a correct mapping.
3. Local mappings to nodes of size smaller than half of the largest local mapping. It is preferred to extend from those larger local mappings, since those are more reliable.

These filters are linearly relaxed with each pass of the algorithm, in the last iteration all local mappings are considered. After the local alignments are computed for the current pass, the next phase begins: chaining the alignments between different nodes by following unique paths in the graph. During this phase every local alignment is extended by considering the possible paths in the graphs. Both directions of the alignments are extended in the same manner, as follows:

1. If there is a unique edge, this edge must be correct and the local alignment is extended along this edge.
2. If there are several edges, the lengths of the end nodes are considered. Since the extension takes place between two regions of the read, certain estimates can be made for the maximal distance between the alignments, edges that are too long are then not considered.
3. If at any point there are no suitable edges to extend along, a mistake was made at some point. Either the graph is incorrect or the original local chaining was erroneous. In either case the erroneous region is reprocessed in a new local chaining step.

In the rest of this section the distance between corrected regions on a read is denoted as n and the estimated insertion and deletion rates of the data are denoted as i and d. After the unique-extension step, the resulting chains may overlap in the graph, in which case they can be linked together to make one consecutive path. Overlapping chains are however not a sufficient condition for linking, the sequences represented by the path and the read need to be compared. If the sequence on the path is smaller than $(1 + 2i)^{-1}n$, the shortest cycle at the common point is considered. If this shortest cycle can not adequately fill the gap, then the paths are not joined and the gap is left for the next pass. To determine whether the shortest cycle is a good fit a local alignment is performed between the sequence dictated by the path and the read, using a Smith-Waterman approach [22]. Likewise, if the resulting chains do not meet, the shortest path between both end points is considered. If this shortest path can not adequately fill the gap, the gap is again left for the next iteration.

By building from seeds and only using shortest path algorithms to chain the nodes, computationally expensive path searching can be avoided, however, this can not be avoided indefinitely. After the final iteration a bounded path search is performed between consecutive corrected regions, in an attempt to fill the remaining gaps. This search looks for paths that contain a sequence with length bounded by the interval $[(1 + 2i)^{-1}n, (1 + 2d)n]$.

Read Ends. The parts of the read beyond the extremal seeds, i.e., the ends of the read, are not corrected. Such a correction can be trivial, if the end is completely contained within the same node as the extremal seed, or in the unique path flowing from that node. In this case correcting the end would not provide any information about the genome that is not already contained in the graph. The correction can also be far from trivial, if there are several possible paths. In this case the correction of the end requires aligning the possible paths against the read. This can be done by looking for seeds with a lower threshold, or by performing direct global alignments with a dynamic programming approach. This is not done in Jabba, since it is a relatively expensive operation for a small gain.

Settings. Jabba takes several parameters that can affect the results. Most importantly the minimal length l of MEMs for the initial search can be specified, the standard value is $l = 20$, but this should be chosen based on the discussion in Sect. 3.1 in function of the data. The maximal number p of iterations of the algorithm can be specified, the standard value is $p = 5$. A third parameter allows Jabba to trim the reads beyond the extremal corrected positions.

3   Results Concerning Maximal Exact Matches

In this section the occurrence of maximal exact matches in reads is investigated. Insertions and deletions have a different effect on the size of maximal exact matches than substitutions. A substitution error puts a firm stop to any running exact matches, while an insertion or deletion may allow for the exact match to continue, effectively looking like an error at a further point in the read. In the following this difference is ignored and all errors are treated like they were substitutions. Because of this, the size of MEMs is slightly underestimated for sequences that contain insertions or deletions. It is also assumed that errors are uniformly distributed in the sequences, as is the case for Pacific Biosciences SMRT reads.

3.1   Coverage by Exact Regions

In this section the expected ratio of a long read that should be covered by MEMs larger than a given size is explored, under the assumption that the reference contains no errors. Variations on this topic have been explored in [23–25]. In the following n is the length of the read, p is the error-rate and m the threshold for maximal exact matches. An *exact region of size* k on a read is defined as k correct consecutive bases in that read. The *coverage by exact regions* is the ratio of bases that are contained in exact regions.

The expected number of exact regions (including those of length 0) is the expected number of errors, i.e., np. The expected coverage of a read by exact regions of size k is then the product of (i) the coverage of the read by one exact region of size k: k/n, (ii) the expected number of exact regions: np, and (iii) the probability that an exact region has size k: $(1 - p)^k p$. This results in:

$$k(1-p)^k p^2 \ . \tag{1}$$

Summing (1) over all $k \geq m$ gives the expected coverage of the read by exact regions of size $k \geq m$:

$$\sum_{k=m}^{\infty} k(1-p)^k p^2 = (1-p) - \sum_{k=0}^{m-1} k(1-p)^k p^2 \ , \tag{2}$$

the right hand side provides a finite formula to compute this expected coverage. Figure 2 shows the expected coverage by exact regions larger than m, for error-rate $p = 15\%$ and $p = 35\%$. The maximum $1 - p$ is obtained at $\{0, 1\}$ since every correct base is contained in an exact region of size ≥ 1. It can be seen that increasing p leads to a steeper descent near the inflection point. While it was a priori clear that a lower error rate leads to larger exact regions, this also shows that the equilibrium between a sufficient amount of seeds and a sufficiently large minimal seed length, is less stable for higher error rates.

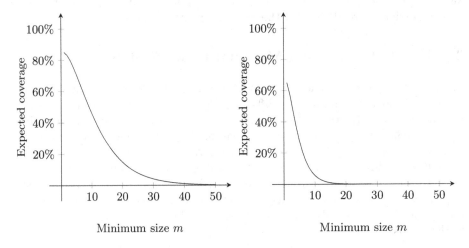

Fig. 2. Expected coverage by exact regions of size $k \geq m$ for reads of size 10000 with 15 % errors (*left*) and 35 % errors (*right*), expressed as percentages of the whole read as a function of the minimal size of the exact regions.

3.2   Occurrence of Exact Regions

The expected length of the longest exact region in a read of size n is denoted by $ER_p(n)$. If $np(1-p)^m \geq 1$ then at least one exact region of size $k \geq m$ is expected in a read of size n, hence the expected length of the longest run can be approximated by solving $np(1-p)^m = 1$ for m:

$$ER_p(n) \approx -\log_{1-p} np. \tag{3}$$

The distribution around this average can be approximated by the complement of a Gumbel distribution with cumulative distribution function

$$F(x) = \exp -(1-p)^{x+1}; \tag{4}$$

the probability that a read of length n will have an exact region of size $k \geq m$ is then approximated by

$$P(n, p, m) = 1 - F(m + ER_p(n)) = 1 - \exp\left(-np(1-p)^{m+1}\right). \tag{5}$$

These approximations are highly accurate when p and n are sufficiently large. Figure 3 shows the ratio of reads of length n that are expected to have an exact region of size m. For sufficiently large values of n, replacing n by $n' > n$ shifts the graph to the right by a term $\log_{1-p} n/n'$, replacing p by $p' < p$ shifts the graph to the left and steepens the descent near the inflection point. This again shows that larger error rates make the determination of a proper seed size threshold less stable.

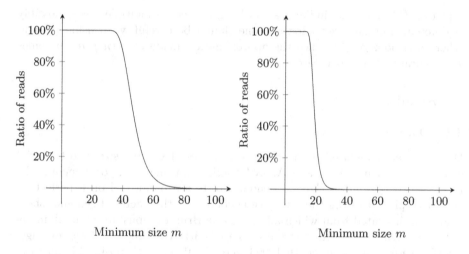

Fig. 3. Expected percentage of reads of size 10000 that contain at least one exact region of size $k \geq m$, for reads with 15 % errors (*left*) and 35 % errors (*right*).

3.3 Applications

During the local chaining step from Sect. 2.3 one can apply the results of Sect. 3.1 to decide whether a local mapping is plausible or not. For each mapping the coverage by exact regions can easily be computed by counting seed sizes. The resulting number can then be compared to the expected coverage that can be obtained from Sect. 3.1. If there is a significant deviation in either direction, the local mapping gets a lower rating.

When computing mappings it is required to have at least 1 seed available, hence the results from Sect. 3.2 propose good upper bounds for the minimum length of seeds, depending on the read size and error rates. To a certain extent this result can also be used to estimate the probability of a read containing several exact regions of a minimal size. If a read of size n contains a MEM of size $k \geq m$, then this MEM divides the read in two pieces, one of size n' and the other of (approx.) size $n - n'$. This approximation of the piece-sizes is made since typically k is significantly smaller than n, and k is not known a priori. The conditional probability of the read containing a second MEM of size larger than m then becomes $1 - (1 - P(n', p, m))(1 - P(n - n', p, m))$, with P as in (5). Since n' depends on the read, it is a priori not known and integrating over n' is required. The distribution of the size of n' can be approximated by the uniform distribution on $\{0, \ldots, n\}$, and because of symmetry this leads to the following estimate of the a priori probability of a read of size n containing at least 2 exact regions with size larger than m:

$$P(n, p, m) \frac{2}{n} \sum_{n'=0}^{n/2} \left(1 - \left(1 - P(n', p, m)\right)\left(1 - P(n - n', p, m)\right)\right). \qquad (6)$$

Equation (6) can in a similar fashion be extended to multiple seeds, possibly of different minimal sizes. However one should be careful when using (6) and other extensions of (5), since the approximation made by $P(n, p, m)$ becomes less accurate when n decreases.

4   Results

4.1   Data

Datasets were simulated from reference genomes of varying size, two bacterial genomes: N. meningitidis and A. hydrophila; and one eukaryotic genome: D. melanogaster (fruit fly). For all genomes tests are performed on a perfect de Bruijn graph built from the reference genome. For the bacterial genomes short reads are simulated from which additional de Bruijn graphs are built. Illumina paired-end reads of length 100 were simulated with ART [26] with 100x coverage. PacBio reads of average length 10000 were simulated with pbsim [27] and 10x coverage, with 15 % errors distributed as 60 % insertions, 30 % deletions and 10 % substitutions. Real Illumina and PacBio datasets were used for E. coli. For D. melanogaster real PacBio data was mapped on the de Bruin graph built from the reference genome. The sources of the data are specified in Table 1.

4.2   Parameters

LoRDEC was run with $k = 19$ for the bacterial data sets, as suggested in [6]. For the larger fruit fly the values $k = 20 \ldots 23$ were tested. The best results were obtained for $k = 21$ and these results have been included in the results tables. For Jabba the de Bruijn graphs were built with $k = 31$ and the minimum MEM size was 20.

4.3   Evaluation Metrics

In [6] it is demonstrated that LoRDEC performs better than both LSC [13] and PacBioToCA [14]. Hence, Jabba is only compared to LoRDEC. By using

Table 1. The data sets and reference genomes. The D. melanogaster reference genome can be accessed on http://www.fruitfly.org/sequence/release5genomic.shtml.

N. meningitidis	Reference genome NC_003116
A. hydrophila	Reference genome NC_008570
D. melanogaster	Reference genome Release 5
	PacBio data from http://datasets.pacb.com.s3.amazonaws.com/2014/Drosophila/reads/list.html.
E. coli	Reference genome NC_000913
	Illumina data accession number ERR022075
	PacBio data from https://github.com/PacificBiosciences/DevNet/wiki/E-coli-K12-MG1655-Hybrid-Assembly

simulated data, the corrected read can be aligned to the original sequence from which the read was simulated. This way a multiple alignment of the original read, the corrected read and the genomic region is created. In this alignment each position is analyzed separately, in order to obtain a confusion matrix as follows:

- True Positive, an erroneous position was corrected.
- False Positive, a new error was introduced at a correct position.
- True Negative, a correct position remains unchanged.
- False Negative, an erroneous position remains unchanged.

To interpret this confusion matrix the following statistics are computed:

- Sensitivity $= TP/(TP + FN)$. This expresses the relative amount of errors that were corrected.
- Specificity $= TN/(TN + FP)$. This expresses the relative amount of correct positions that were recognized as such.
- Precision $= TP/(TP + FP)$. This expresses how reliable a correction is, if one is made.
- Gain $= (TP - FP)/(TP + FN)$. This expresses the quality of the corrected reads compared to the original read.

For real data the reads are aligned to the reference genome with BLASR [28] and the identity of the mapping is computed. All experiments were run on dual-socket octa-core Intel Xeon Sandy Bridge computing nodes at 2.6 GHz and 64 GB of memory. The runtimes and memory usage are measured using the standard Linux time command. The runtime includes only the actual mapping of long reads and does not include the generation and correction of the de Bruijn graph.

4.4 Evaluation

Table 2 shows the results for LoRDEC and Jabba as they were run on each of the simulated data sets. The results on the bacterial genomes suggest that a

Table 2. Results on simulated data for LoRDEC and Jabba. The subscript $_p$ indicates the usage of a perfect de Bruijn graph built from the reference genome. The absence of the subscript indicates the usage of a de Bruijn graph built from simulated short reads. Both real time and CPU time are averages over all reads, with 16 threads in parallel.

	Sensitivity	Specificity	Precision	Gain	Real time	CPU Time	memory
Neisseria meningitidis							
LoRDEC	95.6 %	99.0 %	94.4 %	89.9 %	266.5 ms	1732.5 ms	63 MB
LoRDEC$_p$	98.7 %	99.3 %	96.0 %	94.6 %	141.7 ms	921.1 ms	58 MB
Jabba	92.3 %	99.1 %	94.4 %	86.9 %	126.1 ms	1596.5 ms	126 MB
Jabba$_p$	92.3 %	99.1 %	94.5 %	87.0 %	125.0 ms	1583.3 ms	115 MB
Aeromonas hydrophila							
LoRDEC	99.2 %	99.7 %	98.4 %	97.6 %	84.1 ms	1093.4 ms	75 MB
LoRDEC$_p$	99.8 %	99.9 %	99.1 %	99.0 %	70.5 ms	597.3 ms	46 MB
Jabba	95.7 %	99.4 %	96.4 %	92.1 %	180.0 ms	2315.2 ms	161 MB
Jabba$_p$	95.3 %	99.4 %	96.1 %	91.4 %	178.5 ms	2311.3 ms	168 MB
Drosophila melanogaster							
LoRDEC$_p$	91.6 %	98.5 %	91.1 %	82.7 %	435.1 ms	2828.6 ms	538 MB
Jabba$_p$	93.9 %	99.2 %	95.3 %	89.3 %	246.4 ms	2285.7 ms	1181 MB

corrected de Bruijn graph yields comparable results to using a perfect de Bruijn Graph based on a reference genome. For the bacteria, LoRDEC performs better than Jabba, since the MEMs contain the same information as k-mers, this is most likely due to shortcomings in the chaining algorithm in Jabba. A noticeable drop in the performance of LoRDEC can be observed when using the frequency-based de Bruijn graph compared to the perfect de Bruijn graph.

The performance of Jabba does not drop when moving from the bacterial genomes to the fruit fly. On the other hand, our evaluation of LoRDEC shows that LoRDEC obtains a lower gain for this larger genome, to the point where Jabba outperforms it. It can be seen in Table 2 that Jabba is slower than LoRDEC on A. hydrophila, but faster on the other two genomes, and that Jabba consistently requires 2–4 times more memory than LoRDEC.

The results for the real data can be found in Table 3. The higher performance of Jabba on D. melanogaster when compared to LoRDEC might be explained by the use of MEMs. In LoRDEC the seed size and the k-mer size of the graph are identical. Since the seed size must be kept relatively low as shown in Sect. 3.2,

Table 3. Results on real data for LoRDEC and Jabba, as obtained by mapping the reads to the reference genome.

	Original Identity	LoRDEC Identity	LoRDEC Gain	Jabba Identity	Jabba Gain
Escherichia coli	83.8 %	98.6 %	91.9 %	98.5 %	90.8 %
Drosophila melanogaster	86.7 %	96.6 %	74.3 %	98.0 %	85.3 %

the de Bruijn graph has to be built with a small value of k, leading to a tangled de Bruijn graph. This is expected to become more apparent for larger and more complex genomes. When using MEMs however, the seed size and k-mer size are independent of each other and optimal values of k can be chosen to construct the de Bruijn graph.

Acknowledgments. The computational resources (Stevin Supercomputer Infrastructure) and services used in this work were provided by the VSC (Flemish Supercomputer Center), funded by Ghent University, the Hercules Foundation and the Flemish Government – department EWI. We acknowledge the support of Ghent University (Multidisciplinary Research Partnership "Bioinformatics: From Nucleotides to Networks").

References

1. Salzberg, S.L., et al.: GAGE: a critical evaluation of genome assemblies and assembly algorithms. Genome Res. **22**, 557–567 (2012)
2. Yang, X., Chockalingam, S.P., Aluru, S.: A survey of error-correction methods for next-generation sequencing. Briefings Bioinform. **14**(1), 56–66 (2013)
3. Kelley, D.R., Schatz, M.C., Salzberg, S.L.: Quake: quality-aware detection and correction of sequencing errors. Genome Biol. **11**, R116 (2010)
4. Greenfield, P., et al.: Blue: correcting sequencing errors using consensus and context. Bioinformatics **30**(19), 2723–2732 (2014)
5. Zerbino, D.R., Birney, E.: Velvet: algorithms for de novo short read assembly using de Bruijn graphs. Genome Res. **18**, 821–829 (2008)
6. Salmela, L., Rivals, E.: LoRDEC: accurate and efficient long read error correction. Bioinformatics **30**(24), 3506–3514 (2014)
7. Schröder, J., et al.: SHREC: a short-read error correction method. Bioinformatics **25**(17), 2157–2163 (2009)
8. Ilie, L., Fazayeli, F., Ilie, S.: HiTEC: accurate error correction in high-throughput sequencing data. Bioinformatics **27**(3), 295–302 (2011)
9. Salmela, L., Schröder, J.: Correcting errors in short reads by multiple alignments. Bioinformatics **27**(11), 1455–1461 (2011)
10. Myers, G.: Efficient local alignment discovery amongst noisy long reads. In: Brown, D., Morgenstern, B. (eds.) WABI 2014. LNCS, vol. 8701, pp. 52–67. Springer, Heidelberg (2014)
11. Berlin, K., et al.: Assembling large genomes with single-molecule sequencing and locality sensitive hashing. Nat. Biotech. **33**, 623–630 (2015)
12. Boetzer, M., Pirovano, W.: SSPACE-longread: scaffolding bacterial draft genomes using long read sequence information. BMC Bioinform. **15**(1), 211 (2014)
13. Au, K.F., et al.: Improving pacbio long read accuracy by short read alignment. PLoS ONE **7**(10), e46679 (2012)
14. Koren, S., et al.: Hybrid error correction and de novo assembly of single-molecule sequencing reads. Nat. Biotechnol. **30**, 693–700 (2012)
15. Hackl, T., et al.: proovread: large-scale high-accuracy pacbio correction through iterative short read consensus. Bioinformatics **30**(21), 3004–3011 (2014)
16. Liu, Y., Schmidt, B.: Long read alignment based on maximal exact match seeds. Bioinformatics **28**(18), i318–i324 (2012)
17. Vyverman, M., et al.: A long fragment aligner called ALFALFA. BMC Bioinform. **16**, 159 (2015)

18. Li, H.: Aligning sequence reads, clone sequences and assembly contigs with BWA-MEM (2013). arXiv:1303.3997 [q-bio.GN]
19. Li, H., Durbin, R.: Fast and accurate long-read alignment with Burrows-Wheeler transform. Bioinformatics **26**(5), 589–595 (2009)
20. Langmead, B., Salzberg, S.L.: Fast gapped-read alignment with bowtie 2. Nat. Methods **9**(4), 357–359 (2012)
21. Vyverman, M., et al.: essaMEM: finding maximal exact matches using enhanced sparse suffix arrays. Bioinformatics **29**(6), 802–804 (2013)
22. Zhao, M., et al.: SSW library: an SIMD Smith-Waterman C/C++ library for use in genomic applications. PLoS ONE **8**(12), e82138 (2013)
23. Arratia, R., Gordon, L., Waterman, M.S.: An extreme value theory for sequence matching. Ann. Stat. **14**(3), 971–993 (1986)
24. Gordon, L., Schilling, M.F., Waterman, M.S.: An extreme value theory for longest head runs. Zeitschrift fur Wahrscheinlichkeitstheories verwandt Gebeite (Probability Theory and Related Fields) **72**, 279–287 (1986)
25. Schilling, M.F.: The surprising predictability of long runs. Math. Assoc. Am. **85**(2), 141–149 (2012)
26. Huang, W., et al.: ART: a next-generation sequencing read simulator. Bioinformatics **28**(4), 593–594 (2012)
27. Ono, Y., Asai, K., Hamada, M.: PBSIM: pacbio reads simulator-toward accurate genome assembly. Bioinformatics **29**(1), 119–121 (2013)
28. Chaisson, M.J., Tesler, G.: Mapping single molecule sequencing reads using basic local alignment with successive refinement (BLASR): theory and application. BMC Bioinform. **238**, 13 (2012)

Optimizing Read Reversals for Sequence Compression

(Extended Abstract)

Zhong Sichen[2], Lu Zhao[2], Yan Liang[2], Mohammadzaman Zamani[1],
Rob Patro[1(✉)], Rezaul Chowdhury[1], Esther M. Arkin[2],
Joseph S.B. Mitchell[2], and Steven Skiena[1]

[1] Department of Computer Science, Stony Brook University, Stony Brook, USA
`rob.patro@cs.stonybrook.edu`
[2] Department of Applied Mathematics and Statistics,
Stony Brook University, Stony Brook, USA

Abstract. New generation sequencing technologies produce massive data sets of millions of reads, making the compression of sequence read files an important problem. The sequential order of the reads in these files typically conveys no biologically significant information, providing the freedom to reorder them so as to facilitate compression. Similarly, for many problems the orientation of the reads (original or reverse complement) are indistinguishable from an information-theoretic perspective, providing the freedom to optimize the orientation of each read.

In this paper, we introduce a class of algorithmic problems concerned with optimizing read ordering and orientation for sequence compression. We show that most of the interesting variants are hard, but provide heuristics yielding strong approximation guarantees. In particular, we give a linear time 2-approximation algorithm for the optimal ordering/orientation under the prefix match criteria. Further, through experiments on a number of data sets, we demonstrate that this heuristic works well in practice. A prototype implementation of this 2-factor approximation is available at https://github.com/LaoZZZZZ/prefixMatching.

1 Introduction

New generation sequencing technologies produce massive data sets of 10 s or even 100s of millions of reads. These data sets are not processed and then discarded, but rather are stored, often in public databases, to allow replication of results and future analysis with new methods and tools and in light of new Biological discoveries. The huge size and growing numbers of such data sets makes sequence read compression an important problem — since these experimental results must be efficiently stored and transferred.

The sequential order in which the reads appear in a typical file generally conveys no biological significance. Specifically, the order in which reads are recorded in a file is usually random and, even when it is not, almost all downstream processing tools are agnostic to this order. This provides the freedom to

© Springer-Verlag Berlin Heidelberg 2015
M. Pop and H. Touzet (Eds.): WABI 2015, LNBI 9289, pp. 189–202, 2015.
DOI: 10.1007/978-3-662-48221-6_14

reorder sequencing reads so as to facilitate compression. Similarly, for double-stranded sequence data, and un-stranded sequencing protocols, the orientation of the reads (original or reverse complement) are typically indistinguishable from an information-theoretic perspective, providing the freedom to optimize the orientation of each read.

The problem of compressing sequencing data has spawned its own small field of research, which can be roughly divided into two categories; reference-based compression, which seeks to compress the information represented by a set of sequencing reads by making use of the reference genome of the corresponding organism and *de novo* compression, which seeks to compress the sequencing reads without using any "external" information. While both strategies are important, *de novo* compression is the more broadly applicable, as we do not have reference genomes for most species. Typically, these approach try to exploit the redundancy present within the set of sequencing reads resulting from an experiment [1–4,7,8,10,11,14,16]. We expect such redundancy to be large since the reads are drawn from the same underlying sequence (i.e. genome or transcriptome), which is usually sequenced at considerable coverage.

The idea of boosting the compression of sequencing reads by altering the order in which they appear in the file was proposed by Hach et al. [10]. Read re-ordering is also one of the basic strategies taken in the recent Mince compressor of Patro and Kingsford [14], who also demonstrate that selectively reverse-complementing certain reads can further boost compression. However, the observation that re-ordering the underlying information improves the compression of sequencing data was not completely new. For example, it is conventional wisdom that re-ordering a BAM file by the position of the alignments therein — which has the effect of placing similar alignments nearby — can significantly reduce the size of the resulting file. While compression is a motivating factor for our work, we focus here only on the related problem of ordering the reads to maximize different types of objectives in the hope that these results might prove useful to new techniques that attempt to boost compression, in part, by re-ordering reads.

Algorithmic string ordering problems have been previously considered under different distance measures, without the complexity of orientation. In particular, the Hamming distance Travelling Salesman (TSP) problem has been shown to be Max SNP-hard [13,18] (and thus hard to approximate within a constant $r > 1$) [17].

Our major contributions here include:

- *A Taxonomy of Read Ordering/Orientation Problems* – We introduce a suite of read ordering/orientation problems, which differ according to the measure of similarity between adjacent reads.

  In particular, our algorithmic and complexity-theoretic results are summarized in Table 1 below. Here, we discuss the reversal of strings, but we note that the operation of interest to us is actually reverse complementation. This distinction doesn't affect the theoretical results and our experimental results. The approximation results hold under reverse complementation since our analysis

Table 1. A taxonomy of the complexity of different variants of the string ordering problem.

	Distance	Reversals	Complexity	Apx. factor (if any)
1	Prefix	No	Sort	Optimal
2	Prefix	Yes	NP-Hard	2-apx
2'	Prefix with alphabet size 2	Yes	NP-Hard	2-apx
3	Max prefix or suffix	No	?	9/7-apx
4	Max prefix or suffix	Yes	?	9/7-apx
5	Max prefix and suffix	No	?	9/7-apx
6	Max prefix and suffix	Yes	?	9/7-apx
7	Max substring	No	NP-Hard	9/7-apx
8	Max substring	Yes	NP-Hard	9/7-apx
9	Levenshtein distance	No	NP-Hard	3/2-apx
10	Levenshtein distance	Yes	NP-Hard	3-apx

treats the prefix and suffix as independent, non-interacting strings. There is a bijective mapping between strings and their reverse complement. Since our analysis applies to all input strings, either interpretation is valid.

- *A Linear-time 2-Approximation Algorithm for Maximizing the Sum of Prefix Matching with Reversals* – Our most interesting result is a heuristic for sum of prefix matching which runs in linear-time (and hence is practical for large data sets), yet provides a 2-factor approximation to optimality – exploiting structural properties of tries to show that a greedy matching is in fact an optimal matching.
- *Experimental Results in Read Ordering/Orientation* – We demonstrate the performance of our 2-approximation algorithm for maximum sum of prefix matching by comparing it to other ordering approaches.

This paper is organized as follows. We define our problem taxonomy in Sect. 2, demonstrating the hardness of most variants in Sect. 3 and presenting approximation algorithms in Sect. 4. Experimental results are presented in Sect. 5.

2   Problem Taxonomy

The major problem variants we consider are illustrated in Fig. 1 and described below:

- **Prefix Matching** - Let $S = \{s_1, s_2, ..., s_N\}$ be a finite and ordered set of strings where each string is the same length l. Denote $s_i[k]$ as the k-th character in string s_i. Then the *prefix match* for an arbitrary pair of strings s_i and s_j is defined as the largest integer $n \le l$ such that $s_i[m] = s_j[m] \; \forall \; 1 \le m \le n$. If no characters match, then the prefix match of s_i and s_j is defined to be zero. The *reverse* of a string s_i is a new string s_i' such that $s_i'[k] = s_i[l - k + 1]$

Distance Function	Fixed Orientation		Reversals Permitted	
Max Prefix				
Max Prefix **Or** Suffix				
Max Prefix **And** Suffix				
Max Substring				

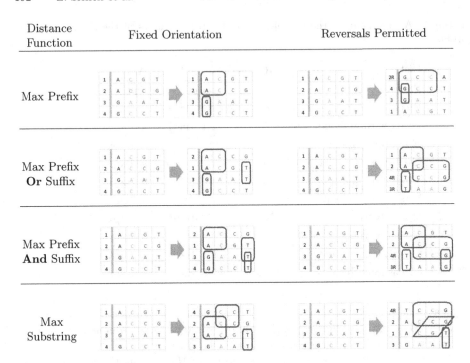

Fig. 1. Taxonomy of read ordering/reversal problems.

$\forall \ 1 \leq k \leq l$. The prefix matching problem is then as follows: Allowing either the forward or reverse orientation of each string in S, find a reordering S^* of S such that the sum of the prefix matches between <u>sequential</u> pairs of strings is maximized.

– **Suffix Matching** - The *suffix match* for an arbitrary pair of strings s_i and s_j is defined as the largest integer $n \leq l$ such that $s_i[m] = s_j[m] \ \forall \ l - n + 1 \leq m \leq l$. The suffix matching problem is the same as the prefix matching problem except we are now maximizing suffixes instead of prefixes.

– **Prefix Matching with Alphabet Size 2** - The proof used for the hardness of Prefix Matching has an implicit assumption. Namely, if all of our strings are length two, and if our strings have infinite alphabet size, then the Prefix Matching problem is already hard.

– **Max Prefix <u>or</u> Suffix** - Similar to the prefix matching and suffix matching problems defined above, we are interested in maximizing either the prefix or suffix. More specifically, for any pair of sequential strings in S, we are allowed to select either the suffix matching or prefix matching, but not both.

– **Max Prefix <u>and</u> Suffix** - This version of the problem is the same as above except instead of picking either the suffix or prefix for a consecutive pair of strings in the ordering, we take the sum of both values.

– **Max Substring** - Let s_i and s_j be two arbitrary strings in S. Then the *max substring* between s_i and s_j is defined as the largest integer n such that for some $0 \leq k \leq l - n$, $s_i[m] = s_j[m] \ \forall \ k + 1 \leq m \leq k + n$. In the max

substring problem, we wish to maximize the sum of the max substring between sequential pairs of strings (with or without reversals).
- **Edit Distance** - We define the distance function here between two strings as the minimum Levenshtein distance. Instead of maximizing like before, we wish to minimize the total Levenshtein distance.

3   Hardness Results

We begin with the problem of ordering strings so as to maximize the sum of the length of prefix matches between successive strings. We note that for strings of fixed orientation, sorting the strings lexicographically suffices for optimality. Assume an optimal ordering that didn't agree with the lexicographic ordering. Then moving any maximal block A of consecutive strings sharing a prefix P next to a non-adjacent block B with the same prefix (to which A would have been adjacent in the lexicographic order) adds a prefix match of length $|P|$ while breaking a shorter one, resulting in a larger total sum of prefix matches.

Allowing sequence reversals as well as reordering makes the problem considerably more difficult:

Theorem 1. *Finding the maximum prefix matching with reversals is NP-Hard.*

Proof. We reduce the classical minimum Vertex Cover Problem to prefix matching. Given a simply connected graph $G(V, E)$ on N vertices and M edges, label each of the vertices a unique character. An edge between any two vertices corresponds to a length two string that is the concatenation of their respective labels. The number of length two strings we construct is exactly equal to M. We may write each length two string either in its forward or reverse orientations.

Let PM denote the maximum/optimal number of prefix matchings and let VC denote the minimum number of vertices needed to cover G. Then, the maximum number of prefix matchings is achieved by the following equation:

$$PM + VC = M \tag{1}$$

Hence, it follows that if we can find PM in polynomial time, then we can also find the minimum vertex cover(VC) in polynomial time.   □

Theorem 2. *Maximum prefix matching with reversals remains NP-Hard even if the alphabet size is 2.*

Proof. Our sequence of reductions is as follows: we first reduce a given instance of the vertex cover problem on N vertices and M edges to an instance of the prefix matching problem on M strings, each with $L = 2$ characters. Call this the $L = 2$ *prefix matching problem*. This first reduction is accomplished via the transformation given in the proof of Theorem 1. Next, we reduce the $L = 2$ prefix matching problem to an instance of the prefix matching problem where the alphabet size is $A = 2$ and each string is no longer restricted to being just length 2. Call this the $A = 2$ *prefix matching problem*. We focus on the second reduction.

Given a set of M strings of length $L = 2$, convert each length two string into its binary representation. Within each length two string's binary representation, reverse the binary representation of the <u>second</u> letter. This is to ensure that the binary representation of the first character will match when strings are reversed. Proceed to insert M^2 zeros between the binary representation of the first character and the reversal of the binary representation of the second character.

Now suppose we can calculate the optimal ordering and objective to the $A = 2$ prefix matching problem in polynomial time. Let OPT denote the optimal solution to the prefix matching problem and define VC as the minimum vertex cover number. We claim the validity of the following inequality:

$$M - VC + 1 \geq \frac{OPT}{\log 2M + M^2} \geq M - VC \tag{2}$$

Note that $\frac{OPT}{\log 2M + M^2} = M - VC + \frac{\epsilon}{\log 2M + M^2}$, where $M - VC$ is the maximum number of prefix matchings in the $L = 2$ problem, and ϵ is the sum of "residual" matchings. To clarify further on the meaning of residual matchings, consider the following two cases. If 2 strings in the $L = 2$ problem have only one matching prefix, then in the $A = 2$ problem, the number of matchings is <u>greater than</u> or equal to $\log 2M + M^2$. This is because for M strings each with two characters, the total number of digits needed to represent all the characters is $\log 2M$, so if there is one character match, then in $A = 2$ problem there are at least $\log(2M) + M^2$ matches. On the other hand, if 2 strings in the $L = 2$ problem have no matching prefixes, then in the $A = 2$ problem, the number of matching prefixes is greater than or equal to 0 but <u>less than</u> or equal to $\log(2M) - 1$.

Define $OPT = \sum_{i=1}^{M-1} m_i$, where m_i is the number of matchings from the ith string to the $(i + 1)$th string in the $A = 2$ problem. For any index k which falls into case one, we can define r_k as $r_k = m_k - (log2M + M^2)$. For indices l which fall into case two, we define $r_l = m_l$. We can see then that the exact definition of ϵ can be written as $\epsilon = \sum_{i=1}^{M-1} r_i$. Since $r_i \leq \log(2M) - 1$ for all i, it follows that $\frac{\epsilon}{log2M + M^2} \leq 1$, and the inequality follows. Immediately, we can see if OPT is found in polynomial time, then we can solve for VC in polynomial time.   □

Theorem 3. *The maximum substring problem without reversals is NP-Hard.*

Proof. We reduce Hamiltonian cycle to max substring matching without reversals. Let $G(V, E)$ be a simply connected graph. Define the vertex set as $V = \{v_1, v_2, v_3, ...v_N\}$ and label each edge in E a unique character. We construct N strings using the following procedure. For each vertex i, list the characters which have v_i as one of its endpoints. Now repeat for all vertices i. Since no two pairs of strings constructed in this manner has more than one letter in common, it follows that finding a solution to max substring matching will also give us a valid Hamilton cycle.   □

The next corollary follows immediately from Theorem 3. If we allow reversals, the optimal solution obtained remains the same. Since every pair of strings has one unique character in common, reversing strings gives no incremental benefit when our distance function is max substring.

Corollary 1. *Max substring matching with reversals is NP-Hard.*

4   Approximation Algorithms

For each of the value functions discussed above, we give a 2-approximation algorithm that takes into account reversals; the method is independent of the particular value function. For a given set, $S = \{s_1, s_2, \ldots, s_N\}$, of N strings, each of length m, we consider the complete graph, K_N whose nodes are the strings, and the weight, $w_{i,j}$, of edge (s_i, s_j) is given by $w_{i,j} = \max\{p(s_i, s_j), p(\overline{s_i}, s_j), p(s_i, \overline{s_j}), p(\overline{s_i}, \overline{s_j})\}$, where $\overline{s_i}$ is the reversal of string s_i, and $p(s_i, s_j)$ is the value function (e.g., length of the common prefix) associated with strings s_i and s_j. A 2-approximation is obtained by finding a maximum weight matching in this weighted complete graph.

Lemma 1. *A maximum weight matching in K_N yields a 2-approximation for finding an optimal value permutation.*

Proof. Consider an optimal ordering of the strings; without loss of generality, it is $(s_1, s_2, s_3, \ldots, s_N)$. Assume N is odd; the case for N is even is handled similarly. The value of this optimal ordering is

$$\sum_i^{N-1} w_{i,i+1} = (w_{1,2} + w_{3,4} + \cdots + w_{N-2,N-1}) + (w_{2,3} + w_{4,5} + \cdots + w_{N-1,N}).$$

We know that a maximum-weight matching has value at least $(w_{1,2} + w_{3,4} + \cdots + w_{N-2,N-1})$; also, a maximum-weight matching has value at least $(w_{2,3} + w_{4,5} + \cdots + w_{N-1,N})$. Thus, the value of an optimal permutation is at most twice the weight of a maximum-weight matching, implying that any permutation that respects the ordering implied by the edges of a maximum-weight matching is within a factor 2 of being optimal.   □

Actually constructing the complete graph K_N explicitly, and computing an optimal matching thereof, would be impractical for large datasets. Thus, we analyze the structure of this problem further to obtain a fast and practical approximation algorithm for the prefix matching variant of our problem.

Our algorithm first builds a *trie* over all N input strings, s_i and their reversal (or reverse-complement), $\overline{s_i}$. This construction is a linear-time operation. We note, here, that the fact that what is, essentially, a greedy algorithm (below) yields a 2-approximation for the problem of finding a maximum prefix matching with reversals has some interesting (perhaps deep) analogies to the recent work of Cazaux and Rivals [5], who prove the utility of greedy algorithms in achieving good approximation ratios for problems in overlap graphs of finite words.

Using this trie, we apply a greedy algorithm, running in time $O(N)$, to compute a matching. Specifically, we scan through our leaf nodes and pick pairs of strings that share the largest depth parent node in monotonically decreasing order. As we search through all $2N$ leaves, we mark and forbid a leaf's reverse if we have already picked a leaf's forward orientation.

If the value function is prefix matching, then the greedy algorithm above actually gives us an optimal solution to pairwise matching. In other words, the optimal value obtained by finding a maximum-weight matching on K_n will be equivalent to the objective value obtained by applying the greedy algorithm on the trie. This is summarized in the following lemma.

Lemma 2. *Suppose the value function is common prefix length. Then, the greedy algorithm on the trie of the N strings gives an optimal matching.*

Proof. It is sufficient to show if strings a and b are two strings of S with the largest prefix match, then there exists an optimal matching of S in which a and b are adjacent. If this always holds, then we can apply this rule iteratively to show there exists an optimal matching that is given by our greedy algorithm.

The proof is by an exchange argument. Suppose, to the contrary, that an optimal matching matches string a to $a' \neq b$, and string b to $b' \neq a$. There are three cases; refer to Fig. 2. Let $d(U)$ denote the depth in the trie of node U; i.e., $d(U)$ is the number of edges in the path from the root, R, to U. In the figure, we let C be the node of the trie that is the common ancestor of strings a and b, D be the common ancestor of strings a and a', and E be the common ancestor of strings b and b'; by assumption, C has the largest depth, $d(C)$, of any non-leaf node of the trie. Thus, node E must be either on the path from the root R to D (case (1)), or on the path from D to the leaf associated with a' (case (2)), or on the path from D to C (case (3)).

In case (1), if we replace the matched edges (a, a') and (b, b') with the edges (a, b) and (a', b'), the objective function goes up by $d(C)+d(E)-d(D)-d(E) \geq 1$. In case (2), if we replace the matched edges (a, a') and (b, b') with the edges (a, b) and (a', b'), the objective function goes up by $d(C)+d(E)-d(D)-d(D) \geq 2$. In case (3), if we replace the matched edges (a, a') and (b, b') with the edges (a, b) and (a', b'), the objective function goes up by $d(C)+d(D)-d(D)-d(E) \geq 1$.

Thus, in all cases, the swap improves the matching, proving the claim.

4.1   Approximations Based on TSP

As long as the benefit, $p(s_i, s_j)$, of edge (s_i, s_j) does not affect the benefit of any other edge (in particular, of edge (s_j, s_k)), the problem of ordering the strings S to maximize total benefit is exactly the Maximum-TSP problem — which seeks to find a cycle in a graph that visits all vertices and maximizes the sum of the weights of the traversed edges — for which there are known 9/7-approximation algorithms [12]. The corresponding entries in the table include "Max Prefix or Suffix" (without reversals), "Max Prefix and Suffix" (without reversals), and "Max Substring" (without reversals). Other cases also can be addressed with Maximum-TSP.

One can alternatively model the optimal ordering problem as a minimization problem, in which the goal is to minimize the total cost of the permutation. Here,

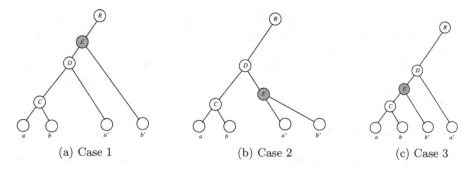

(a) Case 1   (b) Case 2   (c) Case 3

Fig. 2. The three cases to consider for the greedy matching algorithm.

$p(s_i, s_j)$ now reflects a cost (e.g., edit distance) between s_i and s_j. Provided the cost function obeys the triangle inequality, then, we obtain a 3/2-approximation, using the Christofides approximation [6] for metric TSP.

In the case of minimization allowing reversals, we obtain a 3-approximation, since this is now a one-of-a-set TSP, with sets of size 2 (each set consists of a string, together with its reversal), so Slavik's approximation gives a factor $(3/2) \cdot 2 = 3$ [15].

4.2   Ordering with Reversals for Palindromic Value Functions

In this section we describe a simple Maximum-TSP based $\frac{9}{7}$-approximation algorithm for string ordering with reversals when the value function satisfies a *palindromic* property as defined below.

Definition 1 [Palindromic Value Function (PVF)]. A value function $p(\cdot, \cdot)$ is *palindromic* w.r.t. a given set of strings S, provided $p(s_i, s_j) = p(\overline{s_i}, \overline{s_j})$ holds for all $s_i, s_j \in S$, where $\overline{s_i}$ (resp. $\overline{s_j}$) is the reversal of s_i (resp. s_j).

Examples of PVF include $\max\{prefix\text{-}length(s_i, s_j), suffix\text{-}length(s_i, s_j)\}$, $prefix\text{-}length(s_i, s_j) + suffix\text{-}length(s_i, s_j)$, $max\text{-}substring\text{-}length(s_i, s_j)$, $max\text{-}subsequence\text{-}length(s_i, s_j)$, $complement\text{-}of\text{-}Hamming\text{-}distance(s_i, s_j)$, etc.

Given a set $S = \{s_1, s_2, \ldots, s_N\}$ of strings as input we start by creating a weighted complete graph K_N with the strings in S as nodes, and setting the weight of the undirected edge between every pair of distinct vertices s_i and s_j to $w_{ij} = \max\{p(s_i, s_j), p(s_i, \overline{s_j})\}$. We then find a $\frac{9}{7}$-approximation of a symmetric Maximum-TSP path in K_N maximized over choices of starting and ending vertices. W.l.o.g. let $P = \langle s_1, s_2, \ldots, s_N \rangle$ be the resulting path. Observe that each string in P is still in its original input orientation. But if an edge weight $w_{k,k+1}$ on this path is set to $p(s_k, \overline{s_{k+1}})$ then those two strings s_i and s_{k+1} must appear in one of the following two orientations in our output to achieve value $p(s_k, \overline{s_{k+1}})$: $(s_k, \overline{s_{k+1}})$, $(\overline{s_k}, s_{k+1})$. On the other hand, if $w_{k,k+1} = p(s_k, s_{k+1})$ then the strings must be in one of these two orientations: (s_k, s_{k+1}), $(\overline{s_k}, \overline{s_{k+1}})$. So we may need to reverse some of the strings in P so the orientations of every two consecutive strings on this path are consistent with the weight of the edge between them. We do this reorientation by keeping s_1 fixed in its input orientation and then starting from $k = 1$ for each $k < N$, we reorient s_{k+1}, if needed, so that the orientations of s_k and s_{k+1} are consistent with $w_{k,k+1}$. This reorientation step ensures that the total value of our string ordering matches the total weight of path P.

5   Experimental Results

Here, we explore a practical implementation of our linear-time 2-approximation algorithm for the prefix matching problem with reversals. We show that read ordering alone can have significant results on compression and allowing reversals further boosts this compression. We experimentally measure the running-time of our algorithm on a varied selection of different data sets, and we also examine how well our algorithm is able to optimize the average prefix length, defined as $\frac{1}{n-1} \sum_{i=1}^{n-1} f(s_i, s_{i+1})$, where $f(s_i, s_{i+1})$ is the length of the prefix shared by the reads appearing at positions i and $i+1$ in the computed order and with the prescribed orientation. We compare the average prefix length obtained by the order and orientation computed using our 2-approximation algorithm to the average prefix length of the order in the original input file, the order obtained by sorting the reads (without allowing reversals), and the order obtained by running the reads through SCALCE [10] (by compressing then decompressing them). So as to not confuse the SCALCE-induced order with the actual compression results of the full SCALCE algorithm, we refer to it below as "SO". We also consider the compressed file sizes for these different read orderings and orientations and the time required to compute them.

We note that we consider only the compression of read sequences, leaving the conceptually and practically very different problems of read name and quality score compression to be addressed by other methods. Most existing approaches for read name and quality score compression do not assume an underlying ordering for the reads (e.g. the Quartz software of Yu et al. [19] for quality score compression) and, hence, can be used along with reordering approaches, such as ours, for read sequence compression.

Table 2 shows the average prefix length obtained under different read ordering algorithms mentioned above. The first column gives an index to each data set and the second lists the data set's accession IDs. The third column shows read lengths of each dataset. The last three columns contain the average prefix length obtained by the ordering (and potentially reversals) imposed by each algorithm. The datasets chosen reflect a range of DNA sequences from different species. Some of the data sets presented below are from paired-end experiments. To avoid the complexities involved in handling such data when computing a single ordering, we choose only the first sequence out of each pair, and thus use only half of all reads. Another approach to this problem is to merge the paired-end reads as is done in [14], wherein the authors demonstrate that this approach seems to work well in practice. This problem could also be solved by determining a good joint-ordering, or efficiently encoding the permutation between two separate orderings — but we are unaware of tools which take such approaches. All datasets are obtained from NCBI's database.

As expected, the 2-approximation, allowing for reversals, obtains an often substantially larger average prefix length than either sorting or SO; a result that holds across all data sets. It must be noted that SCALCE does *not* attempt to optimize for prefix matches, and so the SO results are not directly comparable to our algorithm under this metric. On the other hand, sorting without taking into account reverse complements actually provides an optimal solution to the prefix matching without reversals problem, though it has no constant factor guarantee when reversals are allowed. Thus, the difference between the 5th and 6th columns is illustrative of the benefit one can expect to achieve in the average prefix length objective by allowing reverse complementation.

Table 3 gives a summary of how different read orderings and orientations affect downstream compression. Column 3 gives the original sequence file size in megabytes, while columns 4–7 list the post-compression file size among the original ordering and the orderings produced by the three algorithms we consider. To obtain these results we compress the re-ordered files from the 2-approximation, sorting and the SCALCE-induced ordering with gzip [9]. We consider here only the sizes of the read sequences, and discard the read names and quality values. The resulting file sizes from this procedure are given in column 7.

Table 3 demonstrates that grouping sequences by common prefixes improves — sometimes substantially — the compression performance of gzip. In fact, in all datasets but 5, all of the re-orderings improve the compressed file size, with the ordering and orientation produced by the 2-approximation usually meeting or surpassing that produced by simple sorting. For datasets 11, 14, and 15, sorting achieves a better compression result than our 2-apx. For these datasets, the 2-approximation algorithm reverses a small proportion of the total number of strings. If the number of reversals done on a dataset is small, then sorting may obtain a better result. For these 3 datasets, the proportion of strings reversed is 22.8 %, 8.5 %, and 10.6 % respectively.

As might be expected, the SCALCE ordering, being optimized for boosting gzip compression, often produces the smallest file sizes. Nonetheless, it is surprising how much of the gap between the size of the original compressed file and the impressive compression rate of the SCALCE ordering can be accounted for simply by re-ordering the reads, since the more local ordering optimization computed by SCALE — essentially bucketing reads sharing similar substrings — might seem a more natural fit for boosting dictionary-based compression tools like gzip. However, the simple prefix-matching score we attempt to optimize captures a substantial percentage of the file compression achieved by the SCALCE ordering, even achieving better compression on data sets 2, 4, 11, and 13. This suggests that it may be an interesting objective to study further.

Finally, we observe that our 2-approximation algorithm demonstrates comparable, practical runtime compared to the other re-ordering methods. We compared the time taken to run each of the re-ordering algorithms on the data followed by the time required to compress the results with gzip. On average, the

Table 2. Average prefix-length under different ordering and reversal criteria.

	Dataset	Read length	Original	2-apx	Sorting	SO
1	Typhi 15098S solexa	51	0.710	13.753	12.479	1.694
2	Typhi 404ty solexa	26	0.359	14.962	13.575	5.201
3	ERR009815	146	0.407	17.962	15.563	2.764
4	DRR000003	36	0.754	14.263	13.456	2.466
5	SRR390728	72	2.723	20.095	17.725	4.252
6	DRR000002	72	0.867	22.614	19.127	2.955
7	ERR233147	69	0.400	27.333	26.327	4.976
8	DRR000001	72	0.865	23.079	19.882	2.972
9	ERR019065	152	0.638	34.953	25.909	9.265
10	ERR406998	152	0.493	64.317	60.576	19.877
11	ERR406999	152	1.105	123.149	121.004	101.914
12	SRR034939	200	0.372	19.244	15.866	0.703
13	ERR018601	152	0.369	30.378	25.595	0.369
14	SRR062421	200	92.987	157.569	156.963	92.987
15	SRR346694	200	91.739	156.168	153.964	156.168

Table 3. The compressed file sizes under various orderings and orientations of the reads; all units are reported in megabytes. All read files of the sequences for each dataset are in the FASTQ format. File sizes represent only read sequences, with quality and ID discarded. We extract the sequence reads after running SCALCE on the original dataset and apply gzip to this ordering. These compression results are given under the SO column.

	Dataset	Sequence size	Gzip only	2-apx	Sorting	SO
1	Typhi 15098S solexa	85	26	22	22	18
2	Typhi 404ty solexa	127	39	24	26	25
3	ERR009815	226	67	60	61	54
4	DRR000003	239	72	56	56	59
5	SRR390728	524	98	125	128	114
6	DRR000002	607	182	142	147	109
7	ERR233147	627	188	132	134	106
8	DRR000001	740	222	171	177	133
9	ERR019065	1004	259	217	230	125
10	ERR406998	1759	508	266	270	188
11	ERR406999	2356	609	114	109	123
12	SRR034939	3638	1066	971	982	921
13	ERR018601	4532	1248	1007	1028	985
14	SRR062421	4966	504	310	308	276
15	SRR346694	5422	536	333	331	333

linear time 2-apx algorithm takes $\approx 92\%$ of the time required to simply sort the file. We note that we have focused here on providing a correct implementation of our algorithm, and there is much room for optimizing the specific implementation, which would surely lead to marked performances in runtime improvement. Obtaining the SCALCE ordering takes longer than either of the other methods, but it is also doing significantly more work (e.g. compressing quality values), so we do not discuss its runtime here. All timing results were computed on a shared server with 4×6-core 2.6 GHz Xeon CPUs and 256 GB of RAM.

Acknowledgments. This research was partially supported by NSF Grants DBI-1355990 and IIS-1017181, and a Google Faculty Research Award to Steven Skiena. E. Arkin and J. Mitchell acknowledge support from the National Science Foundation (CCF-1540890) and the US-Israel Binational Science Foundation (BSF project 2010074). Rezaul Chowdhury was supported by NSF grant CCF-1439084. Rob Patro would like to acknowledge Carl Kingsford and Geet Duggal for useful discussions and for helping to initially pose and explore the prefix matching with reversal problem. Finally, we would like to thank the anonymous reviewers for suggestions and comments which greatly improved the manuscript.

References

1. Adjeroh, D., Zhang, Y., Mukherjee, A., Powell, M., Bell, T.: DNA sequence compression using the Burrows-Wheeler transform. In: Proceedings on Bioinformatics Conference, 2002, pp. 303–313. IEEE Computer Society (2002)
2. Bhola, V., Bopardikar, A.S., Narayanan, R., Lee, K., Ahna, T.: No-reference compression of genomic data stored in FASTQ format. In: IEEE International Conference on Bioinformatics and Biomedicine (BIBM 2011), pp. 147–150. IEEE (2011)
3. Bonfield, J.K., Mahoney, M.V.: Compression of FASTQ and SAM format sequencing data. PLoS One **8**(3), e59190 (2013)
4. Brandon, M.C., Wallace, D.C., Baldi, P.: Data structures and compression algorithms for genomic sequence data. Bioinformatics **25**(14), 1731–1738 (2009)
5. Cazaux, B., Rivals, E.: Approximation of greedy algorithms for max-ATSP, maximal compression, maximal cycle cover, and shortest cyclic cover of strings. In: PSC 2014: Prague Stringology Conference, pp. 148–161. Czech Technical University in Prague, Czech Republic (2014)
6. Christofides, N.: Worst-case analysis of a new heuristic for the travelling salesman problem. Technical report, DTIC Document (1976)
7. Cox, A.J., Bauer, M.J., Jakobi, T., Rosone, G.: Large-scale compression of genomic sequence databases with the Burrows-Wheeler transform. Bioinformatics **28**(11), 1415–1419 (2012)
8. Deorowicz, S., Grabowski, S.: Data compression for sequencing data. Algorithms Mol. Biol. **8**(1), 25 (2013)
9. Gailly, J., Adler, M.: Gzip program (2014). http://www.gnu.org/software/gzip/. Accessed 16 June 2014
10. Hach, F., Numanagić, I., Alkan, C., Sahinalp, S.C.: SCALCE: boosting sequence compression algorithms using locally consistent encoding. Bioinformatics **28**(23), 3051–3057 (2012)

11. Jones, D.C., Ruzzo, W.L., Peng, X., Katze, M.G.: Compression of next-generation sequencing reads aided by highly efficient de novo assembly. Nucleic Acids Res. **40**, e171 (2012)
12. Paluch, K., Mucha, M., Madry, A.: A 7/9 - approximation algorithm for the maximum traveling salesman problem. In: Dinur, I., Jansen, K., Naor, J., Rolim, J. (eds.) Approximation, Randomization, and Combinatorial Optimization. LNCS, vol. 5687, pp. 298–311. Springer, Heidelberg (2009)
13. Papadimitriou, C., Yannakakis, M.: Optimization, approximation, and complexity classes. In: Proceedings of the Twentieth Annual ACM Symposium on Theory of Computing, pp. 229–234. ACM (1988)
14. Patro, R., Kingsford, C.: Data-dependent bucketing improves reference-free compression of sequencing reads. Bioinformatics, btv248 (2015)
15. Slavik, P.: Approximation Algorithms for Set Cover and Related Problems. Ph.D. thesis, Buffalo, NY, USA, AAI9833643 (1998)
16. Tembe, W., Lowey, J., Suh, E.: G-SQZ: compact encoding of genomic sequence and quality data. Bioinformatics **26**(17), 2192–2194 (2010)
17. Trevisan, L.: When hamming meets euclid: the approximability of geometric TSP and steiner tree. SIAM J. Comput. **30**, 475–485 (2000)
18. Trevisan, L.: When hamming meets euclid: the approximability of geometric TSP and MST. In: Proceedings of the Twenty-ninth Annual ACM Symposium on Theory of Computing, pp. 21–29. ACM (1997)
19. Yu, Y.W., Yorukoglu, D., Peng, J., Berger, B.: Quality score compression improves genotyping accuracy. Nat. Biotechnol. **33**(3), 240–243 (2015)

Circular Sequence Comparison with q-grams

Roberto Grossi[1], Costas S. Iliopoulos[2], Robert Mercaş[2,3], Nadia Pisanti[1],
Solon P. Pissis[2(✉)], Ahmad Retha[2], and Fatima Vayani[2]

[1] Department of Computer Science, University of Pisa and Erable Team, INRIA,
Pisa, Italy
[2] Department of Informatics, King's College London, London, UK
solon.pissis@kcl.ac.uk
[3] Department of Computer Science, Kiel University, Kiel, Germany

Abstract. Sequence comparison is a fundamental step in many important tasks in bioinformatics. Traditional algorithms for measuring approximation in sequence comparison are based on the notions of distance or similarity, and are generally computed through sequence alignment techniques. As *circular* genome structure is a common phenomenon in nature, a caveat of specialized alignment techniques for circular sequence comparison is that they are computationally expensive, requiring from super-quadratic to cubic time in the length of the sequences. In this paper, we introduce a new distance measure based on q-grams, and show how it can be computed efficiently for circular sequence comparison. Experimental results, using real and synthetic data, demonstrate orders-of-magnitude superiority of our approach in terms of efficiency, while maintaining an accuracy very competitive to the state of the art.

1 Introduction

Circular molecular structures are present, in abundance, in all domains of life: bacteria, archaea, and eukaryotes; and in viruses. They can be composed of both amino and nucleic acids. The following is a superficial description of such occurrences. Exhaustive reviews can be found in [10] (proteins) and [19] (DNA).

Circular genomes and plasmids are found in bacteria and archaea. Whole-genome comparison is a very useful tool in identifying bacterial strains, as well as inferring phylogenies. The extended benefit of aligning plasmids is the ability to identify important genes, such as antibiotic resistance genes, thereby enabling their study and enhancement by genetic engineering techniques [12].

The most familiar examples of such structures in eukaryotes are mitochondrial (MtDNA) and plastid DNA. However, there exist other structures, called extra-chromosomal circular DNA, which are described as one of the characteristics of genomic plasticity in eukaryotes [9]. MtDNA is generally conserved from parent

R. Mercaş—Supported by the P.R.I.M.E. programme of DAAD co-funded by BMBF and EU's 7th Framework Programme (grant 605728).
S.P. Pissis—Supported by a Research Grant (#RG130720) awarded by the Royal Society.

© Springer-Verlag Berlin Heidelberg 2015
M. Pop and H. Touzet (Eds.): WABI 2015, LNBI 9289, pp. 203–216, 2015.
DOI: 10.1007/978-3-662-48221-6_15

to offspring, and so it can be used as an indicator of evolutionary relationships among species. The absence of recombination in these sequences allows them to be used as simple tests of phylogenetic evolution, and their high mutation rate is a powerful discriminating feature [17,33].

It is common knowledge that many viral genomes are circular. Multiple sequence alignment of viral genomes can be useful in the elucidation of novel sites of interest [4], as well as the inference of evolutionary relationships [3]. Viroids are plant pathogens that comprise very small single-stranded circular RNA. Their multiple sequence alignment could prove useful in the analysis of their secondary structures and pathogenicity [24].

Ribosomally synthesized circular proteins occur in both prokaryotes and eukaryotes [10]. An interesting phenomenon known to occur naturally in genes encoding linear protein structures is circular permutation [34]. This can be exemplified by swaposins: highly-similar proteins resulting from circularly permuted linear peptide sequences [29]. The ability to align linear sequences from circular proteins can significantly speed-up and enhance their analyses, and could also lead to the discovery of novel pairs of circularly permuted proteins.

Conventional tools to align circular sequences could yield an incorrectly high genetic distance between closely-related species. Indeed, when sequencing molecules, the position where a circular sequence starts can be totally arbitrary. Due to this arbitrariness, a suitable rotation of one sequence would give much better results for a pairwise alignment. A practical example of the benefit this can bring to sequence analysis is the following. Linearized human (NC_001807) and chimpanzee (NC_001643) MtDNA sequences do not start in the same region. Their pairwise sequence alignment using EMBOSS Needle [31] gives a similarity of 85.1 % and consists of 1,195 gaps. However, taking different rotations of these sequences into account yields a much more significant alignment with a similarity of 91 % and only 77 gaps. This example motivates the design of efficient algorithms that are devoted to the specific comparison of circular sequences, as they can be relevant in the analysis of organisms containing these structures.

Our Problem. We consider the pairwise circular sequence comparison problem. Under the edit distance model, it consists in finding an optimal linear alignment of two circular strings. This problem for two strings x and y of length m and $n \geq m$, respectively, can be solved under the edit distance model in time $\mathcal{O}(nm \log m)$ [21]. Several other super-quadratic [23] and *approximate* quadratic-time [5] algorithms exist. Trivially, for molecular biology applications, the same problem can be solved in time $\mathcal{O}(nm^2)$ with scoring matrices and affine gap penalty scores. A *direct* application of pairwise circular sequence comparison is for multiple circular sequence alignment [1,14,24]. The latter has also been considered in [20] under the Hamming distance model.

To the best of our knowledge, there is no fast (that is, with sub-quadratic time complexity) and exact (or at least very accurate) algorithm for circular sequence comparison under some realistic model (that is, allowing *indels*). Taking into account edit distance rather than Hamming distance is computationally challenging as the search space for seeking similarity is wider. Moreover, filters that work for Hamming distance do not work in general for edit distance [28] as well. An exception to this are the q-gram filtering techniques [32] that have successfully been used for string matching under the edit distance model (e.g. [7,26,30]), as well as for multiple local alignments both under the Hamming [27] and edit [26] distance model.

Our Contribution. We present new efficient q-gram-based methods for pairwise circular sequence comparison. Specifically, our contribution is threefold.

1. We introduce the *β-blockwise q-gram distance* between two strings x and y, that is, a more powerful generalization of the q-gram distance introduced as a string distance measure in [32]. Intuitively, and similarly to [7,26,30], this generalization comprises partitioning x and y in $β$ *blocks* each, as evenly as possible, computing the q-gram distance between the corresponding block pairs, and then summing up the distances computed blockwise.
2. We present an algorithm based on the suffix array [22] that finds the rotation of x such that the $β$-blockwise q-gram distance between the rotated x and y is minimal, in time and space $\mathcal{O}(βm + n)$, where $m = |x|$ and $n = |y|$, thereby solving *exactly* the circular sequence comparison problem under the $β$-blockwise q-gram distance measure. We also present a simple heuristic algorithm to solve an *approximate* version of the problem.
3. We present an experimental study, using real and synthetic data, which demonstrates orders-of-magnitude superiority of our approach, in terms of efficiency, while maintaining an accuracy very competitive to the *optimal* obtained after considering all rotations of x against y using EMBOSS Needle.

The paper is organized as follows. Section 2 gives some preliminary definitions, notation, and properties. Section 3 describes two algorithms, one is a heuristic approach and the other is an exact algorithm for circular sequence comparison. Section 4 shows the experimental results of performance and accuracy of our algorithms. Section 5 gives some concluding remarks and future proposals.

2   Definitions and Properties

We begin with a few definitions, following [11]. We think of a *string* x of *length* m as an array $x[0 \mathinner{.\,.} m-1]$, where every $x[i]$, $0 \le i < m$, is a *letter* drawn from some fixed *alphabet* $Σ$ of size $|Σ| = \mathcal{O}(1)$. By a *q-gram* we refer to any string $x \in Σ^q$. The *empty string* of length 0 is denoted by $ε$. A string x is a *factor* of a string y if there exist two strings u and v, such that $y = uxv$. Let x be a non-empty string of length m and y be a string. We say that there is an *occurrence* of x in y, or, simply, that x *occurs in* y, when x is a factor of y. The *Parikh vector* associated with a string $w \in Σ^*$ is denoted by $\mathcal{P}(w)$ and represents a vector of size $|Σ|$, where each component denotes the number of occurrences in w of the corresponding letter from $Σ$.

Consider the strings x, y, u, and v, such that $y = uxv$. If $u = ε$, then x is a *prefix* of y. If $v = ε$, then x is a *suffix* of y. We denote by SA the *suffix array* of y of length n, that is, an integer array of size n storing the starting positions of all (lexicographically) sorted suffixes of y, i.e. for all $1 \leq r < n$, we have $y[SA[r-1]..n-1] < y[SA[r]..n-1]$ [22]. Let $\mathsf{lcp}(r, s)$ denote the length of the longest common prefix between $y[SA[r]..n-1]$ and $y[SA[s]..n-1]$, for all positions r, s on y, and 0 otherwise. We denote by LCP the *longest common prefix* array of y defined by $\mathsf{LCP}[r] = \mathsf{lcp}(r-1, r)$, for all $1 \leq r < n$, and $\mathsf{LCP}[0] = 0$. The inverse iSA of the array SA is defined by $\mathsf{iSA}[SA[r]] = r$, for all $0 \leq r < n$. SA, iSA, and LCP of y can be computed in time and space $\mathcal{O}(n)$ [15].

A circular string of length m can be viewed as a traditional linear string which has the left- and right-most letters wrapped around and glued together in some way. Under this notion, the same circular string can be seen as m different linear strings, which would all be considered equivalent. Given a string x of length m, we denote by $x^i = x[i..n-1]x[0..i-1]$, $0 < i < m$, the ith *rotation* of x and $x^0 = x$. Consider, for instance, the string $x = x^0 = \mathtt{ababbbc}$; this string has the following rotations: $x^1 = \mathtt{babbbca}$, $x^2 = \mathtt{abbbcab}$, and so on.

We give some further definitions following [32]. The *q-gram profile* of a string x of length m is the vector $G_q(x)$, where $q > 0$ and $G_q(x)[v]$ denotes the total number of occurrences of $v \in \Sigma^q$ in x. The *q-gram distance* between two strings x and y is defined as

$$D_q(x, y) = \sum_{v \in \Sigma^q} |G_q(x)[v] - G_q(y)[v]|. \qquad (1)$$

Note that D_q is a *pseudo-metric* as $D_q(x, y)$ can be 0 even if $x \neq y$. D_q has the following properties [32] for all $x, y, z \in \Sigma^*$ of length at least q.

1. Positivity: $D_q(x, y) \geq 0$
2. Symmetry: $D_q(x, y) = D_q(y, x)$
3. Triangular inequality: $D_q(x, y) \leq D_q(x, z) + D_q(z, y)$
4. $|(|x| - |y|)| \leq D_q(x, y) \leq |x| + |y| - 2q - 2$
5. $D_q(x_1 x_2, y_1 y_2) \leq D_q(x_1, y_1) + D_q(x_2, y_2) + 2(q - 1)$
6. $D_q(h(x), h(y)) \leq D_q(x, y)$, for a non-length-increasing morphism h on Σ^*.

For a given integer parameter $\beta \geq 1$, we define a generalization of the q-gram distance in (1) by partitioning x and y in β *blocks* as evenly as possible, and using the q-gram distance within each pair of blocks, one from x and one from y. The rationale is to enforce *locality* in the resulting distance. For the sake of presentation in the rest of the paper, we assume that the lengths $|x| = m$ and $|y| = n$ are both multiples of β, so that x and y are conceptually partitioned into β blocks, each of size m/β for x and n/β for y.

Definition 1. *Given strings x of length m and y of length $n \geq m$ and integers $\beta \geq 1$ and $q > 0$, the β-blockwise q-gram distance $D_{\beta,q}(x, y)$ is defined as*

$$D_{\beta,q}(x, y) = \sum_{j=0}^{\beta-1} D_q\left(x\left[\frac{jm}{\beta}..\frac{(j+1)m}{\beta} - 1\right], y\left[\frac{jn}{\beta}..\frac{(j+1)n}{\beta} - 1\right]\right). \qquad (2)$$

In this paper, we consider the following problem, where we search for the ith rotation of x that minimizes its blockwise distance from y as defined in (2). Ties are broken arbitrarily.

CIRCULAR SEQUENCE COMPARISON (CSC)
Input: strings x and y of lengths m and $n \geq m$, respectively, and integers $\beta \geq 1$ and $q < m$
Output: i such that $D_{\beta,q}(x^i, y)$ is minimal

---

3 Algorithms

We use the following result to first give a naïve solution to the CSC problem.

Lemma 2. ([32]). *If we have space $\mathcal{O}(|\Sigma|^q)$ available, then the q-gram distance $D_q(x,y)$ can be computed in time $\mathcal{O}(m+n)$ and extra space $\mathcal{O}(m+n)$, where $m = |x|$ and $n = |y|$.*

We then apply Lemma 2 to each pair of blocks of x and y separately.

Lemma 3. *If we have space $\mathcal{O}(|\Sigma|^q)$ available, then the β-blockwise q-gram distance $D_{\beta,q}(x,y)$ can be computed in time $\mathcal{O}(m+n)$ and extra space $\mathcal{O}(\frac{m+n}{\beta})$, where $m = |x|$ and $n = |y|$.*

The naïve algorithm, denoted by nCSC, computes for $x' = xx$ the values

$$\delta_i = D_{\beta,q}(x'[i \mathbin{..} i+m-1], y),$$

for all $0 \leq i < m$; we report position i such that δ_i is minimal. This requires the application of Lemma 3, m times. Therefore, we obtain the following.

Lemma 4. *If we have space $\mathcal{O}(|\Sigma|^q)$ available, then algorithm nCSC solves the CSC problem in time $\mathcal{O}(m(m+n))$ and extra space $\mathcal{O}(\frac{m+n}{\beta})$.*

3.1 Algorithm hCSC: a Heuristic Algorithm

Here we give a simple heuristic algorithm, denoted by hCSC, to solve the CSC problem faster than nCSC, and return an approximation of the best rotation.

Step 1: We split $x' = xx$ in 2β non-overlapping string *blocks* of length m/β. We obtain strings $x_0, x_1, \ldots, x_{2\beta-1}$, such that $x_i = x'[\frac{im}{\beta} \mathbin{..} \frac{(i+1)m}{\beta} - 1]$, for all $0 \leq i < 2\beta$. We split y in β non-overlapping string blocks of length n/β. We obtain strings $y_0, y_1, \ldots, y_{\beta-1}$, such that $y_i = y[\frac{in}{\beta} \mathbin{..} \frac{(i+1)n}{\beta} - 1]$, for all $0 \leq i < \beta$.

Step 2: For a given sequence $x_j, \ldots, x_{j+\beta-1}$ of strings and y, we compute the β-blockwise q-gram distance as follows

$$\delta_j = D_{\beta,q}(x'[\frac{jm}{\beta} \mathbin{..} \frac{jm}{\beta} + m - 1], y) = \sum_{i=0}^{\beta-1} D_q(x_{j+i}, y_i).$$

We compute δ_j, for all $0 \leq j \leq \beta$. We choose $j_{best} = j$ such that δ_j is minimal, for all $0 \leq j \leq \beta$. In other words, we have found a *window* of length m starting at position j_{best}, such that $(j_{best} + 1)\bmod(m/\beta) = 0$, consisting of β blocks of length m/β each, that minimizes its β-blockwise q-gram distance from y.

Step 3: To perform a refinement on the position of the window, we consider all starting positions included in the two blocks starting at positions j_{best} and $j_{best} - m/\beta$. This includes $2m/\beta - 1$ starting positions in total—we do not need to consider position $j_{best} - m/\beta$ as this was already considered by another window in Step 2. Similarly to Step 2, we obtain the β-blockwise q-gram distance δ_i between the window starting at position i and y, for all $j_{best} - m/\beta < i \leq j_{best} + m/\beta - 1$. We report position $i_{best} = i$ such that δ_i is minimal, for all $j_{best} - m/\beta < i \leq j_{best} + m/\beta - 1$.

Analysis. Step 1 can be done trivially in time $\mathcal{O}(m+n)$. If we have space $\mathcal{O}(|\Sigma|^q)$ available, then, by Lemma 2, $D_q(x_{j+i}, y_i)$ can be computed in time $\mathcal{O}(\frac{m+n}{\beta})$. By Lemma 3, δ_j can be computed in time $\mathcal{O}(\beta(\frac{m+n}{\beta})) = \mathcal{O}(m+n)$. Hence, Step 2 can be done in time $\mathcal{O}(\beta(m+n))$. In Step 3, the blockwise q-gram distance δ_i between a single window and y can be computed in time $\mathcal{O}(\beta(\frac{m+n}{\beta})) = \mathcal{O}(m+n)$. There exist $2m/\beta - 1$ such windows. Hence, Step 3 can be done in time $\mathcal{O}(\frac{m(m+n)}{\beta})$. Overall, the algorithm requires time $\mathcal{O}(\beta(m+n) + \frac{m(m+n)}{\beta})$ and space $\mathcal{O}(|\Sigma|^q + m + n)$.

For practical purposes, setting $\beta = \mathcal{O}(\sqrt{m})$ and $q = \mathcal{O}(\log_{|\Sigma|} m)$ gives an algorithm with time complexity $\mathcal{O}(\sqrt{m}(m+n))$ and space complexity $\mathcal{O}(m+n)$.

3.2   Algorithm saCSC: an Exact Suffix-Array-based Algorithm

The above heuristics hCSC does not guarantee to find the exact value i, for which $\delta_i = D_{\beta,q}(x^i, y)$ is minimal. In particular, when we identify in Step 2 j_{best}, that is, the j for which δ_j is minimal, we take into account only the values of j such that $(j+1)\bmod(m/\beta) = 0$. Thus, Step 3 cannot guarantee that i_{best}, the local minimum obtained by shifting the window m/β positions to the right and left of j_{best}, is minimal for all $0 \leq i < m$. In this section, we give a fast and exact algorithm, denoted by saCSC, to find i such that $\delta_i = D_{\beta,q}(x^i, y)$ is minimal, based on the suffix array (see Sect. 2).

We partially follow the idea from [13]. This work investigates the string matching problem in the setting of k-abelian equivalences: two strings are considered k-abelian equivalent for some positive integer k, if they have the same length and share the same factors of length at most k, including multiplicities. Note that if k is greater than or equal to the string's length, then the strings must be equal. A version of this result, called extended k-abelian equivalence, focuses only on the factors of length k. By setting $k = q$, it is quite straightforward to notice the equivalence with q-grams. Therefore, in order to avoid confusion we will refer to the former notion from now on as *q-abelian equivalence*.

In [13], the authors propose a linear-time algorithm to solve the string matching problem when looking at q-abelian equivalent strings: given a string x of length m, a string y of length $n \geq m$, and a positive integer $q < m$, all factors of y that are q-abelian equivalent to x can be found in time and space $\mathcal{O}(m + n)$. The idea of the algorithm in [13] consists in constructing the suffix array of the string xy, and ranking sets of identical q-length prefixes of suffixes in the suffix array in the order of their appearance. Then it constructs new strings based on this ranking, and solves the problem as in the *jumbled matching* case [6], i.e. identifying all factors of y that have the same Parikh vector as x.

Basic Algorithm for $\beta = 1$. We construct the suffix array of the string xxy and assign a *rank* to the prefix with length q of each suffix with length at least q, based on its order in the suffix array. That is, the first i_0 suffixes in the suffix array, all sharing the same prefix of length at least q, will get rank 0; the next i_1 suffixes sharing the same prefix of length at least q, different from the previous one, will get rank 1, and so on. Next, based on this ranking, we construct two new strings x' of length $2m - q + 1$ and y' of length $n - q + 1$, such that $x'[i] = j$, if j is the rank of the q-length prefix of the $(i + 1)$th suffix of xx in the suffix array of xxy (the same goes for y). It is not difficult to see that the ranks go up at most to value $m + n - q + 1$. However, we can reduce this value to $m + 2$ by introducing two new ranks a_x and a_y: we can conceptually replace by a_x every letter of x' that does not occur in y', and by a_y every letter of y' that does not occur in x'. Hence we can consider that the new strings x' and y' are defined over an integer alphabet of size *at most* $\min(n - q + 1, m) + 2 \leq m + 2$.

Example 5. Let $x = $ GAGTCTA, $y = $ TCTAGCG, and $q = 3$. We denote xxy by z.

i	0	1	2	3	4	5	6	7	8	9	10	11	12	13	14	15	16	17	18	19	20
$z[i]$	G	A	G	T	C	T	A	G	A	G	T	C	T	A	T	C	T	A	G	C	G
$SA[i]$	6	17	1	8	13	19	4	15	11	20	0	7	18	2	9	5	16	12	3	14	10
$LCP[i]$	0	2	2	6	1	0	1	4	3	0	1	7	1	1	5	0	3	2	1	5	4
$x'[i]$	a_x	a_x	a_x	2	a_x	1	a_x	a_x	a_x	a_x	a_x	0									
$y'[i]$	2	0	1	a_y	a_y																

$x'[3] = y'[0] = 2$ denotes that $x[3 . . 5] = y[0 . . 2] = $ TCT. $x'[0] = a_x$ denotes that $x[0 . . 2] = $ GAG does not occur in y. □

We observe that when identifying the q-gram distance between two blocks, we can apply the idea in [13], with the only difference that we should also maintain a Parikh vector that stores the *differences* between the number of occurrences of q-grams in the current block of xx and y (in fact the new letters given by the ranks). Moreover, at the time of the construction of y', we also construct a Parikh vector $\mathcal{P}(y')$, storing, for each letter of y', the number of its occurrences in y'. Notice that $|\mathcal{P}(y')| \leq m + 2$. Later on, when computing the q-gram distances, we can construct another vector diff to store the letter differences between $\mathcal{P}(y')$ and the Parikh vector covering the $m - q + 1$ letters of x' associated with a window of length m on the string xx. This gives us the current Parikh difference and, in fact, represents the q-gram distance between the two analyzed blocks, where $|\text{diff}| \leq m + 2$. Apart from these, we only need another vector δ of size m, which stores at each position i the actual q-gram distance δ_i between y and the window starting at position i in xx, which is the ith rotation x^i of x.

We use a sliding window of length m to maintain the above information. When the window is shifted one position to the right, we have to add to the difference-vector diff the previous first element of the window, and deduct from it the current last element of it. The distance δ_i between y' and the factor of x' starting at position i is thus updated using, in addition, the value of the q-gram distance δ_{i-1} as follows. If, after adding the previous first element to the vector, we have a non-positive value at this position, we update the distance by decreasing the previous value by 1; otherwise, we increase it by 1. If, after deducting the current last element to the vector, we have a non-negative value at this position, we update the distance by decreasing the previous value by 1; otherwise, we increase it by 1. The distance will never be less than the number of occurrences of a_y. Furthermore, if the previous first element was a_x, the new distance decreases by 1, and for every newly added a_x, it increases by 1. As these operations require constant time, after going once through x' with y', we obtain the list of distances δ_i from y to each rotation x^i in linear time.

We are now able to give a more formal description of the steps to solve the CSC problem for $\beta = 1$, which follow a dynamic programming scheme.

Step 1: Construct the SA, iSA, and LCP of xxy. Rank the q-length prefixes of suffixes using LCP-array queries. Construct x' and y', as well as $\mathcal{P}(y')$, the Parikh vector storing, for each letter of y', the number of its occurrences in y'; make proper use of letters a_x and a_y, the ranks that do not occur in either y' or x', respectively. Further, create diff $= \mathcal{P}(y')$ and $\delta_0 = \sum_{i=0}^{|\mathcal{P}(y')|-1} \mathcal{P}(y')[i]$.

Step 2: Read the first $m - q + 1$ letters of x', which constitute our sliding window of length m on the string xx. When reading letter $x'[i]$, update diff by decreasing by 1 the value of the newly read letter, and update δ_0, by either increasing the current value of the distance when there were read too many of the current letters, or decreasing it, when more of these letters still occur in y'

$$\text{diff}[x'[i]] = \text{diff}[x'[i]] - 1 \quad \text{and} \quad \delta_0 = \begin{cases} \delta_0 - 1, & \text{if diff}[x'[i]] \geq 0 \\ \delta_0 + 1, & \text{if diff}[x'[i]] < 0. \end{cases}$$

Step 3: Let i be the current position in x' and repeat this step, one position at a time. Shift the window to the right, update the information for diff

$$\text{diff}[x'[i]] = \text{diff}[x'[i]] + 1 \quad \text{and} \quad \text{diff}[x'[i+m]] = \text{diff}[x'[i+m]] - 1,$$

and calculate δ_{i+1}, based on this information, sequentially applying the two following rules

$$\delta_{i+1} = \begin{cases} \delta_i - 1, & \text{if diff}[x'[i]] \leq 0 \\ \delta_i + 1, & \text{if diff}[x'[i]] > 0 \end{cases}$$

$$\delta_{i+1} = \begin{cases} \delta_{i+1} - 1, & \text{if diff}[x'[i+m]] \geq 0 \\ \delta_{i+1} + 1, & \text{if diff}[x'[i+m]] < 0. \end{cases}$$

Correctness. Steps 1 and 2 are trivially correct as at the end of them we have that diff is the difference between $\mathcal{P}(y')$ and the vector corresponding to the window. These operations follow directly from the definitions of SA and LCP, and are followed by a simple traversal of the suffix array in order to obtain the ranks and create the $\mathcal{P}(y')$ and diff vectors. Also, δ_0, which was initially the number of letters in y', is decreasing as long as the difference between the vectors for a specific letter is non-negative (thus, we still have more occurrences of that letter in y' compared to the window), and increasing otherwise. In Step 3, we update the difference vector by increasing the value at position $x'[i]$ and decreasing that of the new letter $x'[i+m]$ added to the difference. The q-gram distance at that position is based on the values of the newly obtained difference vector, as well as the q-gram distance at the previous position: if $\mathsf{diff}[x'[i]] \leq 0$, then obviously there were more letters $x'[i]$ in y' than in the window, thus we need to decrease, while, if $\mathsf{diff}[x'[i]] > 0$, then there were at least as many letters $x'[i]$ in the window as in y', and taking one out increases the distance. The complementary reasoning applies to the newly added letter $x'[i+m]$. The value of δ_i never goes below the number of occurrences of a_y in y' (it is equal to that, when all other elements of diff are 0) and represents the q-gram distance between y and x^i, the corresponding window of length m starting at position i in xx.

Analysis. In Step 1, constructing SA, iSA, and LCP of xxy can be done in time and extra space $\mathcal{O}(m+n)$ (Sect. 2). Furthermore, the construction of x', y', $\mathcal{P}(y')$, diff, and δ_0 is done with the same time and space cost. In Step 2, updating diff and δ_0 after reading each letter takes constant time, as we execute two operations, thus $\mathcal{O}(m)$ in total. Constant time is required for each iteration in Step 3 to compute the value of δ_i, $1 \leq i < m$, and update diff, since a constant number of operations are executed, thus $\mathcal{O}(m)$ in total. Hence, we can solve the CSC problem for $\beta = 1$ in time and space $\mathcal{O}(m+n)$.

General Algorithm for $\beta \geq 1$. We can now generalize this algorithm to solve the CSC problem for any $\beta \geq 1$, which gives algorithm saCSC. We maintain a Parikh vector for each block, and apply the above basic algorithm for each pair of blocks, computing their q-gram distance. If we denote by $\mathcal{P}_j(y')$ and diff_j, for all $0 \leq j < \beta$, the β Parikh vectors of y' and of the q-gram distances, respectively, as well as by $\delta_{i,j}$ the q-gram distance between the jth block of y and x^i, then the updates will be given by the formulae below. Hence, at each position $i < m$, we can update all of the β Parikh vectors corresponding to the blocks, as previously described, in time $\mathcal{O}(\beta)$. As an example, see here the modification of the previous Step 3, with the other two steps being easily adapted in a similar fashion.

Step 3': When shifting the window one position to the right from position i, update the information for every diff_j, where $0 \leq j < \beta$, as follows

$$\mathsf{diff}_j[x'[i + \frac{jm}{\beta}]] = \mathsf{diff}_j[x'[i + \frac{jm}{\beta}]] + 1$$

$$\mathsf{diff}_j[x'[i + \frac{(j+1)m}{\beta}]] = \mathsf{diff}_j[x'[i + \frac{(j+1)m}{\beta}]] - 1,$$

and calculate $\delta_{i+1,j}$, based on this information, sequentially applying the two following rules

$$\delta_{i+1,j} = \begin{cases} \delta_{i,j} - 1, & \text{if } \mathsf{diff}_j[x'[i + \frac{im}{\beta}]] \leq 0 \\ \delta_{i,j} + 1, & \text{if } \mathsf{diff}_j[x'[i + \frac{im}{\beta}]] > 0 \end{cases}$$

$$\delta_{i+1,j} = \begin{cases} \delta_{i+1,j} - 1, & \text{if } \mathsf{diff}_j[x'[i + \frac{(j+1)m}{\beta}]] \geq 0 \\ \delta_{i+1,j} + 1, & \text{if } \mathsf{diff}_j[x'[i + \frac{(j+1)m}{\beta}]] < 0. \end{cases}$$

Theorem 6. *Algorithm* saCSC *solves the CSC problem in* $\mathcal{O}(\beta m + n)$ *time and space.*

4   Experimental Results

We implemented algorithms nCSC, hCSC, and saCSC as the program CSC. Given one of the three methods, two sequences x and y in (Multi)FASTA format, the number β of blocks, and the length q of the q-grams, CSC finds the rotation of x (or an approximation of it) that minimizes its β-blockwise q-gram distance from y. The implementation is distributed under the GNU General Public License (GPL), and it is available at http://github.com/solonas13/csc. For comparison purposes, we also implemented a naïve algorithm that compares all rotations of x against y using the Needleman-Wunsch algorithm [25] with substitution matrices and affine gap penalty scores [18]; we denote this by cNW.

The following experiments were conducted on a desktop computer using one core of Intel® Core™ i7-2600 CPU at 3.4 GHz and 12 GB of RAM under 64-bit GNU/Linux. All programs were compiled with gcc version 4.7.3. We used both synthetic data (Sects. 4.1–4.2) and real data (Sect. 4.3). All input datasets referred to in this section are publicly maintained at the same web-site.

4.1   Accuracy

We began with simulating three DNA sequence datasets using INDELible [16], with each dataset consisting of 12 sequences, each of length approximately 2,500 base pairs (bp). INDELible produces linear sequences with substitutions, insertions, and deletions at rates defined by the user. Three unique substitution rates were set per dataset using the substitution model JC69 (Jukes-Cantor, 69): 5 %, 20 %, and 35 %. The insertion and deletion rates were set, respectively, to 4 % and 6 %, relative to substitution rate of 1, similar to those observed in MtDNA in primates and mammals [14]. We refer to these datasets as *Original*.

To allow for comparison of the performance of the algorithms in realigning randomly-rotated sequences, which should be similar to those obtained from sequencing circular DNA structures, such as MtDNA, one random rotation was generated from each sequence in all datasets, creating new datasets which will be referred to as *Random*. Using the three *Random* datasets allowed us to test the accuracy of hCSC and saCSC; notice that nCSC and saCSC always return the same rotation. For each *Random* dataset, an all-against-all sequence comparison

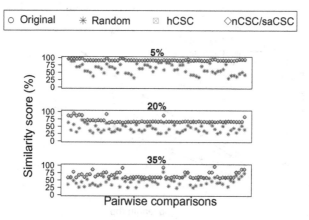

Fig. 1. Accuracy comparison for substitution rates 5%, 20%, and 35%; the black, green, and blue points coincide implying that algorithms hCSC, nCSC, and saCSC return the rotation maximizing the similarity score for all pairwise comparisons (Colour figure online)

was performed. That is, all possible pairs, 66 in total, of sequences in each dataset were input to both hCSC and saCSC. β was set to 50 and q was set to 6. The resultant re-rotated sequences were aligned using EMBOSS Needle [31] and the similarity scores were compared to those of the *Original* and *Random* datasets, which were input directly to EMBOSS Needle. The results can be found in Fig. 1.

The results show that: (a) hCSC and saCSC yield significantly improved similarity scores compared to those obtained from inputting *Random* datasets directly to EMBOSS Needle; and (b) hCSC and saCSC yield similarity scores that are identical or almost identical—notice that the black (Original), green (hCSC), and blue (nCSC/saCSC) points coincide—to those obtained from inputting *Original* datasets directly to EMBOSS Needle. This implies that algorithms hCSC, nCSC, and saCSC return the rotation maximizing the similarity score for all pairwise comparisons.

Hence what we establish here is that the introduced distance measure coupled with the respective algorithms consistently yield a very high accuracy, compared to the standard measure [18, 25, 31], for both *low* and *high* substitution rates.

4.2 Time Performance

We then compared the time performance of the algorithms. Each algorithm was given a pair of randomly generated sequences starting from $m = n = 50$ bp and doubling 8 times to a length of $m = n = 12,800$ bp. It was expected that the slowest algorithm would be cNW which runs in time $\mathcal{O}(nm^2)$. Then it would be algorithm nCSC which runs in time $\mathcal{O}(m(m + n))$, then algorithm hCSC, which runs in time $\mathcal{O}(\beta(m+n)+\frac{m(m+n)}{\beta})$, and lastly algorithm saCSC, which runs in time $\mathcal{O}(\beta m+n)$.

Initially, β was set to $\lceil \sqrt{m} \rceil$ and q was set to $\lceil \log m / \log |\Sigma| \rceil$. The results in Fig. 2 demonstrate orders-of-magnitude superiority of saCSC compared to

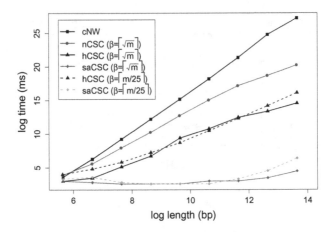

Fig. 2. Elapsed-time comparison

cNW and nCSC, confirming our theoretical findings. hCSC is the second fastest. Although β was set to $\lceil\sqrt{m}\rceil$, saCSC clearly outperforms hCSC, due to the use of a highly optimized implementation of the suffix-array construction, thus highlighting the importance of suitably implemented data structures such as suffix arrays.

Since the time complexities of hCSC and saCSC depend on β, we repeated the same experiment with these two algorithms setting β to $\lceil m/25 \rceil$ and q to $\lceil \log m / \log |\Sigma| \rceil$—notice that q does not affect the time efficiency of the algorithms. The results in Fig. 2 show that hCSC and saCSC are still the fastest, even though $m = \mathcal{O}(\beta)$, and that saCSC is clearly the fastest of all. As expected for $m = \mathcal{O}(\beta)$, we observe that hCSC and saCSC become gradually slower when m grows.

More algorithms could have been included in the comparison but their (at least) quadratic time complexity [5,23] prevents them to compete with saCSC.

4.3 Application to Real Data

As the input dataset, we used two real sequences from GenBank [2]: human (NC_001807) and chimpanzee (NC_001643) MtDNA sequences. The MtDNA genome size for human is 16,571 bp and for chimpanzee is 16,554 bp. Their pairwise sequence alignment using EMBOSS Needle with the default parameters (Gap opening penalty 10.0 and Gap extension penalty 0.5) gives a similarity of 85.1 %. We used saCSC to obtain the rotation of NC_001807 that minimizes its β-blockwise q-gram distance from NC_001643, for $\beta = 850$ and $q = 5$. We obtained rotation 578 of NC_001807 and used EMBOSS Needle to align this rotation with NC_001643. EMBOSS Needle gave a significantly improved similarity of 91 %. This rotation is exactly the rotation we obtained after naïvely searching for the rotation of NC_001807 that maximizes similarity using cNW. Finding this rotation took approximately 28 hours for cNW and only a quarter of a second for saCSC. We repeated this experiment with the human and gorilla (NC_011120) MtDNA sequences. The MtDNA genome size for gorilla is $16,412$ bp. Their pairwise sequence alignment using EMBOSS Needle with the default parameters gives a similarity of 83.5 %. After using saCSC to rotate sequence NC_001807, EMBOSS Needle gave a significantly improved similarity of 88.4 %.

5 Final Remarks

In this paper, we introduced a new distance measure for sequence comparison based on q-grams, and showed how it can be applied *effectively* and computed *efficiently* for circular sequence comparison. Furthermore, we presented an experimental study, using both real and synthetic data, demonstrating orders-of-magnitude superiority of our approach, in terms of efficiency, while maintaining an accuracy which is very competitive to the state of the art.

Our immediate target is twofold: (a) implement algorithm saCSC in BEAR [1], a state-of-the-art tool for improving multiple circular sequence alignment; and (b) evaluate alternative methods for circular sequence comparison based on local alignment heuristics [8].

References

1. Barton, C., Iliopoulos, C.S., Kundu, R., Pissis, S.P., Retha, A., Vayani, F.: Accurate and efficient methods to improve multiple circular sequence alignment. In: Bampis, E. (ed.) SEA 2015. LNCS, vol. 9125, pp. 247–258. Springer, Heidelberg (2015)
2. Benson, D.A., Karsch-Mizrachi, I., Lipman, D.J., Ostell, J., Rapp, B.A., Wheeler, D.L.: GenBank. Nucleic Acids Res. **28**(1), 15–18 (2000)
3. Bray, N., Pachter, L.: MAVID: constrained ancestral alignment of multiple sequences. Genome Res. **14**(4), 693–699 (2004)
4. Brodie, R., Smith, A.J., Roper, R.L., Tcherepanov, V., Upton, C.: Base-By-Base: single nucleotide-level analysis of whole viral genome alignments. BMC Bioinform. **5**(1), 96 (2004)
5. Bunke, H., Buhler, U.: Applications of approximate string matching to 2D shape recognition. Pattern Recogn. **26**(12), 1797–1812 (1993)
6. Burcsi, P., Cicalese, F., Fici, G., Lipták, Z.: Algorithms for jumbled pattern matching in strings. Int. J. Found Comput. Sci. **23**(2), 357–374 (2012)
7. Burkhardt, S., Crauser, A., Ferragina, P., Lenhof, H.P., Rivals, E., Vingron, M.: q-gram based database searching using a suffix array (QUASAR). In: 3rd RECOMB, pp. 77–83 (1999)
8. Chao, K.M., Zhang, J., Ostell, J., Miller, W.: A tool for aligning very similar DNA sequences. CABIOS **13**(1), 75–80 (1997)
9. Cohen, S., Houben, A., Segal, D.: Extrachromosomal circular DNA derived from tandemly repeated genomic sequences in plants. Plant J. **53**(6), 1027–1034 (2008)
10. Craik, D.J., Allewell, N.M.: Thematic minireview series on circular proteins. J. Biol. Chem. **287**(32), 26999–27000 (2012)
11. Crochemore, M., Hancart, C., Lecroq, T.: Algorithms on Strings. Cambridge University Press, New York (2007)
12. del Castillo, C.S., Hikima, J.I., Jang, H.B., Nho, S.W., Jung, T.S., Wongtavatchai, J., Kondo, H., Hirono, I., Takeyama, H., Aoki, T.: Comparative sequence analysis of a multidrug-resistant plasmid from Aeromonas hydrophila. Antimicrob. Agents Chemother. **57**(1), 120–129 (2013)

13. Ehlers, T., Manea, F., Mercaş, R., Nowotka, D.: k-Abelian pattern matching. In: Shur, A.M., Volkov, M.V. (eds.) DLT 2014. LNCS, vol. 8633, pp. 178–190. Springer, Heidelberg (2014)

14. Fernandes, F., Pereira, L., Freitas, A.T.: CSA: an efficient algorithm to improve circular DNA multiple alignment. BMC Bioinform. **10**(1), 1–13 (2009)

15. Fischer, J.: Inducing the LCP-array. In: Dehne, F., Iacono, J., Sack, J.-R. (eds.) WADS 2011. LNCS, vol. 6844, pp. 374–385. Springer, Heidelberg (2011)

16. Fletcher, W., Yang, Z.: INDELible: a flexible simulator of biological sequence evolution. Mol. Biol. Evol. **26**(8), 1879–1888 (2009)

17. Goios, A., Pereira, L., Bogue, M., Macaulay, V., Amorim, A.: mtDNA phylogeny and evolution of laboratory mouse strains. Genome Res. **17**(3), 293–298 (2007)

18. Gotoh, O.: An improved algorithm for matching biological sequences. J. Mol. Biol. **162**(3), 705–708 (1982)

19. Helinski, D.R., Clewell, D.B.: Circular DNA. Annu. Rev. Biochem. **40**(1), 899–942 (1971)

20. Lee, T., Na, J.C., Park, H., Park, K., Sim, J.S.: Finding consensus and optimal alignment of circular strings. Theor. Comput. Sci. **468**, 92–101 (2013)

21. Maes, M.: On a cyclic string-to-string correction problem. IPL **35**(2), 73–78 (1990)

22. Manber, U., Myers, E.W.: Suffix arrays: a new method for on-line string searches. SIAM J. Comput. **22**(5), 935–948 (1993)

23. Marzal, A., Barrachina, S.: Speeding up the computation of the edit distance for cyclic strings. In: 15th ICPR, vol. 2, pp. 891–894 (2000)

24. Mosig, A., Hofacker, I.L., Stadler, P.F.: Comparative analysis of cyclic sequences: viroids and other small circular RNAs. In: GCB. LNI, vol. 83, pp. 93–102. GI (2006)

25. Needleman, S.B., Wunsch, C.D.: A general method applicable to the search for similarities in the amino acid sequence of two proteins. J. Mol. Biol. **48**(3), 443–453 (1970)

26. Peterlongo, P., Sacomoto, G.T., do Lago, A.P., Pisanti, N., Sagot, M.F.: Lossless filter for multiple repeats with bounded edit distance. Algorithm Mol. Biol. **4**(3), 1–20 (2009)

27. Peterlongo, P., Pisanti, N., Boyer, F., do Lago, A.P., Sagot, M.F.: Lossless filter for multiple repetitions with Hamming distance. JDA **6**(3), 497–509 (2008)

28. Pisanti, N., Giraud, M., Peterlongo, P.: Filters and seeds approaches for fast homology searches in large datasets. In: Elloumi, M., Zomaya, A.Y. (eds.) Algorithms in computational molecular biology, chap. 15, pp. 299–320. John Wiley & sons (2010)

29. Ponting, C.P., Russell, R.B.: Swaposins: circular permutations within genes encoding saposin homologues. Trends Biochem. Sci. **20**(5), 179–180 (1995)

30. Rasmussen, K., Stoye, J., Myers, E.: Efficient q-gram filters for finding all epsilon-matches over a given length. J. Comput. Biol. **13**(2), 296–308 (2006)

31. Rice, P., Longden, I., Bleasby, A.: EMBOSS: the european molecular biology open software suite. Trends Genet. **16**(6), 276–277 (2000)

32. Ukkonen, E.: Approximate string-matching with q-grams and maximal matches. Theor. Comput. Sci. **92**(1), 191–211 (1992)

33. Wang, Z., Wu, M.: Phylogenomic reconstruction indicates mitochondrial ancestor was an energy parasite. PLoS ONE **10**(9), e110685 (2014)

34. Weiner, J., Bornberg-Bauer, E.: Evolution of circular permutations in multidomain proteins. Mol. Biol. Evol. **23**(4), 734–743 (2006)

Bloom Filter Trie – A Data Structure for Pan-Genome Storage

Guillaume Holley[1,2,3(✉)], Roland Wittler[1,2,3], and Jens Stoye[1,2,3]

[1] Genome Informatics, Faculty of Technology, Bielefeld University,
Bielefeld, Germany
`gholley@cebitec.uni-bielefeld.de`
[2] Center for Biotechnology, Bielefeld University, Bielefeld, Germany
[3] International Research Training Group 1906, Bielefeld University,
Bielefeld, Germany

Abstract. High throughput sequencing technologies have become fast and cheap in the past years. As a result, large-scale projects started to sequence tens to several thousands of genomes per species, producing a high number of sequences sampled from each genome. Such a highly redundant collection of very similar sequences is called a pan-genome. It can be transformed into a set of sequences "colored" by the genomes to which they belong. A colored de-Bruijn graph (C-DBG) extracts from the sequences all colored k-mers, strings of length k, and stores them in vertices. In this paper, we present an alignment-free, reference-free and incremental data structure for storing a pan-genome as a C-DBG: the Bloom Filter Trie. The data structure allows to store and compress a set of colored k-mers, and also to efficiently traverse the graph. Experimental results prove better performance compared to another state-of-the-art data structure.

1 Introduction

A *string* x is a sequence of characters drawn from a finite, non-empty set, called the *alphabet* $\mathcal{A}$. Its *length* is denoted by $|x|$. The character at position i is denoted by $x[i]$, the substring starting at position i and ending at position j by $x[i..j]$. Strings are concatenated by juxtaposition. If $x = ps$ for (potentially empty) strings p and s, then p is a *prefix* and s is a *suffix* of x.

A *genome* is the collection of all inheritable material of a cell. Ideally it can be represented as a single string over the DNA alphabet $\mathcal{A} = \{a, c, g, t\}$ (or as a few strings in case of species with multiple chromosomes). In practice, however, genomes in databases are often less perfect, either left unchanged in form of the raw data as produced by sequencing machines (millions of short sequences called *reads*), or after some incomplete assembly procedure in form of contiguous chromosome regions (hundreds of *contigs* of various lengths). We are interested in the problem of storing and compressing a set of multiple highly similar genomes, e.g. the pan-genome of a bacterial species, comprising hundreds, or even thousands of strains that share large sequence parts, but differ

© Springer-Verlag Berlin Heidelberg 2015
M. Pop and H. Touzet (Eds.): WABI 2015, LNBI 9289, pp. 217–230, 2015.
DOI: 10.1007/978-3-662-48221-6_16

by individual mutations from one another. An abstract structure that has been proposed for this task is the *colored de-Buijn graph* (C-DBG) [13]. It is a directed graph $G = (V_G, E_G)$ in which each vertex $v \in V_G$ represents a k-mer, a string of length k over $\mathcal{A}$, associated with a set of colors representing the genomes in which the k-mer occurs. A directed edge $e \in E_G$ from vertex v to vertex v', respectively from k-mer x to k-mer x', exists if $x[2..k] = x'[1..k-1]$. Each k-mer x has $|\mathcal{A}|$ possible successors $x[2..k]c$ and $|\mathcal{A}|$ possible predecessors $cx[1..k-1]$ with $c \in \mathcal{A}$. An implementation of such a graph does not have to store edges since they are implicitly given by vertices overlapping on $k-1$ characters.

In this paper, we propose a new data structure for indexing and compressing a pan-genome as a C-DBG, the Bloom Filter Trie (BFT). It allows any format for the input genomes (completely sequenced, set of contigs, set of reads, and even mixtures of them), is alignment-free, reference-free and incremental, i.e., it does not need to be entirely rebuilt when a new genome is inserted. BFTs provide insertion and look-up operations for strings of fixed length associated with an annotation.

In the next section, existing data structures and software for pan-genome representation are reviewed. Section 3 presents the BFT and Sect. 4 the operations it supports. Then, Sect. 5 describes the traversal of a C-DBG stored as a BFT. Finally, Sect. 6 contains experimental results showing the performance of the data structure. Section 7 concludes. Our implementation of the BFT is available at https://github.com/GuillaumeHolley/BloomFilterTrie.

2   Existing Approaches

The BFT, as well as existing tools for pan-genome storage, uses a variety of basic data structures reviewed in the following. Existing tools for pan-genome storage will then be discussed in Sect. 2.2.

2.1   Data Structures

One common way to index and compress a set of strings is to use as a first step the *Burrows-Wheeler Transform* (BWT) [2] that rearranges the input data to enable better compression by aggregating characters with similar context. For multiple sets of strings, a disk-based approach [4] or different terminator characters must be used. The *FM-Index* [9] allows to count and locate the occurrences of a substring in the BWT.

Introduced by Bloom [1], a *Bloom filter* (BF) records the presence of elements in a set. Based on the hash table principle, look-up and insertion times are constant. The BF is composed of a bit array $B[1..m]$, initialized with 0s, in which the presence of n elements is recorded. A set of f hash functions $h_1, ..., h_f$ is used, such that for an element e, $h_i(e) : e \rightarrow \{1, .., m\}$. Inserting an element into B and testing for its presence are then

$$\mathsf{Insert}(e, B) : B[h_i(e)] \leftarrow 1 \text{ for all } i = 1, ..., f$$

and

$$\text{MayContain}(e, B) : \bigwedge_{i=1}^{f} B[h_i(e)],$$

respectively, where $\bigwedge$ is the logical conjunction operator. The BF does not generate false negatives but may generate false positives, as MayContain can report the presence of elements which are not present but a result of independent insertions.

The *Sequence Bloom Tree* (SBT) [21] is a binary tree with BFs as vertices. An internal vertex is the union of its two children BFs, i.e., a BF where a cell is set to 1 if the cell at the same position in at least one of the two children is set to 1.

A *trie* [10] is a rooted edge-labeled tree $T = (V_T, E_T)$ storing a set of strings. Each edge $e \in E_T$ is labeled with a character. A path from the root to a leaf represents the string obtained by concatenating all the characters on this path. The depth of a vertex v in T is denoted by $depth(v, T)$ and is the number of edges between the root of T and v. The height of T, denoted by $height(T)$, is the number of edges on the longest path from the root of T to a leaf. The *burst trie* [11] is an efficient implementation of a trie which reduces its number of branches by compressing sub-tries into leaves. Its internal vertices are labeled with multiple prefixes of length 1, linked to children. The leaves are labeled with multiple suffixes of arbitrary length. A leaf has a limited capacity of suffixes and is *burst* when this capacity is exceeded. A burst splits suffixes of a leaf into prefixes of length 1, linked to new leaves representing the remaining suffixes.

2.2   Software for Pan-Genome Storage

Existing tools for pan-genome storage are mostly alignment-based or reference-based and take a set of assembled genomes as input. Alignments naturally exhibit shared and unique regions of the pan-genome but are computationally expensive to obtain. In addition, misalignments can lead to an inaccurate estimation of the pan-genome regions [7]. PanCake [8] is an extension of string graphs, known from genome assembly [17], which achieves compression based on pairwise alignments. Experiments showed compression ratios of 3:1 to 5:1. Nguyen *et al.* [18] formulated the pan-genome construction problem as an optimization problem of arranging alignment blocks for a set of genomes partitioned by homology. The complexity of the problem has been shown to be NP-hard, and a heuristic using Cactus graphs [19] was provided. A multiple sequence alignment is required for creating the blocks, another NP-hard problem.

Among the reference-based tools, Huang *et al.* [12] proposed to build a pan-genome by adding all the variants detected between a set of genomes to a reference genome. The BWT of the augmented reference is then computed and can be used by an aligner based on the FM-Index. While being more accurate with the augmented reference genome than BWA [14] with the reference alone, the aligner is between 10 to 100 times slower, uses significantly more memory and can introduce false positive alignments. RCSI [22] (Referentially Compressed

Search Index) uses referential compression with a compressed suffix tree to store a pan-genome and to search for exact or inexact matches. The inexact matching allows a limited number of edit distance operations. 1,092 human genomes totaling 3.09 TB of data were compressed into an index of 115 GB, offering a compression ratio of about 28:1. Yet, the index is built for a maximum length query and a maximum number of edit operations.

Close to our approach is SplitMEM [16], which uses a C-DBG to build a pan-genome made of assembled genomes and to extract the shared regions. Although the C-DBG is directly constructed in a compressed way, where a non-branching path is stored in a single vertex, the resulting size of the data structure is larger than the sum of the original sizes of the input sequences, due to the use of an augmented suffix tree.

Recently, the authors of Khmer [5] introduced in their software library a de-Bruijn graph labeling method. Khmer provides a lightweight representation of de-Bruijn graphs [20] based on Bloom filters and a graph labeling method based on graph partitioning. Unfortunately, this functionality was made available only a few days before submission.

The SBT [21] is an alignment-free, reference-free and incremental data structure that allows to label sequences with their colors. The proposed tool is designed to index and compress data from sequencing experiments for effective query of full-length genes or transcripts by separation into k-mers. A leaf of an SBT is used to represent a sequencing experiment by extracting all its k-mers and storing them in the BF of the leaf. SBTs do not represent exactly the set of k-mers of the sequencing experiments they contain, though, due to the inexact nature of BFs.

3   The Bloom Filter Trie

The Bloom Filter Trie (BFT) that we propose in this paper is an implementation of a C-DBG. It is based on a burst trie and is used to store k-mers associated with a set of colors. For the moment we may assume that colors are represented by a bit array $color$ initialized with 0s. Each color has an index i_{color} such that $color_x[i_{color}] = 1$ records that k-mer x has color i_{color}. Sets of colors will later be compressed as explained in Sect. 4.3. All arrays in a BFT are dynamic: An insertion at position pos in an array reallocates it and shifts every cell having an index $\geq pos$ by one position.

In the following, let $t = (V_t, E_t)$ be a BFT created for a certain value of k where we assume that k is a multiple of an integer l such that k-mers can be split into $\frac{k}{l}$ equal-length substrings. The maximum height of t is $height_{max}(t) = \frac{k}{l} - 1$. The alphabet we consider is the DNA alphabet $\mathcal{A} = \{a, c, g, t\}$, and because $|\mathcal{A}| = 4$, each character can be stored using two bits. A vertex in a BFT is a list of containers, zero or more of which are *compressed*, plus zero or one *uncompressed* container. In the following, we will explain how the containers are represented and how an uncompressed container is burst when its capacity is exceeded.

3.1   Uncompressed Container

An uncompressed container of a vertex v in a BFT is a limited capacity set of tuples $<s, color_{ps}>$ where s is a suffix and p is the prefix represented by the path from the root to v. Uncompressed containers are burst when the number of suffixes stored exceeds their capacity $c > 0$. Then, each suffix s of the uncompressed container is split into a prefix s_{pref} of length l and a suffix s_{suf} of length $|s| - l$ such that $s = s_{pref}s_{suf}$. Prefixes are stored in a new compressed container. Suffixes, attached with their colors, are stored in new uncompressed containers, themselves stored in the children of the compressed container. An example of a BFT and a bursting is given in Fig. 1.

3.2   Compressed Container

A bursting replaces an uncompressed container by a compressed one, used to:

- store q suffix prefixes in compressed form (in Fig. 1(b), $q = 4$),
- store links to children containing the suffixes,
- reconstruct suffix prefixes and find the corresponding children.

In the following, each suffix prefix s_{pref} is split into a prefix a and a suffix b with respective binary representations α and β. A compressed container is composed of four structures $quer$, $pref$, suf and $clust$, where:

- $quer$ is a BF represented as a bit array of length m and f hash functions, used to record and filter for the presence of q suffix prefixes;
- $pref$ is a bit array of $2^{|\alpha|}$ bits initialized with 0s and used to record prefix presence exactly. Here the binary representation α of a prefix a is interpreted as an integer such that $pref[\alpha]$ set to 1 records the presence of a;
- suf is an array of q suffixes b sorted in ascending lexicographic order of the original suffix prefixes they belong to;

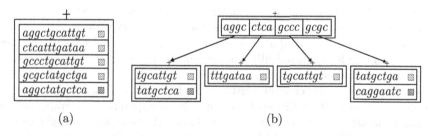

(a)   (b)

Fig. 1. Insertion of six suffixes (that are here complete k-mers) with different colors (boxes with diagonal lines) into a BFT with $k = 12$, $l = 4$ and $c = 5$. In (a), the first five suffixes are inserted at the root into an uncompressed container. When a sixth suffix $gcgccaggaatc$ is inserted, the uncompressed container exceeds its capacity and is burst, resulting in the BFT structure shown in (b).

- *clust* is an array of q bits, one per suffix of array *suf*, that represents cluster starting points. A cluster is a list of consecutive suffixes in array *suf* that share the same prefix. It has an index $i_{cluster}$ with $1 \leq i_{cluster} \leq 2^{|\alpha|}$ and a start position $pos_{cluster}$ in the array *suf* with $i_{cluster} \leq pos_{cluster} \leq q$. Position *pos* in array *clust* is set to 1 to indicate that the suffix in $suf[pos]$ starts a cluster because it is the lexicographically smallest suffix of its cluster. A cluster contains $n \geq 1$ suffixes and, therefore, position i in array *clust* is set to 0 for $pos < i < pos + n$. The end of a cluster is indicated by the beginning of the next cluster or if $pos \geq q$.

For example, the internal representation of the compressed container shown in Fig. 1(b) with $|a| = 2$ and $|b| = 2$ would be:

| *quer* | 0 | 0 | 1 | 0 | 1 | 1 | 0 | 0 | 0 | 1 | 1 | 1 |

| *suf* | gc | ca | cc | gc |

| *pref* | 0 | 0 | 1 | 0 | 0 | 0 | 0 | 1 | 0 | 1 | 0 | 0 | 0 | 0 | 0 | 0 |

| *clust* | 1 | 1 | 1 | 0 |

The size required by a set of q substrings in a compressed container is $m + 2^{|\alpha|} + q \cdot (|\beta| + 1)$ bits. A bursting minimizes this size by choosing a prefix length $|a|$ and a BF size m such that the set of substrings stored in a compressed container does not occupy more memory than their original representation in an uncompressed container, i.e., $m + 2^{|\alpha|} \leq q \cdot (|\alpha| - 1)$. Each suffix prefix inserted after a bursting costs only $|\beta| + 1$ bits. When the average size per suffix prefix stored is close to $|\beta| + 1$ bits, arrays *pref*, *suf* and *clust* can be recomputed by increasing $|a|$ and decreasing $|b|$, such that $2^{|\alpha'|} + q \cdot |\beta'| < 2^{|\alpha|} + q \cdot |\beta|$, where α' and β' are the values of α and β, respectively, after resizing.

4   Operations Supported by the Bloom Filter Trie

The BFT supports all operations necessary for storing, traversing and searching a pan-genome, as well as to extract the relevant information of the contained genomes and subsets thereof. Here we describe the most basic ones of them, Look-up (Sect. 4.1) and Insertion (Sect. 4.2), as well as how the sets of colors are compressed (Sect. 4.3). Traversal of the graph is discussed in Sect. 5.

The algorithms use two auxiliary functions. HammingWeight(α, *pref*) counts the number of 1s in $pref[1..\alpha]$ and corresponds to how many prefixes represented in array *pref* are lexicographically smaller than or equal to an inserted prefix a with binary representation α. This requires $\mathcal{O}(2^{|\alpha|})$ time. The second function, Rank(i, *clust*), iterates over array *clust* from its first position until the i-th entry 1 is found and returns the position of this entry. It corresponds to the start position of cluster i in array *clust*. If the entry is not found, the function returns $|clust| + 1$ as a position. While Rank could be implemented in $\mathcal{O}(1)$ time [9], we use a more naive but space efficient $\mathcal{O}(q)$ time implementation.

4.1   Look-Up

The function that tests whether a suffix prefix $s_{pref} = ab$ with binary representation $\alpha\beta$ is stored in a compressed container cc is given in Algorithm 1. Line 1 uses MayContain to filter for presence of s_{pref} inside cc by querying the BF $quer$ in $\mathcal{O}(f)$ time. If present as a true or false positive, the presence of the prefix a is verified in the array $pref$ in $\mathcal{O}(1)$ time. If a is not present, this was clearly a false positive and nothing else has to be done. If a is present, line 2 computes in $\mathcal{O}(2^{|\alpha|})$ time the Hamming weight i of a, i.e., the index of the cluster in which suffix b is possibly situated. Line 3 locates the rank of i, i.e., the start position of the cluster, and lines 4–7 compare the suffixes of the cluster to b. Lines 3–7 are computed in $\mathcal{O}(q)$ time. Algorithm 1 has therefore a worst case running time of $\mathcal{O}(f + 2^{|\alpha|} + q)$.

Algorithm 1. Contains(ab, cc)

1: **if** MayContain($ab, cc.quer$) **and** $cc.pref[\alpha] = 1$ **then**
2:     $i \leftarrow$ HammingWeight($\alpha, cc.pref$)
3:     $start \leftarrow$ Rank($i, cc.clust$)
4:     $pos \leftarrow start$
5:     **while** $pos \leq |suf|$ **and** ($pos = start$ **or** $cc.clust[pos] = 0$) **do**
6:         **if** $cc.suf[pos] = b$ **then return** $true$
7:         $pos \leftarrow pos + 1$
8: **return** $false$

The function that tests whether a k-mer x is present in a BFT $t = (V_t, E_t)$ is given in Algorithm 2. Each vertex $v \in V_t$ represents k-mer suffixes possibly stored in its uncompressed container or rooted from its compressed containers. The look-up traverses t and, for a vertex v, queries its containers one after the other for suffix $x_{suf} = x[l \cdot depth(v, t) + 1 .. k]$. If the queried container is a compressed container, its BF $quer$ is queried for $x_{suf}[1..l]$ and, in case of a positive answer, the function Contains is used for an exact membership of $x_{suf}[1..l]$. If it is found, the traversing procedure continues recursively on the corresponding child. The absence of $x_{suf}[1..l]$ indicates the absence of x in t since $x_{suf}[1..l]$ cannot be in another container of v. If the container is an uncompressed container, its suffixes are compared to x_{suf}. As an uncompressed container has no children, a match indicates the presence of the k-mer. Algorithm 2 is initially called as TreeContains($x, 1, l, root$). In the worst case, all vertices on a traversed path represent all possible suffix prefixes and the BFs $quer$ have a false positive ratio of 0. In such case, each traversed vertex contains $\lceil \frac{|A|^l}{c} \rceil$ containers. The longest path of a BFT has $height_{max}(t) + 1$ vertices. Therefore, the worst case time of TreeContains is $\mathcal{O}(height_{max}(t) \cdot \lceil \frac{|A|^l}{c} \rceil \cdot (f + 2^{|\alpha|} + q))$.

4.2 Insertion

Prior to any k-mer insertion into a BFT t, a look-up verifies if the k-mer is already present. If it is, only its set of colors is modified. Otherwise, the look-up stops the trie traversal on a container $cont$ of a vertex v where the searched suffix prefix or k-mer suffix is not present. If $cont$ is an uncompressed container, the

---

Algorithm 2. TreeContains(x, i, l, v)

1: **for each** container $cont$ in v **do**
2:     **if** $cont$ is compressed **and** MayContain($x[i..i+l-1]$, $cont.quer$) **then**
3:         **if** Contains($x[i..i+l-1]$, $cont$) **then**
4:             $v \leftarrow$ child associated with $x[i..i+l-1]$ in $cont.suf$
5:             **return** TreeContains($x, i+l, l, v$)
6:         **else return** $false$
7:     **else if** $cont$ is uncompressed **then**
8:         **for each** $<s, color_{x[1..i-1]s}>$ in $cont$ **do**
9:             **if** $s = x[i..k]$ **then return** $true$
10: **return** $false$

---

insertion of the k-mer suffix and its color is a simple $\mathcal{O}(c)$ time process. If $cont$ is compressed, the insertion of suffix prefix $s_{pref} = ab$ is a bit more intricate. In fact, it will only be triggered if $cont$ is the first compressed container of v to have s_{pref} as a false positive (MayContain(s_{pref}, $cont.quer$) = $true$ and Contains(s_{pref}, $cont$) = $false$). False positives are therefore "recycled", which is a nice property of BFTs: The BF $quer$ remains unchanged, and only $pref$, suf and $clust$ need to be updated in a way similar to Algorithm 1: The presence of prefix a must first be verified by testing the value of $pref[\alpha]$ where α is the binary representation of a. If $pref[\alpha] = 0$, prefix a is not present and is recorded by setting $pref[\alpha]$ to 1. Then, the index $id_{cluster}$ and start position $pos_{cluster}$ of the new cluster are computed using HammingWeight and Rank. Suffix b is inserted into $suf[pos_{cluster}]$ and a 1 into $clust[pos_{cluster}]$. This takes $\mathcal{O}(2^{|\alpha|} + 2q)$ time. If $pref[\alpha] = 1$ prior to insertion, prefix a is already present, and $id_{cluster}$ and $pos_{cluster}$ have already been computed by Contains(s_{pref}, $cont$). Let n be the number of suffixes in cluster $id_{cluster}$. Suffix b is inserted into $suf[pos]$ such that $pos_{cluster} \leq pos \leq pos_{cluster} + n$ and $suf[pos-1] < suf[pos]$. If $pos = pos_{cluster}$, b starts its cluster: A 1 is inserted into $clust[pos]$ and $clust[pos+1]$ is set to 0. Otherwise, a 0 is inserted into $clust[pos]$. This takes $\mathcal{O}(2q)$ time. The worst case time insertion of a k-mer is $\mathcal{O}(d + 2^{|\alpha|} + 2q)$ with d being the worst case time look-up.

The internal representation of the compressed container shown in Fig. 1(b) after insertion of the suffix prefix $gtat$ is given below (inserted parts are highlighted). The presence of prefix gt is recorded in $pref[12]$. Then, its cluster index and start position are computed as 4 and 5, respectively. Consequently, after reallocation of arrays suf and $clust$, suffix at is inserted in $suf[5]$ and $clust[5]$ is set to 1 to indicate that $suf[5]$ starts a new cluster.

$quer$ | 0 | 0 | 1 | 0 | 1 | 1 | 0 | 0 | 0 | 1 | 1 | 1 |

$pref$ | 0 | 0 | 1 | 0 | 0 | 0 | 0 | 1 | 0 | 1 | 0 | 1 | 0 | 0 | 0 | 0 |

suf | gc | ca | cc | gc | at |

$clust$ | 1 | 1 | 1 | 0 | 1 |

4.3   Color Compression

Remember from Sect. 3 that color sets associated with k-mers in a C-DBG are initially stored as bit arrays in BFTs. However, these can be compressed. To this end, a list of all color sets occurring in the BFT is built and sorted in decreasing order of total size, i.e., the number of k-mers sharing a color set multiplied by its size. Then, by iterating over the list, each color set is added incrementally to an external array if the integer encoding its position in the array uses less space than the size of the color set itself. Finally, each color set present in the external array is replaced in the BFT by its position in the external array.

5   Traversing Successors and Predecessors

Let t be a BFT that represents a C-DBG G. For a k-mer x, visiting all its predecessors or successors in G, denoted $pred(x, G)$ and $succ(x, G)$, respectively, implies the look-up of $|\mathcal{A}|$ different k-mers in t. Such a look-up would visit in the worst case $|\mathcal{A}| \cdot (height_{max}(t) + 1)$ vertices in t. This section describes how to reduce the number of vertices and containers visited in t during the traversal of a vertex in G.

Observation 1. *Let G be a C-DBG represented by a BFT t and x a k-mer corresponding to a vertex of G. All k-mers of $succ(x, G)$ share $x[2..k]$ as a common prefix and therefore share a common subpath in t starting at the root. On the other hand, k-mers of $pred(x, G)$ have different first characters and, therefore, except for the root of t do not share a common subpath. Hence, the maximum number of visited vertices in t for all k-mers of $succ(x, G)$ is $1 + height_{max}(t)$ and for all k-mers of $pred(x, G)$ is $1 + |\mathcal{A}| \cdot height_{max}(t)$.*

Lemma 1. *Let G be a C-DBG represented by a BFT t, x a k-mer in t and v a vertex of t that terminates the shared subpath of the k-mers in $succ(x, G)$. If $depth(v, t) = height_{max}(t)$, $succ(x, t)$ suffixes may be stored in any container of v. If not, they are stored in the uncompressed container of v.*

Proof. A vertex v is the root of a sub-trie storing k-mer suffixes of length $l \cdot (height_{max}(t) - depth(v, t) + 1)$ with $l = \frac{k}{height_{max}(t)+1}$. Let s be a k-mer suffix of $succ(x, t)$ rooted at a vertex $v \in V_t$. If $depth(v, t) \neq height_{max}(t)$ but s is rooted at a compressed container in v, then this compressed container stores $s[1..l]$, and $s[l + 1..|s|]$ is rooted in one of its children. As the divergent character between the k-mer suffixes of $succ(x)$ is in position $|s| - 1$, this character is in $s[l + 1..|s|]$, rooted at one child of this compressed container. Therefore v does not terminate the common subpath shared by $succ(x, t)$ k-mers.   □

Lemma 1 proves that the only two cases where a look-up of $pred(x, G)$ or $succ(x, G)$ must search in different containers of a vertex are:

– searching at the root of t for k-mers of $pred(x, G)$,
– if $depth(v, t) = height_{max}(t)$, searching at vertex v for suffixes of $succ(x, G)$.

Restricting the hash functions used in the compressed containers to take only positions 2 through $l-1$ into account, allows to limit the search space.

Lemma 2. *Let t be a BFT where the f hash functions h_i of quer have the form $h_i(s_{pref}) : s_{pref}[2..l-1] \rightarrow \{1,..,m\}$ for $i = 1,...,f$. Then, for a vertex v of t and a suffix prefix s_{pref}, all possible substrings $s'_{pref} = c_1 s_{pref}[2..l-1]c_2$ are contained in the same container of v.*

Proof. Assume a k-mer suffix s inserted in a vertex v of t. A look-up for s analyzes the containers of v from the head to the tail of the container list. In the worst case, s can be rooted, according to BFs *quer*, in all compressed containers as a true positive or as a false positive. However, a look-up stops either on the first compressed container claiming to contain the suffix prefix $s_{pref} = s[1..l]$, or on the uncompressed container. As the hash functions of *quer* consider only $s_{pref}[2..l-1]$, a look-up will therefore stop on the same container for any substring $s'_{pref} = c_1 s_{pref}[2..l-1]c_2$. □

As a consequence of Lemma 2, each suffix prefix s_{pref} stored or to store in arrays *pref*, *suf* and *clust* is modified such that $s_{pref} = s_{pref}[2..l]s_{pref}[1]$, which guarantees that all $s'_{pref} = s_{pref}[2..l-1]c_2c_1$ are in the same container. Furthermore, suffixes stored in array *suf* are required to have a minimum length of two characters to ensure that characters c_1 and c_2, the variable parts between the different s'_{pref}, are stored in array *suf*. Hence, as all s'_{pref} share $s_{pref}[2..l-1]$ as a prefix, they share the same cluster in arrays *suf* and *clust*. Suffix prefixes $s'_{pref} = s_{pref}[1..l-1]c_2$ also have consecutive suffixes in their cluster.

6   Evaluation

We implemented the BFT in C and compared it to the SBT [21], version 0.3.1, on a mid-class laptop with an SSD hard drive and an Intel Core i5-4300M processor cadenced at 2.6 GHz. All software was run with a single thread. Both data structures were used to represent one real and one simulated pan-genome dataset. The real dataset (NCBI BioProject PRJEB5438) consists of raw sequencing data from 473 clinical isolates of *Pseudomonas aeruginosa*, sampled from 34 patients, resulting in 844.37 GB of FASTQ files. The simulated dataset was generated from 19 strains of *Yersinia pestis*. For each strain, we used Wgsim[1] to create 6,000,000 reads of length 100 with a substitution sequencing error rate of 0.5 %, resulting in 31 GB of FASTQ files. We first used KmerGenie [3] on a subsample of the files for each dataset to estimate the best k-mer length and the minimum number of occurrences for considering a k-mer valid (not resulting from a sequencing error). A length of $k = 63$ with a minimum number of 3 occurrences was selected for the real and a length of $k = 54$ with a minimum of 15 occurrences for the simulated data set.

---

[1] https://github.com/lh3/wgsim.

Table 1. Running time and memory usage for the real (*P. aeruginosa*) and simulated (*Y. pestis*) dataset. The compression ratio is given w.r.t. the original file sizes and (NA) indicates unavailable information.

	P. aeruginosa		Y. pestis	
	BFT	SBT	BFT	SBT
Insertion time	**14 h 34 min**	44 h 4 min	**11 min 29 s**	38 min 6 s
(without *k*-mer counting)	**(8 h 5 min)**	(NA)	**(2 min 18 s)**	(NA)
Uncompressed size	**7.25 GB**	11 GB	**79 MB**	115.2 MB
(compression ratio)	**(116:1)**	(77:1)	**(402:1)**	(276:1)
Compressed size	**2.2 GB**	4.8 GB	**76 MB**	117.2 MB
(compression ratio)	**(384:1)**	(176:1)	**(418:1)**	(271:1)

For the BFT, we used KMC2 [6] to extract all valid *k*-mers from each genome. The capacity c influences the compression ratio as well as the time for insertion and look-up. We chose a value of $c = 248$ as it showed a good tradeoff in practice. The prefix length l determines the size of several internal structures of the BFT and how efficiently they can be stored. We selected $l = 9$ as this limits the internal fragmentation of the memory. The color set compression was applied regularly during the insertion process in order to keep the memory used to build the BFT as low as possible. After insertion of each dataset, the BFT was written to disk.

The SBT employs Jellyfish [15] to extract from each genome all valid *k*-mers. As the size of BFs used in the SBT must be specified prior to the *k*-mer counting and should be the same for all vertices, the authors of the SBT suggested to estimate the number of unique *k*-mers in each dataset to design the size of BFs, at the price of an extra computation time (personal communication). Since we knew the exact number of unique *k*-mers from the BFT construction, we used this instead: 93,202,452 *k*-mers for the real dataset, resulting in a BF size of 11.1 MB. However, our simulated dataset corresponds to a very well conserved species with an average of 4,557,245 unique 54-mers per genome for a total of 5,121,443 unique 54-mers in the pan-genome: Each BF of the SBT would hold a very high false positive ratio, 59 % on average, by choosing 5,121,443 bits for the BFs size. To avoid saturation, we computed a BF size of 24,910,142 bits (2.97 MB) for the simulated dataset to obtain a smaller false positive ratio of approximately 7.2 % – similar to the ratio for the real dataset. We also reused *k*-mer counts computed for the BFT to estimate the number of hash functions: One hash function for the real dataset and four hash functions for the simulated dataset. The SBT counts the *k*-mers and builds the leaves in a one step process: It is not possible to differentiate these two sub-steps nor to extract the valid *k*-mers using a different software. According to the SBT paper and the CPU usage of this step, the insertion time is mainly dominated by the *k*-mer extraction. Note that SBTs are streamed on disk, each vertex being kept in a separate file.

Running times and memory usage of both tools are shown in Table 1.

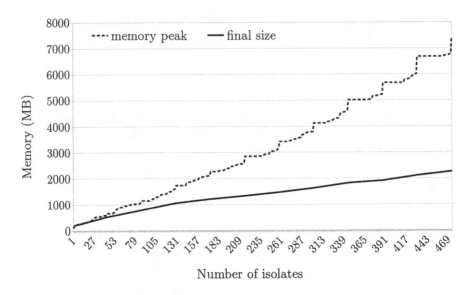

Fig. 2. Memory used by the BFT during the insertion of *P. aeruginosa* isolates.

Suprisingly, the compressed version of the SBT for the simulated dataset takes more disk space than the uncompressed version. Memory usage during the insertion of the real dataset in the BFT is shown in Fig. 2. Note that after storage on disk, the BFT can be compressed further using a standard compressor. We compressed the BFT using 7z[2], resulting in a file of 980 MB for the real dataset and 40.1 MB for the simulated dataset (about 882:1 and 792:1 w.r.t. the original file sizes). We suspect that 7z delivers such compression ratio by taking advantage of the data redundancy among the uncompressed containers.

For each dataset, the set of unique k-mers in the BFT was written to disk in random order and reused as a batch query for the presence of all unique k-mers in both data structures. It was not possible to query the SBT for a single batch query of all 93,202,452 63-mers for the real dataset as the memory used exceeded the 16 GB of memory available on the test machine, even when specifying that BFs could be loaded into memory separately. We suspect this is because k-mers are first loaded into memory before querying and the results are also stored in memory before writing to disk. Therefore, we divided the set of unique 63-mers into ten subsets, the first nine subsets containing 10,000,000 k-mers each and the last subset containing 3,202,452 k-mers. Query times are shown in Table 2.

A second experiment gives an estimation of the time required to traverse the graph represented by a BFT: It verifies for each k-mer whether its corresponding vertex in the graph is branching. This experiment first computes information about the root in a negligible amount of time and memory. Then, the BFT is queried for its branching vertices. For the real dataset, this experiment took 14 min (average time of 8.71 μs per 63-mer), resulting in 14,314,840 branching

---

[2] http://www.7-zip.org/.

Table 2. Query time for the real (*P. aeruginosa*) and simulated (*Y. pestis*) dataset). Real and simulated dataset batch queries contain 93,202,452 63-mers and 5,121,443 54-mers, respectively.

	P. aeruginosa		*Y. pestis*	
	BFT	SBT	BFT	SBT
Total running time	**13 min 9h**	18 min 8 s	**3 min 47 s**	
Running time	**8.04 μs**	359 μs	**1.56 μs**	44.35 μs

vertices. For the simulated dataset, this experiment took 14 s (average time of 2.73 μs per 54-mer), resulting in 6,312 branching vertices.

In summary, in our experiments the BFT was multiple times faster than the SBT on the building time while using about 1.5 times less memory. The BFT was about 30 times faster than the SBT for querying a k-mer.

7   Conclusion

We proposed a novel data structure called the Bloom Filter Trie for storing a pan-genome as a colored de-Bruijn graph. The trie stores k-mers and their colors. A new representation of vertices is proposed to compress and index shared substrings. It uses four basic data structures, which allow to quickly verify the presence of substrings. In the worst case, the compressed strings have a memory footprint close to their binary representation. However, we observe in practice substantial memory savings. Future work concerns the possibility to compress non-branching paths that share the same colors [16], but also the extraction of the different pan-genome regions.

Acknowledgments. The authors wish to thank the anonymous reviewers and the authors of SBT for helpful comments. GH and RW are funded by the International DFG Research Training Group GRK 1906/1.

References

1. Bloom, B.H.: Space/time trade-offs in hash coding with allowable errors. Comm. ACM **13**(7), 422–426 (1970)
2. Burrows, M., Wheeler, D.J.: A block-sorting lossless data compression algorithm. Digital SRC Research Report 124 (1994)
3. Chikhi, R., Medvedev, P.: Informed and automated k-mer size selection for genome assembly. Bioinformatics **30**(1), 31–37 (2014)
4. Cox, A.J., Bauer, M.J., Jakobi, T., Rosone, G.: Large-scale compression of genomic sequence databases with the Burrows-Wheeler transform. Bioinformatics **28**(11), 1415–1419 (2012)
5. Crusoe, M., Edvenson, G., Fish, J., Howe, A., McDonald, E., Nahum, J., Nanlohy, K., Ortiz-Zuazaga, H., Pell, J., Simpson, J., Scott, C., Srinivasan, R.R., Zhang, Q., Brown, C.T.: The khmer software package: enabling efficient sequence analysis figshare (2014)

6. Deorowicz, S., Kokot, M., Grabowski, S., Debudaj-Grabysz, A.: KMC 2: Fast and resource-frugal k-mer counting. Bioinformatics **31**(10), 1569–1576 (2015)
7. DePristo, M.A., Banks, E., Poplin, R., Garimella, K.V., Maguire, J.R., Hartl, C., Philippakis, A.A., del Angel, G., Rivas, M.A., Hanna, M., et al.: A framework for variation discovery and genotyping using next-generation DNA sequencing data. Nat. Genet. **43**(5), 491–498 (2011)
8. Ernst, C., Rahmann, S.: PanCake: A data structure for pangenomes. In: Proceedings of the German Conference on Bioinformatics, vol. 34, pp. 35–45 (2013)
9. Ferragina, P., Manzini, G.: An experimental study of an opportunistic index. In: Proceedings of the 12th ACM-SIAM Symposium on Discrete Algorithms, vol. 1, pp. 269–278 (2001)
10. Fredking, E.: Trie memory. Comm. ACM **3**(9), 490–499 (1960)
11. Heinz, S., Zobel, J., Williams, H.E.: Burst tries: a fast, efficient data structure for string keys. ACM Trans. Inf. Syst. **20**(2), 192–223 (2002)
12. Huang, L., Popic, V., Batzoglou, S.: Short read alignment with populations of genomes. Bioinformatics **29**(13), i361–i370 (2013)
13. Iqbal, Z., Caccamo, M., Turner, I., Flicek, P., McVean, G.: De novo assembly and genotyping of variants using colored de Bruijn graphs. Nat. Genet. **44**(2), 226–232 (2012)
14. Li, H., Durbin, R.: Fast and accurate short read alignment with Burrows-Wheeler transform. Bioinformatics **25**(14), 1754–1760 (2009)
15. Marçais, G., Kingsford, C.: A fast, lock-free approach for efficient parallel counting of occurrences of k-mers. Bioinformatics **27**(6), 764–770 (2011)
16. Marcus, S., Lee, H., Schatz, M.C.: SplitMEM: a graphical algorithm for pangenome analysis with suffix skips. Bioinformatics **30**(24), 3476–3483 (2014)
17. Myers, E.W.: The fragment assembly string graph. Bioinformatics **21**, ii79–ii85 (2005)
18. Nguyen, N., Hickey, G., Zerbino, D.R., Raney, B., Earl, D., Armstrong, J., Haussler, D., Paten, B.: Building a pangenome reference for a population. J. Comput. Biol. **22**(5), 387–401 (2015)
19. Paten, B., Diekhans, M., Earl, D., St. John, J., Ma, J., Suh, B., Haussler, D.: Cactus graphs for genome comparisons. J. Comput. Biol. **18**(3), 469–481 (2011)
20. Pell, J., Hintze, A., Canino-Koning, R., Howe, A., Tiedje, J.M., Brown, C.T.: Scaling metagenome sequence assembly with probabilistic de Bruijn graphs. Proc. Natl. Acad. Sci. U.S.A. **109**(33), 13272–13277 (2012)
21. Solomon, B., Kingsford, C.: Large-Scale Search of Transcriptomic Read Sets with Sequence Bloom Trees. bioRxiv, 017087 (2015)
22. Wandelt, S., Starlinger, J., Bux, M., Leser, U.: RCSI: Scalable similarity search in thousand(s) of genomes. Proc. VLDB Endow. **6**(13), 1534–1545 (2013)

A Filtering Approach for Alignment-Free Biosequences Comparison with Mismatches

Cinzia Pizzi[✉]

Department of Information Engineering, University of Padova, Padova, Italy
`cinzia.pizzi@dei.unipd.it`

Abstract. Alignment-free approaches for sequence similarity based on substring composition are increasingly attracting interest from the scientific community. In fact, in several contexts, with respect to alignment-based approaches, alignment-free techniques are faster but less accurate. Recently, several studies (e.g. [4,8,9]) attempted to bridge the accuracy gap with the introduction of approximate matches in the definition of composition-based distance measures.

In this work we present *MissMax*, an exact algorithm for the computation of the longest common substring with mismatches between each suffix of a sequence x and a sequence y. This collection of statistics is useful for the computation of two similarity distances that have been recently extended to incorporate approximate matching, namely the longest and the average common substring with k mismatches. Our approach is exact, and it is based on a filtering technique that showed, in a set of preliminary experiments, to substantially reduce the size of the set of potential sites of a longest match.

1 Introduction

Sequence similarity has long been playing a crucial role in Computational Biology and Bioinformatics. As a matter of fact, similarity is considered as a key ingredient in the prediction of functional and structural properties, and of evolutionary mechanisms, by comparing new elements with other elements whose properties are known, or by comparing elements that show a similar behaviour to infer the common mechanism that underlies the observed phenomena.

Since the introduction of high throughput techniques, hundreds of fully sequenced genomes of different species have been made available at a fast pace. The increasing number of available sequences makes all kind of sequence analysis, most notably assembly, phylogenetic reconstruction, and multiple alignments, more challenging due to the time consuming and memory-demanding operations that need to be carried out on these huge datasets.

To try to cope with the increasing demand of time efficiency, a wide range of alignment-free (or composition-based) approaches have been proposed. The idea behind compositional approaches is to model each sequence in terms of the

This research was partially supported by PRIN 20122F8B2.

M. Pop and H. Touzet (Eds.): WABI 2015, LNBI 9289, pp. 231–242, 2015.
DOI: 10.1007/978-3-662-48221-6_17

substrings that it contains, and then to devise appropriate similarity measures to compare two sequences based on this model [13].

Traditionally, alignment-free approaches rely on the frequency or presence of L-mers, for a fixed length L, and consider exact matches. Although usually very fast, in several contexts such approaches can be much less accurate than alignment-based counter-parts.

For this reason, within the last decade several approaches have been proposed to improve the ability to better capture the nature of the similarity/dissimilarity between biological sequences with alignment-free techniques. Among the wide literature, we can mention, for example, the introduction of over-representation, rather than simple frequency count, in the definition of the similarity measure for fixed length [11] and maximal length [3] components; and the definition of distances based on average longest shared substrings [12], which frees the analysis from fixing the length of the substrings to analyse.

More recently, several studies proposed to model the intrinsic variability of biosequences by using spaced-words, or by considering approximate matches with a bounded number of mismatches in the characterization of the sequence composition. Several related experiments showed that, in the context of phylogenetic tree reconstruction, the introduction of approximate matches can improve the quality of the detected sequence similarity [4, 8, 9].

Given these premises, we focused our attention on these more involved formulations of the alignment-free approach, in particular on those allowing for approximate matching within a bounded number of mismatches. Within this framework we developed an algorithm for the computation of the longest common substring with mismatches between all the suffixes of a sequence x and another sequence y. This primitive is at the basis of recently developed distance measures for the problem of phylogentic tree reconstruction: the longest and the average common substring distances with mismatches.

The paper is organized as follows: in Sect. 2 we will introduce the basic notation used throughout the paper, and also give a brief overview of recent research on the subject. In Sect. 3 we will describe the proposed algorithm, and in Sect. 4 we will present the results of preliminary experiments on its performances.

2   Preliminaries

Let us consider two sequences $x = x_1 \ldots x_n$ and $y = y_1 \ldots y_m$ defined over an alphabet Σ. Let us indicate with $X_i = x_i \ldots x_n$ and $Y_j = y_j \ldots y_m$ the suffixes of x and y starting at position i and j respectively. In the following, we will assume, without loss of generality, that both sequences have the same length n.

Since we are interested in studying sequence similarity measures based on string composition, and considering maximal and approximate matches, as a warm up we will give a brief overview of recently proposed approaches, and will use some of the presented results to introduce further notation used throughout the paper.

2.1   Related Work

An early result was presented in [10], where an $O(n^2)$ algorithm was proposed to compute the number of occurrences with k mismatches of all the substrings of length L in a string x of length n. The algorithm was proposed within the pattern discovery framework [5], thus the need to count the occurrences within the same string in order to subsequently estimate their over-representation. However, the proposed solution can be easily adapted to compute the number of occurrences of all the substrings of length L of a string x in another string y, leading to the definition of a similarity measure between the two sequences based on the frequency of shared approximate occurrences.

More recently, in [4] several similarity measures based on shared maximal substrings with mismatches were introduced[1]. Let $LCP_k(x,y)$ be the length of the longest common prefix between two strings x and y when k mismatches are allowed. Now, consider the set of $LCP_k(X_i, Y_j)$ defined for all the suffixes $X_i, i = 1, 2, \ldots, n$ of x, and for all the suffixes $Y_j, j = 1, 2, \ldots, n$ of y.

The following measures of cross correlation were defined for a given number of mismatches k:

- $MaxCor_k$, defined as the maximum value attained by $LCP_k(X_i, Y_j)$ over all values of $i \in (1, 2, ..., n)$ in x and $j \in (1, 2, ..., n)$ in y.
- $AvCor_k$, that is, the average value attained by $LCP_k(X_i, Y_j)$ over all values of $i \in (1, 2, ..., n)$ of x and $j \in (1, 2, ..., n)$ in y.
- $MaxCor_k(i), i = 1, 2, ..., n$: the maximum value attained by $LCP_k(X_i, Y_j)$ for each i over all values correspondingly spanned by j.
- $AvCor_k(i), i = 1, 2, ..., n$: the average value attained by $LCP_k(X_i, Y_j)$ for each i over all values correspondingly spanned by j.

For measures such as $MaxCor_k$ and $AvCor_k$ a subquadratic solution $O(\frac{kn^2}{\log n})$ was also proposed in [4].

In [9] kmacs, a greedy heuristic, was proposed to generalize the well known Average Common Substring (ACS) distance [12] so to account for k mismatches when considering the longest common substring between pairs of positions in the two strings. We refer to this variant of the ACS problem as $kACS$. The algorithm proposed in [9] has time complexity $O(nkz)$, where z is the maximum number of occurrences in y of a string of maximal length occurring in both x and y. Being based on a heuristic, this method is very fast, but it does not guarantee to find the optimal solution to the problem. Note that the kACS problem can be described in terms of the measures of cross correlation previously defined as the mean over all positions i in x of $MaxCor_k(i)$.

We end our overview with two results published earlier this year. In [6] the k-LCF problem is introduced as the generalization of the longest common substring (there named "factor" to avoid confusion with LCS as the Longest Common Subsequence problem) as finding the longest common shared match between two sequences when up to k mismatches are allowed. Also this problem can

---

[1] Here we use a slightly different notation than the one used in [4].

be described in terms of the previously defined scheme, as it corresponds to $MaxCor_k$. In [6] an $O(nm)$ time and $O(1)$ space solution is provided for a generic k and two strings of length n and m respectively, and also an $O(n+m)\log(n+m)$ solution for the case $k = 1$. Finally, in [2] an $O(n \log^{k+1} n)$ time and $O(n)$ space algorithm was proposed to provide a subquadratic solution to the kACS problem.

Within such a context, we focussed our attention on the computation of the values of $MaxCor_k(i)$, for all $i = 1\ldots n$, because such vector allows us to compute the values of both $MaxCor_k$ (or equally k-LCF) and of $kACS$, being respectively:

$$MaxCor_k = \max_{i=1\ldots n} MaxCor_k(i) \tag{1}$$

$$kACS = \frac{1}{n} \sum_{i=1\ldots n} MaxCor_k(i) \tag{2}$$

3   The MissMax Algorithm

Our aim is to compute the values of $MaxCor_k(i)$ for each position i in x. The main idea behind the proposed algorithm is to avoid the computation of the $LCP_k(X_i, Y_j)$ for all pairs of positions $i \in x$ and $j \in y$. To this purpose we will compute the value of $MaxCor_k(i + 1)$ starting from the value of the already computed $MaxCor_k(i)$. This procedure will initially give us a candidate longest match L_{max} that is at least equal to $MaxCor_k(i-1) - 1$. We will use this information, among some others that will be discussed in the following subsections, to reduce the cardinality of the set $\mathcal{C}$ of possible candidates for approximate matches longer than L_{max}, and then we will verify them.

Note that, when computing the $MaxCor_k(i)$ for each i, one can either take track of their maximum value to compute $MaxCor_k$, or of their sum to later compute kACS at no extra cost.

3.1   Initial Set Up: $MaxCor_k(1)$

We start with the computation of $MaxCor_k(1)$ as the maximum approximate match of the suffix X_1 against the sequence y. For this purpose we will start from, and exploit, the classical concept of longest common substring (without any mismatch allowed).

Definition 1. *Longest Common Substring*
Given two sequences x and y, of length n, find the maximum length L for which a pair of indexes (i, j) exists such that $x_i \ldots x_{i+L-1} = y_j \ldots j_{j+L-1}$.

The problem of finding the longest common substring between two sequences is a well known problem in pattern matching that can be solved in linear time by the traversal of a generalized suffix tree of the two sequences. More in details, we want to be able to find the longest common match starting at any two positions

i in x and j in y. This problem can be solved through a call to the Lowest Common Ancestor of the corresponding leaves n_i and n_j in the generalized suffix tree. The length of the label of the path from the root to $LCA(n_i, n_j)$ is the length of their longest common prefix. LCA queries can be carried out for any i and j in constant time after a linear-time preprocessing step [7].

In particular, similarly to the routine step in [9], we will perform $k + 1$ *jump-extensions* to compute the longest approximate match between X_1 and the generic Y_j. As after the first jump-extension of length l_1 we know we will have a mismatch, we will call LCA on the nodes corresponding to positions $1 + l_1 + 1$ and $j + l_1 + 1$, and repeat the procedure until the $(k+1)$-th mismatch is found. This is repeated for each $j = 1 \ldots n$, thus taking $O(kn)$ time overall.

3.2 Minimum $MaxCor_k$ from the Previous Step

Assume now we have computed $L = MaxCor_k(i)$, and we want to compute $MaxCor_k(i + 1)$. Let j be the position in y of a longest approximate match of X_i. Two cases may hold, which are illustrated in Figs. 1 and 2:

1. $\mathbf{x_i} = \mathbf{y_j}$: in this case the k mismatches all lie within $x[i + 1, i + L - 1]$ and $y[j+1, j+L-1]$, respectively. Therefore we have $LCP_k(X_{i+1}, Y_{j+1}) = L-1$. Note that this might or might not be the final $MaxCor_k(i + 1)$ over all positions of Y.
2. $\mathbf{x_i} \neq \mathbf{y_j}$: in this case the mismatch between the first characters will be lost when considering the alignment of $i + 1$ and $j + 1$, leading to $k - 1$ mismatches in the following $L - 1$ positions. After L positions we know we must have a mismatch, which is now counted as the k-th. To finally obtain $LCP_k(X_{i+1}, Y_{j+1})$ we need a further call to LCA on the nodes corresponding to the positions $i + L + 1$ and $j + L + 1$ to obtain $LCP_0(X_{i+L+1}, Y_{j+L+1})$ that will end on the $(k + 1)$-th mismatch. In summary: $LCP_k(X_{i+1}, Y_{j+1}) = L + LCP_0(X_{i+L+1}, Y_{j+L+1})$. Again, note that this might or might not be the final $MaxCor_k(i + 1)$ over all positions of y.

Let L_{max} be the candidate value for $MaxCor_k(i + 1)$ obtained either from Case 1 or Case 2.

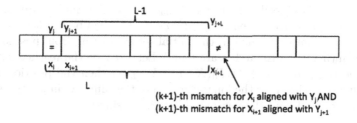

Fig. 1. Candidate $MaxCor_k(i + 1)$ from $MaxCor_k(i)$ when $x_i = y_j$

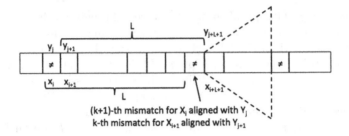

Fig. 2. Candidate $MaxCor_k(i+1)$ from $MaxCor_k(i)$ when $x_i \neq y_j$

3.3   Potential Candidates from the Previous Step

Let us consider now a generic position r in Y. We must have $L' = LCP_k(X_i, Y_r) \leq L$, since L was the absolute maximum found in the step to compute $MaxCor_k(i)$.

If $\mathbf{x_i} = \mathbf{y_r}$ then k mismatches lie between $x[i+1, i+L'-1]$ and $y[r+1, r+L'-1]$, respectively, and $LCP_k(X_{i+1}, Y_{r+1}) = L'-1 \leq L-1$. As a consequence, the pair $(i+1, r+1)$ can be ruled out as one that cannot have an approximate match longer than the one we are currently considering (which is greater or equal than $L-1$). Note that this observation allows us to exclude from the candidate set $\mathcal{C}$ all the positions $r+1$ in y that are preceed by a symbol matching x_i.

The case where $\mathbf{x_i} \neq \mathbf{y_r}$ is more involved. With reference to Fig. 3, the aligment (X_{i+1}, Y_{r+1}) loses the mismatch in the first position of the alignment (X_i, Y_r), and includes the one at position $i+L'$ and $r+L'$, in x and y respectively. To obtain the length of LCP_k for this alignment we should add $LCP_0(X_{i+L'+1}, Y_{r+L'+1})$, which gives the last exact contribution till the $(k+1)$-th match. It may happen that the addition of this term to L' allows one to obtain a match longer than the potential $MaxCor_k(i+1) = L_{max}$ we had from the previously discussed Case 1 or Case 2. The main problem here is that we do not know the value of L'.

We will then proceed by assuming r is indeed the site of a match longer than the current maximum L_{max}. If this is the case, the gap with L_{max} must be closed assuming the $(k+1)$-th mismatch occurs after the positions $i+L$ and $r+L$ in the two strings. As a consequence, $LCP_0(X_{i+L}, Y_{r+L})$ will end exactly where $LCP_0(X_{i+L'}, Y_{r+L'})$ would end (see Fig. 3).

If this value is indeed bigger than L_{max} we need to make sure no further mismatch was present between $i+L'$ and $i+L$. This can be checked by running the jump-extension performed in the initial setup starting from positions $i+1$ and $r+1$ until $k+1$ mismatches are found. Let L_{true} be the reached extension. If its value is equal to $L + LCP_0(X_{i+L+1}, Y_{r+L+1})$ then the position r is the new candidate position for the longest match of the suffix X_i in y, otherwise the position is dropped, and the next candidate is considered.

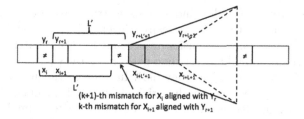

Fig. 3. Guessing the maximum extension between the suffix X_i and a candidate Y_r

3.4 Theoretical and Practical Considerations

We now discuss some theoretical and practical issues emerging from the proposed approach.

Theoretical Observations. The worst case complexity occurs when we inherit from step i an initial candidate that is smaller or equal than $MaxCor_k(i)$. In such a case any position r for which $x_{i+L_{max}+1} = y_{r+L_{max}+1}$ is a possible longer match that we need to verify with the jump-extension. This lead to potential $O(n)$ candidate pairs per position i, and to a worst case time complexity $O(kn^2)$.

However, we observe that even in the worst case, at least the first symbol of the last jump must be a match in order to have a longer match. Therefore, given the generalixed suffix tree we built for x and y, it will suffice to check only the positions $r - L$ where r is a leaf in the subtree of the root with a label matching x_{i+L}. Assuming equal distribution, this means we have to check $\frac{n}{|\Sigma|}$ positions, and can safely ignore the others $n \times \frac{|\Sigma|-1}{|\Sigma|}$ positions that do not match.

Particular care needs to be taken if there are multiple longest matches from the previous steps. All the corresponding positions need to be treated separately. This is because when $L_{max} = MaxCor_k(i) - 1 = L - 1$ and $x_i \neq x_r$ we will have that x_{r+L} lands on the k-th mismatch rather than on the first position after that, thus ending up to erroneously discard position $r + 1$ from further processing. For this reason, we keep track of all ties corresponding to a longest match in a list, and add such positions to the set of candidates C that need to be verified.

Furthermore, we actually need to check only those positions for which the corresponding LCA with X_{i+L} is at a depth greater than $LCP_0(X_{i+L+1}, Y_{j+L+1})$. Statistics about the length of the longest common substring between two strings comes handy at this point. The expected length of the longest common substring is $2\log_\Sigma n$. The number of expected pairs (i, r) for which $LCP_0(X_i, Y_r) = m$ is $n^2 p^m$, if we assume the uniform distribution and $p = \frac{1}{|\Sigma|}$ is the probability of any symbol in Σ to occur in any position. Therefore, given the extension $LCP_0(X_i, Y_j)$ of the candidate pair, only those pairs with a match in the first position after the last mismatch, and with an LCA deeper than $LCP_0(X_i, Y_j)$ should be considered. The larger $LCP_0(X_i, Y_j)$, the smaller the number of occurrences to check.

Practical Observations. Building indexing data structures can be expensive, and so can be operations that are theoretically efficient. For example, it was already observed in [9] that a naive extension to account for k mismatches gave better performances than calling LCA (or than performing the equivalent operation on an Enhanced Suffix Array [1], as they did). In our experiments we experienced the same. Therefore, by keeping the original approach in mind, but avoiding reference to indexing data structures, we developed a tool, implementing our algorithm *MissMax*, in which the extensions are performed naively. Note that in many cases we just need to perform a one-step extension, as the k-step extensions are performed only in the initial step, and whenever we have a candidate with an approximate match longer than L_{max}. Also, to reduce the set $\mathcal{C}$ of candidates we associate to each string a bitvector B. We then set up a position j if $y_{j-1} \neq x_{i-1}$ and $y_{j+L_{max}} = x_{i+L_{max}}$. This way we consider only positions that can be a site of a longer or equal match, and substantially reduce the size of the candidate set $\mathcal{C}$. For DNA applications, when reading the input, we can build 4 bitvectors, one for each symbol to speed up this computation. The position i of the bitvector of the symbol s is set if and only if $x_i = s$. When computing $MaxCor_k(i+1)$, the first filter is given by the complement of the bitvector corresponding to x_i. To get the second filter we take the bitvector corresponding to the symbol x_{i+L} and shift it of L positions to the left. The bitwise AND of the two vectors is the bitvector B mentioned above that holds the positions of C (to avoid a further shift of one position to the right, when considering position j we look at the value of this vector at position $j-1$).

4   Preliminary Experimental Analysis

In this section we present the results of preliminary experiments we run to test MissMax. Here we are not mainly interested in the improvement of the quality of the tree reconstruction with mismatches, that was already discussed in both [4,9]. We are rather interested on the time needed to compute the values of $MaxCor_k(i)$, for all the positions i in a sequence x with respect to a second sequence y, as a function of the input length. As explained in Sec.2, this statistics can be used to compute both the $MaxCor_k$ (i.e. the longest common substring with k mismatches) based distance discussed in [4], and the kACS distance discussed in [9]. Our algorithm is an exact algorithm based on filtering, therefore one of the main aspects to investigate is the filtering power, which in turn affects the time performances. For what concerns the comparison with other algorithms, the algorithm described in [2], which holds the best known asymptotic complexity for the exact computation of the n values of $MaxCor_k(i)$, has no available implementation yet. On the other side, the algorithm behind kmacs [9], which is available, is based on a greedy heuristics, which makes it very fast, although it does not guarantee to find the optimal solution. Despite this major difference, we tried to do some tests with it anyway. However, we did experienced some unexpected behavior of kmacs on some of the input sequences we tried. Because the reasons for the sudden decrease of performances we observed are

Algorithm 1. Computing MaxCor(i) from MaxCor(i-1)

```
1: procedure MISSMAX(string x, string y, int k )
2:     MaxCor(1) ← k-jump extension between X₁ and any Yⱼ
3:     MaxPos(1) ← the corresponding position
4:     if There are more positions with the same MaxCor(1) then
5:         Put them in a special list
6:     end if
7:     for each suffix i in x do
8:         Compute the candidate MaxCorₖ(i) = Lₘₐₓ from MaxCorₖ(i − 1)
9:         Compute the candidate set C of potential sites for longer matches
10:        Add to C the elements of the special list and clear it
11:        for each position r ∈ C do
12:            Compute potential longest match L
13:            if L ≥ MaxCorₖ(i) then
14:                Compute the actual extension L'
15:                if L = L' then
16:                    if L > MaxCorₖ(i) then
17:                        Update MaxCorₖ(i) and MaxPos(i), and reset the special list
18:                    else
19:                        Add r to a special list
20:                    end if
21:                end if
22:            end if
23:        end for
24:    end for
25: end procedure
```

still unclear, we preferred not to report such results. Rather, we plan in the near future to perform further experiments to understand what is going on, and to guarantee a fair comparison.

As a consequence of the above considerations, the only algorithm left to compare to is the naive solution to the problem. As the naive algorithm is quite slow, we run this comparison on relatively short mithocondrial DNA sequences (16k bases) taken from a dataset already used in [4,12]. The experiments were conducted on different values of k and on prefixes of the input sequences of increasing length. We also verified whether there was any difference in terms of performances by comparing two close species as Human and Chimpanzee (Fig. 4 - left), and two divergent species as Human and Wallaroo (Fig. 4 - right). The figures clearly show an increasing gain in performances when the length of the input sequence grows. We also observe that the time needed for the computation depends on number of mismatches k.

To reveal possible dependencies of the filtering on the number of mismatches k or on the input length n, we measured the minimum, maximum, and average size of the candidate set C over all suffixes i. We noticed that such values are constant in terms of percentage of the input length (therefore the longer the

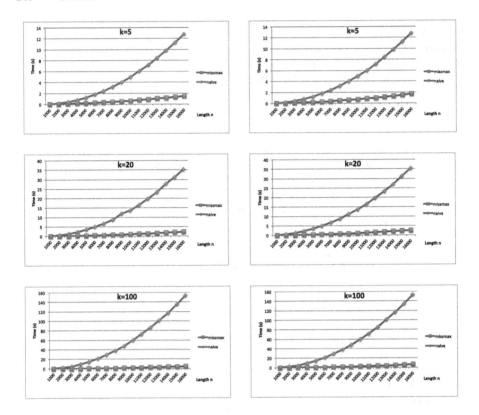

Fig. 4. Time performances as a function of the input size for different values of k. The column of the left refers to two similar species (Human and Chimpanzee) and the column to the right refers to two more different species, as Human and Wallaroo

input the bigger the candidate set), while the number of mismatches does not seem to have much influence.

However, we also noticed that the comparison of similar sequences is faster than the comparison of divergent species. This is somehow expected, as if a long match (which is more likely between similar species than between divergent ones) is found at some point, the filter is expected to work better. This is confirmed by the data we collected. For divergent species the average size of the candidate set for $k = 5$ is 7.14 % of the input size, and for $k = 100$ it is 8.13 %. For similar species the average size of the candidate set is 5.54 % and 5.08 % of the input length, for $k = 5$ and $k = 100$ respectively. For completeness of results, the minimum was about 1 % for both comparison, while the maximum was about 27 % and 28 % for similar and divergent species respectively.

The difference in terms of time performances for the comparison of two similar and two divergent species can be better appreciated in Fig. 5.

To further investigate the scalability of the approach we perform similar tests on longer sequences, avoiding the comparison with the slow naive algorithm.

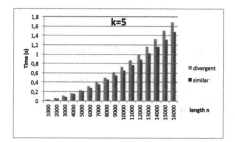
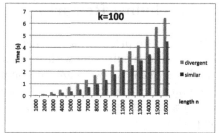

Fig. 5. Time performance of MissMax when comparing similar species and divergent species for $k = 5$ (left) and $k = 100$ (right).

We took two similar Bacteria, the Streptococcus pneumonia and the Streptococcus pyogenes, and an Eukariote, the Saccharomyces cerevisiae, and compute the values of $MaxCor_k(i)$ for all i among them. The results are reported in Fig. 6, where we show the time required for three different values of k.

The average, minimum and maximum size of the candidate set C for this set of experiments is shown in Fig. 7. The dependency of the size of the candidate set on the number of mismatches k is more evident here.

As a final remark, we observe that the variance of the candidate set is quite large as the number of candidates ranges from a minumum of about 2 % to a maximum 25 % of the input size. It might be of interest to further study the distribution of the size of C to detect possible bottlenecks and appropriate ways to overcome them.

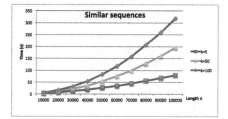
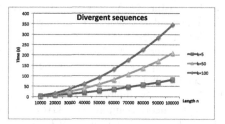

Fig. 6. Time performance of MissMax when comparing similar species (left) and divergent species (right) for different values of the number k of allowed mismatches.

	k=5	k=50	k=100
avg	7.53%	9.51%	9.86 %
min	1.92 %	1.90%	1.89 %
max	26.84 %	26.72%	26.71 %

	k=5	k=50	k=100
avg	7.92%	10.56%	11.05 %
min	1.85 %	1.84%	1.83 %
max	25.76 %	25.76 %	25.71 %

Fig. 7. Average, mimimum and maximum size of the candidate set, as a percentage of the input length, for similar species (left) and divergent species (right)

5   Concluding Remarks

In this work we proposed a filtering-based approach for the computation of the longest common substring with k mismatches between *each* suffix of a sequence x and a sequence y we want to compare to. This statistics is useful for the computation of alignment free distances based on approximate matching, that are a promising approach to improve the quality of alignment free sequence comparison.

References

1. Abouelhoda, M.I., Kurtz, S., Ohlebusch, E.: Replacing suffix trees with enhanced suffix arrays. J. Discrete Algorithms **2**, 53–86 (2004)
2. Aluru, S., Apostolico, A., Thankachan, S.V.: Efficient alignment free sequence comparison with bounded mismatches. In: Przytycka, T.M. (ed.) RECOMB 2015. LNCS, vol. 9029, pp. 1–12. Springer, Heidelberg (2015)
3. Apostolico, A., Denas, O., Dress, A.: Efficient tools for comparative substring analysis. J. Biotechnol. **149**(3), 120–126 (2010)
4. Apostolico A., Guerra, C., Pizzi, C.: Alignment free sequence similarity with bounded hamming distance. In: Data Compression Conference, pp. 183–192. IEEE Press (2014)
5. Apostolico, A., Pizzi, C.: Motif discovery by monotone scores. Discrete Appl. Math. **155**(6–7), 695–706 (2007)
6. Flouri, T., Giaquinta, E., Kobert, K., Ukkonen, E.: Longest common substrings with k mismatches. Inormation Process. Lett. **115**(6–8), 643–647 (2015)
7. Harel, D., Tarjan, R.E.: Fast algorithms for finding nearest common ancestor. SIAM J. Comput. **13**, 338–355 (1984)
8. Leimeister, C.-A., Boden, M., Horwege, S., Lindner, S., Morgenstern, B.: Fast alignment-free sequence comparison using spaced-word frequencies. Bioinformatics **30**(14), 1991–1999 (2014)
9. Leimeister, C.-A., Morgenstern, B.: kmacs: the k-mismatch average common substring approach to alignment-free sequence comparison. Bioinformatics **30**(14), 2000–2008 (2014)
10. Pizzi, C.: K-difference matching in amortized linear time for all the words in a text. Theor. Comput. Sci. **410**(8–10), 983–987 (2009)
11. Qi, J., Wang, W., Hao, B.: Whole proteome prokaryote phylogeny without sequence alignment: a k-string composition approach. Mol. Evol. **58**(1), 1–11 (2004)
12. Ulitsky, I., Burstein, D., Tuller, T., Chor, B.: The average common substring approach to phylogenetic reconstruction. J. Comput. Biol. **13**(2), 336–350 (2006)
13. Vinga, S., Almeida, J.: Alignment-free sequence comparison - a review. Bioinformatics **20**, 206–215 (2003)

Models and Algorithms for Genome Rearrangement with Positional Constraints

Krister M. Swenson[1,2](✉) and Mathieu Blanchette[3]

[1] CNRS, LIRMM, Université de Montpellier, Montpellier, France
swenson@lirmm.fr
[2] Institut de Biologie Computationnelle (IBC), Montpellier, France
[3] McGill University, Montreal, QC, Canada

Abstract. Traditionally, the merit of a rearrangement scenario between two genomes has been measured based on a parsimony criteria alone; two scenarios with the same number of rearrangements are considered equally good. In this paper, we acknowledge that each rearrangement has a certain likelihood of occurring based on biological constraints, *e.g.* physical proximity of the DNA segments implicated, or repetitive sequences. Accordingly, we propose optimization problems with the objective of maximizing overall likelihood, by weighting the rearrangements. We study a binary weight function suitable to the representation of sets of genome positions that are most likely to have swapped adjacencies. We give a polynomial-time algorithm for the problem of finding a minimum weight double cut and join (DCJ) scenario among all minimum length scenarios. In the process, we solve an optimization problem on colored noncrossing partitions which is a generalization of the MAXIMUM INDEPENDENT SET problem on circle graphs.

1 Introduction

A huge body of work exists on modeling the evolution of whole chromosomes [10]. The main difference between such models is the set of rearrangements that they allow. The moves of interest are usually inversion, transposition, translocation, chromosome fission and fusion, deletion, insertion, and duplication.

Almost all versions of the problem are NP-Hard if content modifying operations such at duplication, loss, and insertion are allowed [6,14]. Fortunately, a model that considers genomes with equal content (*i.e.* no duplications or insertions/deletions) is quite pertinent, particularly in eukaryotes, since syntenic blocks of genes can be assigned between genomes so that each block occurs exactly once in each genome. For two genomes with equal content, double cut and join (DCJ) has been the model of choice since it elegantly includes inversion, translocation, chromosome circularization and linearization, as well as chromosome fission and fusion [3,27].

One of the most important problems in comparative genomics is the inference of ancestral gene orders, *i.e.* paleogenetics. Given a realistic model of evolution, one can infer ancestral adjacencies of high confidence from present-day genomes [4,15,20]. However, methods that attempt to infer deeper structure

© Springer-Verlag Berlin Heidelberg 2015
M. Pop and H. Touzet (Eds.): WABI 2015, LNBI 9289, pp. 243–256, 2015.
DOI: 10.1007/978-3-662-48221-6_18

for ancestral species suffer due to the huge number of parsimonious scenarios between genomes [1, 13, 22].

The apparent difficulty of the ancestral inference problem — because of the potentially astronomical number of parsimonious sorting scenarios — highlights the importance of methods that infer scenarios that conform to some extra biological constraints. Yet, aside from methods that weight inversions based on their length [2, 5, 11, 17, 21], to our knowledge no work exists in this direction.

In this paper we use a weight function on rearrangements suitable for modeling *positional* constraints, *i.e.* sets of positions in the genome that are likely to swap adjacencies. Two examples of constraints that fit this paradigm are: (1) the physical 3D location of DNA segments in a nucleus and, (2) repetitive sequences that are the cause or consequence of rearrangement mechanisms. We illustrate the utility of our model with 3D constraints in Sect. 1.4.

We propose a general optimization problem that minimizes the sum of weights over the moves in a scenario. A more constrained version of the problem asks for such a scenario out of all possible unweighted parsimonious scenarios. Our algorithm solves this version of the problem in polynomial time given a binary weight function, despite an exponential growth of the number of parsimonious DCJ scenarios with respect to the distance [7, 19]. The commutation properties of DCJ moves as studied in [19] link certain DCJ scenarios to noncrossing partitions. Our algorithm relies on solving a new optimization problem on *colored* noncrossing partitions, called MINIMUM NONCROSSING COLORED PARTITION. It is a generalization of the MAXIMUM INDEPENDENT SET problem on circle graphs [12, 18, 25].

1.1   Genomes as Sets of Signed Integers

A gene, or more generally a syntenic block of genes, will be represented by a signed integer. A chromosome is a sequence of blocks, and a genome is a set of chromosomes. Thus, we write a genome in list notation where a block is a positive integer if read in one direction in the genome, and a negative integer if read in the opposite direction. For example, a genome A can be written as

$$\{(\circ, 5, -1, -2, 6, -4, -8, \circ), (\circ, -3, 7, \circ), (9, 10)\},$$

where $\circ$ represents a *telomere* at the end of a linear chromosome. Genome A has two linear chromosomes and a circular chromosome $(9, 10)$.

Alternatively, the organization of the blocks on the chromosomes can be given by the set of adjacencies between the extremities of consecutive blocks. A block b has a tail extremity, written b_t, and a head extremity, written b_h. Thus, the adjacency between 5 and -1 in A is $\{5_h, 1_h\}$. A block that is on the end of a linear chromosome implies a *telomeric adjacency*. The first chromosome has two such adjacencies: $\{\circ, 5_t\}$ and $\{8_t, \circ\}$. A circular chromosome has no telomeres, *i.e.* the last block is adjacent to the first. We can write genome A using adjacencies as

$$A = \{\{\{\circ, 5_t\}, \{5_h, 1_h\}, \{1_t, 2_h\}, \{2_t, 6_t\}, \{6_h, 4_h\}, \{4_t, 8_h\}, \{8_t, \circ\}\},$$
$$\{\{\circ, 3_h\}, \{3_t, 7_t\}, \{7_h, \circ\}\},$$
$$\{\{9_h, 10_t\}, \{10_h, 9_t\}\}\}.$$

1.2 DCJ and Sorting DCJs

Double cut and join (DCJ) is an operation on a genome that cuts one or two adjacencies, and glues the resulting ends back together according to the following rules [3]:

1. If a single adjacency is cut, then add new telomeres to the resulting ends (resulting in two new telomeric adjacencies).
2. If two adjacencies are cut, then glue the adjacencies back in one of two new ways.

Application of a single DCJ corresponds to diverse genomic operations such as inversion, chromosome linearization and circularization, transposition, and excision of a circular chromosome.

The DCJ distance between genomes A and B is the minimum number of DCJ moves needed to transform A into B. DCJs that move A closer to B, called *sorting* DCJs, can be found using a graph. The *colored adjacency graph* for A and B is a graph $G(A, B, col)$ whose vertices are the extremities and telomeres of A and B, and whose edges are colored by the color function col. For each adjacency in A or B an *adjacency* edge links the corresponding nodes of the adjacency, and a *cross* edge links non-telomere vertices from A to vertices with the same label in B. The graph for genomes

$$A = \{\{\{\circ, 5_t\}, \{5_h, 1_h\}, \{1_t, 2_h\}, \{2_t, 6_t\}, \{6_h, 4_h\}, \{4_t, 8_h\}, \{8_t, \circ\}\},$$
$$\{\{\circ, 3_h\}, \{3_t, 7_t\}, \{7_h, \circ\}\}\}, \text{ and}$$
$$B = \{\{\{\circ, 1_t\}, \{1_h, 2_t\}, \{2_h, 3_t\}, \{3_h, 4_t\}, \{4_h, 5_t\}, \{5_h, 6_t\}, \{6_h, \circ\}\},$$
$$\{\{\circ, 7_t\}, \{7_h, 8_t\}, \{8_h, \circ\}\}\}$$

is given in Fig. 1. It is easy to confirm that the adjacency and cross edges each form a matching, so that each connected component of the graph will be either a cycle or a path. Note that connected components of the graph are only loosely related to the chromosomes; connected components can span multiple chromosomes.

We denote a cross edge by the label of the vertices that they connect. We denote the connected components of the graph by the set of cross edges that comprise them. The connected components of the graph in Fig. 1 are $\{5_t, 4_h, 6_h\}$, $\{5_h, 6_t, 2_t, 1_h\}$, $\{1_t, 2_h, 3_t, 7_t\}$, $\{8_t, 7_h\}$, and $\{3_h, 4_t, 8_h\}$. The *length* of a path or a cycle is the number of cross edges it has.

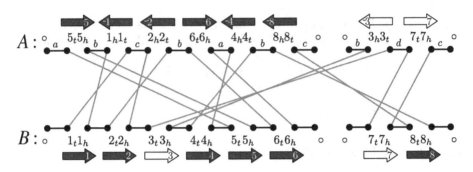

Fig. 1. The colored adjacency graph $G(A, B, col)$. Black edges are adjacency edges and gray edges are cross edges. The color function col maps adjacency edges of genome A to the alphabet $\{a, b, c, d\}$.

To find sorting DCJs, we categorize the connected components by length. In Fig. 1 there is one cycle, two even-length paths, and two odd-length paths. The formula for the DCJ distance is

$$d_{DCJ}(A, B) = N - (C + I/2) \tag{1}$$

where N is the number of blocks, C is the number of cycles, and I is the number of odd-length paths in $G(A, B)$ [3]. Figure 2 depicts a comprehensive list of the possible sorting DCJs on an adjacency graph, and describes the conditions under which they may be applied. See Proposition 1 of [19] for a more thorough treatment. $G(A, A)$, for some genome A, will always have $2M$ paths of length one and $N - M$ cycles of length two, where M is the number of chromosomes and N is the number of blocks.

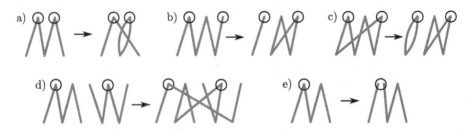

Fig. 2. All possible DCJs that move one genome closer to the other. Adjacency edges are contracted, so that only the cross edges are shown in the connected components. Endpoints that are affected by the DCJ are circled. In the top row, extracting a cycle from (a) an even-length path, (b) an odd-length path, and (c) a cycle are depicted. Even-length paths can be combined to form two odd-length paths if one of the paths has endpoints in genome A and the other in genome B, as depicted in (d). An even-length path can be split into two odd length paths if the split is done in the genome with fewer vertices in the path, as depicted in (e).

1.3   The Minimum Weighted Rearrangements Problem

Consider a genome A_i made of a set of linear or circular chromosomes. Each rearrangement on this genome may have a certain likelihood of occurring. In Sect. 1.5 we will describe a DCJ move on $G(A_i, B)$ as a reconnection of two adjacency edges of $G(A_i, B)$; the resulting graph $G(A_{i+1}, B)$ is identical to $G(A_i, B)$ aside from the connectivity of two adjacency edges. Therefore there is a bijection between edges of $G(A_i, B)$ and edges of $G(A_{i+1}, B)$, so we can weight all pairs of genome adjacencies occurring in a sorting scenario by weighting all pairs of adjacency edges in $G(A, B)$. For the set P of all pairs of adjacency edges in genome A, the weight function for a pair is $w : P \mapsto \mathbb{R}_+$, where $\mathbb{R}_+$ denotes the non-negative real numbers. The higher the value of w the less likely the rearrangement is to occur, $e.g.$ a value of 0 represents a most likely rearrangement.

A sequence of rearrangements $\rho_1, \rho_2, \ldots, \rho_d$ such that $(\cdots ((A\rho_1)\rho_2) \cdots \rho_d) = B$ is called a *sorting scenario*. The weight of a scenario is the sum of the weights of all the rearrangements in the scenario, $i.e.$ $\sum_{i=1}^{d} w(\rho_i)$. The MINIMUM WEIGHTED REARRANGEMENTS problem is the following.

Problem 1. MINIMUM WEIGHTED REARRANGEMENTS

INPUT: Genomes A and B and a weight function w.
OUTPUT: A scenario of rearrangements turning A into B.
MEASURE: The weight of the scenario.

1.4   Positional Constraints as Colored Adjacencies

Although chromosomes are represented as linear or circular sequences of syntenic blocks, in reality they correspond to molecules whose conformation within the nucleus is complex. Recent technological advances, called Hi-C, allow the mapping of chromosome conformation in various cell types and species [8,9,16,23,28]. The positional constraints introduced here are based on the principle that rearrangements (DCJ moves) involving pairs of adjacencies that are close in 3D space are more frequent than others. This model is supported by the pioneering work of Véron, *et al.* [26], who showed that loci that are distant in the linear ordering of the human chromosome yet close in the ordering of the mouse chromosome, are physically close (in 3D) in the human chromosome. Recently we have conducted a study on rearrangement scenarios showing that breakpoint pairs comprising a rearrangement are closer than expected by chance for intrachromosomal and interchromosomal rearrangements. This is true for multiple cell types from multiple laboratories [24]. In this paper, we use the observation that many moves are local to constrain the rearrangement scenarios that we compute. We call this the *positional* constraint.

We incorporate the constraint by grouping adjacencies of the genome into classes that are more likely to swap endpoints. This idea is illustrated in Fig. 3, where the physical (3D) structure of genome A is drawn and the adjacencies are grouped into colored *localities*. According to Véron et al. [26] and our recent results [24], rearrangements are more likely to occur between adjacencies at the same position.

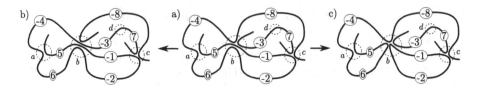

Fig. 3. (a) A 2D cartoon of a possible 3D configuration for genome A. Adjacencies between syntenic blocks are classified by physically close regions, which are marked by dashed circles and labeled by the alphabet $\{a, b, c, d\}$. (b) Genome A after a reciprocal translocation has occurred at position b. (c) Genome A after an excision has occurred at position b.

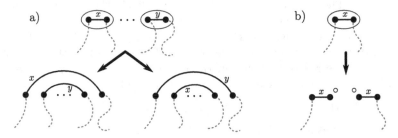

Fig. 4. The update of colors by a DCJ. (a) Adjacency edges with colors x and y are reconfigured in two different ways for the same DCJ operation. In this case the reconfigurations are achieved by swapping either both right-hand endpoints or both left-hand endpoints of the adjacency edges. (b) The adjacency edge with color x is split to make two adjacencies of color x with two new telomeres.

1.5   Locality and the Adjacency Graph

Each adjacency edge in G corresponds to an adjacency in genome A or B. The color of an adjacency is given to the adjacency edge it corresponds to. Figure 1 shows a coloring for the adjacencies of genome A that matches the localities in Fig. 3. The application of a DCJ operation to a genome has the effect of swapping the endpoints of two adjacency edges, or splitting an adjacency edge as in the case of Fig. 4(e).

Throughout a DCJ sorting scenario, adjacency edges always keep the same color. Thus, each DCJ operation corresponds to one of two possible updates of the same pair of adjacency edges, as depicted in Fig. 4(a).

1.6   A Positional Weight Function

Categorize rearrangements into two sets: those that are likely, and those that are not. Such a categorization of rearrangements is powerful enough to encapsulate the positional property discussed earlier.

A DCJ ρ acts on one or two adjacencies. Our model labels each adjacency with some *color* from an alphabet Σ, and weights a DCJ based on the colors that are acted upon. Call i_ρ and j_ρ the adjacencies affected by ρ; $i_\rho = j_\rho$ if the DCJ acts on only a single adjacency, *e.g.* case (e) in Fig. 2. The color of an adjacency i_ρ is written $col(i_\rho)$. Given a DCJ ρ, our weight function is

$$w(\rho) = \begin{cases} 0 \text{ if } i_\rho = j_\rho \text{ or } col(i_\rho) = col(j_\rho) \\ 1 \text{ otherwise.} \end{cases}$$

We call those DCJ moves that have zero weight *likely*, while we call all others *rare*. It is trivial to evaluate our weight function for a given DCJ; simply check the colors of the two adjacency edges that are affected.

Two restricted versions of the general problem are now described. The problem MINIMUM LOCAL SCENARIO is exactly MINIMUM WEIGHTED REARRANGE-MENTS with the positional weight function w.

Problem 2 (MLS). MINIMUM LOCAL SCENARIO

INPUT: Genomes A and B and positional weight function w.
OUTPUT: A scenario of rearrangements turning A into B.
MEASURE: The weight of the scenario.

The problem MINIMUM LOCAL PARSIMONIOUS SCENARIO introduces the con-straint that the scenario output is also a parsimonious scenario, *i.e.* a scenario of minimum length.

Problem 3 (MLPS). MINIMUM LOCAL PARSIMONIOUS SCENARIO

INPUT: Genomes A and B and positional weight function w.
OUTPUT: A parsimonious scenario of rearrangements turning A into B.
MEASURE: The weight of the scenario.

2   Minimum Local Parsimonious Scenario

Since a solution to MINIMUM LOCAL PARSIMONIOUS SCENARIO is limited to sorting moves, most connected components of $G(A, B, col)$ must be sorted inde-pendently of each other, the exception being for even-length paths; all but one DCJ in Fig. 2 act on a single connected component. We first give a method for computing the number of rare operations per connected component when no pair of even-length paths exist, as in Fig. 2(d). We then show in Sect. 2.2 how to solve the problem when such pairs exist.

2.1   Colored Partitions

Consider a connected component C of the graph $G(A, B, col)$. If C is monochro-matic, *i.e.* has adjacency edges of a single color, then the component can be sorted with likely DCJs according to the listed moves in Fig. 2; the move that operates on more than one component in Fig. 2(d) need not be used since each path can be split on its own with a local move, as in Fig. 2(e). If C is polychro-matic then DCJs must be performed to separate the colors, since a fully sorted

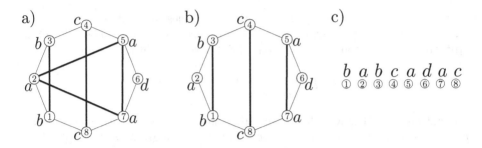

Fig. 5. Colored partitions for the set $[1,8]$ where $col(1) = b$, $col(2) = a$, $col(3) = b$, $col(4) = c$, $col(5) = a$, $col(6) = d$, $col(7) = a$, and $col(8) = c$. Vertices are circles numbered by their order in the set $[1,8]$ and labeled by their color. Thick black lines are drawn between vertices that are in the same class of the partition. (a) The *crossing* partition $\{\{1,3\}, \{2,5,7\}, \{4,8\}, \{6\}\}$. (b) The optimal *noncrossing* partition $\{\{1,3\}, \{2\}, \{4,8\}, \{5,7\}, \{6\}\}$. (c) The instance embedded on a line.

genome has components that each have only a single colored adjacency edge in genome A.

Recall that AA-paths and BB-paths are paths that start and end in the same genome. In this subsection, we assume that there does not exist both an AA-path and a BB-path in the graph (Fig. 2(d)). Ouangraoua and Bergeron established that the DCJs in a sorting scenario can be done in any order for such a graph and that every component will be sorted independently, thereby defining a noncrossing partition on each component (see Sects. 3 and 4 of [19]). Later in this section we show that MINIMUM LOCAL PARSIMONIOUS SCENARIO on a single component is equivalent to the following problem concerning a generalization of noncrossing partitions. A *partition* of a set is a collection of pairwise disjoint subsets whose union is the entire set. The subsets are called *classes*. $[1, n]$ is the set of integers from 1 to n.

Definition 1. *A* noncrossing partition *is a partition* $\mathcal{P}$ *of* $[1, n]$ *such that for any classes* $S_i, S_j \in \mathcal{P}$ *if we have* $p < q < p' < q'$ *for* $p, p' \in S_i$ *and* $q, q' \in S_j$, *then* $S_i = S_j$. *A* noncrossing colored partition *is a noncrossing partition where for any* $p, p' \in S_i$, $col(p) = col(p')$.

Another way to define a noncrossing partition is on a convex polygon. A noncrossing partition is a partition of the vertices of an n-gon with the property that if you draw a line between all pairs of vertices in the same class, for all classes, then no two lines from different classes intersect. A *colored partition* has colored vertices, and respects the property that any pair of vertices in the same class of the partition have the same color (see Figs. 5(a) and (b)).

Problem 4 (MNCP). MINIMUM NONCROSSING COLORED PARTITION

INPUT: Set size n, color set Σ, and color function $col : [1, n] \to \Sigma$.
OUTPUT: A noncrossing colored partition.
MEASURE: The cardinality of the partition.

We present a polynomial-time algorithm for the MINIMUM NONCROSSING COL-ORED PARTITION problem, which according to Lemma 2 (later in this section) gives a solution to MINIMUM LOCAL PARSIMONIOUS SCENARIO on a single component. We describe the algorithm on an instance that has been embedded on a line where the left-most vertex ① represents the smallest element of the set, as shown in Fig. 5(c). For an interval $[i, j]$, let $NCP(i, j)$ be the number of classes in the MNCP on that subproblem. Thus, $NCP(1, n)$ corresponds to the MINIMUM NONCROSSING COLORED PARTITION of $[1, n]$.

For any interval $[i, j]$ we have $NCP(i, i) = 1$, and the following recurrence.

$$NCP(i,j) = \min \begin{cases} NCP(i, j-1) + 1 & \text{for } i < j, \\ NCP(i, j-1) & \text{for } i < j \text{ and } col(i) = col(j) \\ NCP(i, k-1) + NCP(k, j) & \text{for all k where i < k < j} \end{cases}$$

The first case corresponds to the creation of a new class with the single element j. The second case is applicable when element j is the same color as element i; in this case i and j become part of the same class, all the other classes staying the same. The third case tests combinations of subproblems; this case is pertinent when the $col(i) = col(k-1)$ or $col(k) = col(j)$. It is easy to confirm that any feasible solution to MNCP is scored by the recurrence. This dynamic program runs in $O(n^3)$ time in the worst case.

We now show the link between MLPS and MNCP. Consider component C to be sorted. Pick an arbitrary vertex of C if it is a cycle, or either endpoint of C if it is a path, and consider an ordering of the vertices of genome A based on a traversal of the edges of C from that vertex. Embed the vertices of the component on a circle with respect to that ordering, and the edges so that they remain inside the circle. Call this a *circular embedding* of the component. Consider a sorting scenario for C that corresponds to a sequence of adjacency graphs $C_0, C_1, \ldots, C_d$ ($C = C_0$). Call C_i° the graph C_i with vertices embedded according to the circular embedding of C_0.

Lemma 1 ([19]). *C_i° has no pair of crossing adjacency edges for any i.*

Proof. By construction, all adjacency edges in C_0° connect adjacent vertices on the circle, so none of them cross. Assume that C_j° has crossing adjacency edges and C_{j-1}° does not. This implies that the jth DCJ did not split a component. This is a contradiction since every sorting move on C splits a component, never creating both an AA-path and BB-path.   □

Lemma 2. *Given a connected component C, MINIMUM LOCAL PARSIMONIOUS SCENARIO on C can be solved by MINIMUM NONCROSSING COLORED PARTITION.*

Proof. First, transform an instance of MLPS on a single component to an instance of MNCP. Given a cycle C representing genomes A and B, map the set of elements $[1, n]$ from the set of adjacency edges of A ordered according to a circular embedding of C. The color function col maps each element to its corresponding adjacency edge's color.

Now transform an optimal solution of MNCP into an optimal solution for MLPS. Clearly, any partition of $[1, n]$ corresponds to a partition of adjacency edges of genome A. We show that there always exists a scenario of DCJs whose prefix separates C into connected components according to the partition. Any two edges of the same component can be chosen for a DCJ [19] and the DCJs on a cycle can be done in any order (Lemma 1). Since the ordering of the edges on the cycle corresponds to the ordering on $[1, n]$, an edge partition of size k can be achieved with $k - 1$ DCJs. Since k is minimum over all feasible partitions and the remaining DCJs of the scenario are likely, the constructed scenario has a minimum number of rare DCJs.   □

In fact, the two problems are equivalent. We omit the reduction in the other direction since it is out of the scope of this paper.

2.2 Even-length Paths

A MINIMUM NONCROSSING COLORED PARTITION can be computed in polynomial time for a single component independent of all others. Yet it is possible to merge components in a parsimonious DCJ scenario. As described in Fig. 2, the only parsimonious DCJs that *merge* components are those that act on one edge from an AA-path and one edge from a BB-path. Call AA (BB respectively) the set of AA-paths (BB-paths respectively) in the adjacency graph. The key observation is that once a path has been merged with another, the result is always two odd-length paths which subsequently cannot be merged with any other. Thus we devote this section to the computation of which pairs $(a, b) \in AA \times BB$ will be merged in an optimal solution, and which paths will remain unmerged.

Any pair (a, b) can be merged in several ways. For all possible DCJs that merge them, we compute the MNCP on the resulting components. The minimum MNCP over all merges is the cost in rare moves for merging the two paths. To compute the pairs of paths to be merged in an optimal solution, we use the inverse of these costs — the number of likely moves — as weights in a bipartite graph.

Take the elements of AA and BB as vertices in a complete bipartite graph, and label each edge (a, b) with the maximum number of likely DCJs for the merge of paths a and b. Any even-length path could alternatively be used independently of any other, so there is a vertex q' for each $q \in AA \cup BB$ with a single edge (q, q') labeled by the number of likely moves on q alone (computed using the MNCP on that component). Algorithm 1 computes the minimum number of rare DCJs in a parsimonious scenario. It is easy to modify the algorithm to give the list of DCJs.

The function $MNCPonComp(a, col)$ computes the MINIMUM NONCROSSING COLORED PARTITION on the given component a. In other words it builds the color function col according to the component a and then calls $MNCP(1, n, col)$ where n is the number of adjacency edges on the A side of the component a. The function $maxMerge(a, b)$ computes the maximum number of likely DCJs over all possible DCJs that use one edge from a and one edge from b. The function $d(AA)$ computes the sum of DCJ distances from each component in AA using Formula 1. The function $maxMatching(V_A, V_B, w)$ builds the bipartite graph with vertices V_A on one side and vertices V_B on the other, and the edges described by the weight function w.

Algorithm 1. $MLPS(A, B)$

Input: genomes A and B.
Output: cost of parsimonious scenario with a minimum number of rare DCJs.
$\qquad\qquad\qquad\qquad\qquad\qquad\qquad$ ▷ Sort the graph components by type:
$C \leftarrow$ set of cycles in $G(A, B, col)$
$P \leftarrow$ set of odd-length paths in $G(A, B, col)$
$AA \leftarrow$ set of AA-paths in $G(A, B, col)$
$BB \leftarrow$ set of BB-paths in $G(A, B, col)$
$\qquad\qquad\qquad\qquad\qquad$ ▷ Compute the cost of the cycles and odd-length paths:
$cost \leftarrow 0$
for $c \in C$ **do**
$\quad cost \leftarrow cost + MNCPonComp(c, col) - 1$
end for
for $p \in P$ **do**
$\quad cost \leftarrow cost + MNCPonComp(p, col) - 1$
end for

$\qquad\qquad\qquad\qquad\qquad\qquad$ ▷ Compute the cost of the even-length paths:
for $a \in AA$ **do** $\qquad\qquad\qquad$ ▷ Compute weights for not merging AA vertices:
$\quad V_A \leftarrow V_A \cup \{a, a'\}$
$\quad w(a, a') \leftarrow MNCPonComp(a, col) - 1$
end for
for $b \in BB$ **do** $\qquad\qquad\qquad$ ▷ Compute weights for not merging BB vertices:
$\quad V_B \leftarrow V_B \cup \{b, b'\}$
$\quad w(b, b') \leftarrow MNCPonComp(b, col) - 1$
end for
for $a \in AA$ **do** $\qquad\qquad\qquad\qquad\qquad$ ▷ Compute weights for merges:
$\quad$ **for** $b \in BB$ **do**
$\quad\quad w(a, b) \leftarrow maxMerge(a, b)$
$\quad$ **end for**
end for
$\qquad\qquad\qquad\qquad$ ▷ Build the bipartite graph and compute the matching:
$cost \leftarrow cost + d(AA) + d(BB) - maxMatching(V_A, V_B, w)$
return $cost$

To summarize, any path can be merged at most once in a parsimonious scenario. Potential merges, as well as potential non-merges, are encoded into a bipartite graph with edges weighted by the cost of a merge. A maximum weight matching in this graph corresponds to a scenario that minimizes the number of rare moves on the paths. All other connected components of the graph are sorted using the MINIMUM NONCROSSING COLORED PARTITION on the component.

The running time of our algorithm is dominated by the weighting of the edges on the bipartite graph. Consider all merges done between elements of AA and elements of BB. A particular adjacency edge e from a given path $a \in AA$ will take part in exactly one DCJ with every edge f from a path $b \in BB$ throughout the weighting process. Therefore for each pair (e, f), e being an edge from a path in AA and f being an edge from a path in BB, we will compute the MNCP on the resulting merge. If the number of edges in the paths AA (respectively BB) is $n(AA)$

(respectively $n(BB)$), then the running time of our algorithm is $O(n(AA)n(BB)n^3)$. In the worst case, half of the edges are used in AA-paths and half in BB-paths, yielding a running time of $O(n^5)$. We conjecture that in practice even-length paths are rare, yielding a running time of $O(n^3)$.

3   Conclusion

The number of parsimonious DCJ scenarios between two genomes is exponential in the distance between them. However, many of the scenarios are probably unrealistic in the biological sense. This paper takes a step towards modeling realistic scenarios by posing optimization problems that take into account positional constraints. An example of such a positional constraint is the 3D proximity of genome segments given by Hi-C experiments.

An $O(n^5)$ algorithm is proposed for computing a parsimonious DCJ scenario that is most likely, given a function that classifies DCJ as "likely" or "unlikely". In practice the algorithm will be $O(n^3)$ since we expect long even-length path to be rare in nature. For example, the adjacency graph for the mouse/human syntenic map built by Véron, et al. [26] from one-to-one orthologs in Biomart has only 182 edges in even-length paths out of a total of 13302 edges. The largest connected component has 35 edges.

From a biological perspective, a solution to MINIMUM LOCAL PARSIMONIOUS SCENARIO corresponds to finding a maximum likelihood scenario in a situation where likely and unlikely scenarios are both rare, and the difference between the likelihoods of likely and unlikely moves is not very large. In this situation, a most parsimonious scenario made of k unlikely moves is more likely than a non-parsimonious scenario made of $k + 1$ likely moves. Thus the maximum likelihood scenario is the most parsimonious scenario that involves the smallest number of unlikely moves.

We introduce the MINIMUM NONCROSSING COLORED PARTITION problem — a generalization of the MAXIMUM INDEPENDENT SET problem on circle graphs — for weighting the edges of a bipartite graph, on which we obtain a maximum matching. While this technique is essential to our algorithm for finding DCJ scenarios, we believe it will also come in handy for an algorithm that finds likely *inversion* scenarios (*e.g.* for handling the infamous "hurdles"). A multitude of biologically relevant variations on this problem exist, including variations on the model of genome rearrangement, a variant where edges have multiple colors, and a bidirectional sorting variant where edges are weighted on both genomes according to the chromatin conformation on each. Models that incorporate uncertainty or evolution in the Hi-C data would also be relevant. We hope that this work provokes further study from both the algorithmic and the biological perspectives.

Acknowledgments. We would like to thank Anne Bergeron for her helpful comments during the preparation of this manuscript. This work was funded in part by a grant from the Fonds de Recherche du Québec en Nature et Technologies.

References

1. Aganezov, S., Alekseyev, M.: On pairwise distances and median score of three genomes under DCJ. BMC Bioinform. **13**(suppl. 19), S1 (2012)
2. Bender, M.A., Ge, D., He, S., Hu, H., Pinter, R.Y., Skiena, S., Swidan, F.: Improved bounds on sorting by length-weighted reversals. J. Comput. Syst. Sci. **74**(5), 744–774 (2008)
3. Bergeron, A., Mixtacki, J., Stoye, J.: A unifying view of genome rearrangements. In: Bücher, P., Moret, B.M.E. (eds.) WABI 2006. LNCS (LNBI), vol. 4175, pp. 163–173. Springer, Heidelberg (2006)
4. Bertrand, D., Gagnon, Y., Blanchette, M., El-Mabrouk, N.: Reconstruction of ancestral genome subject to whole genome duplication, speciation, rearrangement and loss. In: Moulton, V., Singh, M. (eds.) WABI 2010. LNCS, vol. 6293, pp. 78–89. Springer, Heidelberg (2010)
5. Blanchette, M., Kunisawa, T., Sankoff, D.: Parametric genome rearrangement. Gene **172**(1), GC11–GC17 (1996)
6. Blin, G., Fertin, G., Sikora, F., Vialette, S.: The EXEMPLAR BREAKPOINT DISTANCE for non-trivial genomes cannot be approximated. In: Das, S., Uehara, R. (eds.) WALCOM 2009. LNCS, vol. 5431, pp. 357–368. Springer, Heidelberg (2009)
7. Braga, M.D.V., Stoye, J.: The solution space of sorting by DCJ. J. Comput. Biol. **17**(9), 1145–1165 (2010)
8. Dixon, J.R., Selvaraj, S., Yue, F., Kim, A., Li, Y., Shen, Y., Hu, M., Liu, J.S., Ren, B.: Topological domains in mammalian genomes identified by analysis of chromatin interactions. Nature **485**(7398), 376–380 (2012)
9. Duan, Z., Andronescu, M., Schutz, K., McIlwain, S., Kim, Y.J., Lee, C., Shendure, J., Fields, S., Blau, C.A., Noble, W.S.: A three-dimensional model of the yeast genome. Nature **465**(7296), 363–367 (2010)
10. Fertin, G., Labarre, A., Rusu, I., Tannier, E., Vialette, S.: Combinatorics of Genome Rearrangements. MIT Press, Cambridge (2009)
11. Galvão, G.R., Dias, Z.: Approximation algorithms for sorting by signed short reversals. In: Proceedings of the 5th ACM Conference on Bioinformatics, Computational Biology, and Health Informatics, pp. 360–369. ACM (2014)
12. Gavril, F.: Algorithms for a maximum clique and a maximum independent set of a circle graph. Networks **3**, 261–273 (1973)
13. Haghighi, M., Sankoff, D.: Medians seek the corners, and other conjectures. BMC Bioinform. **13**(suppl. 19), S5 (2012)
14. Jiang, M.: The zero exemplar distance problem. J. Comput. Biol. **18**(9), 1077–1086 (2011)
15. Jones, B.R., Rajaraman, A., Tannier, E., Chauve, C.: Anges: reconstructing ancestral genomes maps. Bioinformatics **28**(18), 2388–2390 (2012)
16. Le, T.B.K., Imakaev, M.V., Mirny, L.A., Laub, M.T.: High-resolution mapping of the spatial organization of a bacterial chromosome. Science **342**(6159), 731–734 (2013)
17. Lefebvre, J.-F., El-Mabrouk, N., Tillier, E.R.M., Sankoff, D.: Detection and validation of single gene inversions. In: Proceedings of the 11th International Conference on Intelligent Systems for Molecular Biology (ISMB 2003). Bioinformatics, vol. 19, pp. i190–i196. Oxford University Press (2003)
18. Nash, N., Gregg, D.: An output sensitive algorithm for computing a maximum independent set of a circle graph. Inform. Process. Lett. **110**(16), 630–634 (2010)
19. Ouangraoua, A., Bergeron, A.: Combinatorial structure of genome rearrangements scenarios. J. Comput. Biol. **17**(9), 1129–1144 (2010)

20. Ouangraoua, A., Tannier, E., Chauve, C.: Reconstructing the architecture of the ancestral amniote genome. Bioinformatics **27**(19), 2664–2671 (2011)
21. Pinter, R.Y., Skiena, S.: Genomic sorting with length-weighted reversals. Genome Inform. **13**, 103–111 (2002)
22. Rajan, V., Xu, A.W., Lin, Y., Swenson, K.M., Moret, B.M.E.: Heuristics for the inversion median problem. BMC Bioinform. **11**(suppl. 1), 54 (2010)
23. Sexton, T., Yaffe, E., Kenigsberg, E., Bantignies, F., Leblanc, B., Hoichman, M., Parrinello, H., Tanay, A., Cavalli, G.: Three-dimensional folding and functional organization principles of the *Drosophila* genome. Cell **148**(3), 458–472 (2012)
24. Swenson, K.M., Blanchette, M.: Large-scale mammalian rearrangements preserve chromatin conformation (2015) (in preparation)
25. Valiente, G.: A new simple algorithm for the maximum-weight independent set problem on circle graphs. In: Ibaraki, T., Katoh, N., Ono, H. (eds.) ISAAC 2003. LNCS, vol. 2906, pp. 129–137. Springer, Heidelberg (2003)
26. Veron, A., Lemaitre, C., Gautier, C., Lacroix, V., Sagot, M.-F.: Close 3d proximity of evolutionary breakpoints argues for the notion of spatial synteny. BMC Genom. **12**(1), 303 (2011)
27. Yancopoulos, S., Attie, O., Friedberg, R.: Efficient sorting of genomic permutations by translocation, inversion and block interchange. Bioinformatics **21**(16), 3340–3346 (2005)
28. Zhang, Y., McCord, R.P., Ho, Y.-J., Lajoie, B.R., Hildebrand, D.G., Simon, A.C., Becker, M.S., Alt, F.W., Dekker, J.: Spatial organization of the mouse genome and its role in recurrent chromosomal translocations. Cell **148**(5), 908–921 (2012)

Sparse RNA Folding Revisited: Space-Efficient Minimum Free Energy Prediction

Sebastian Will[1] and Hosna Jabbari[2](✉)

[1] Bioinformatics/IZBI, University Leipzig, Leipzig, Germany
swill@csail.mit.edu
[2] Ingenuity Lab, National Institute of Nanotechnology, and Department of Chemical and Materials Engineering, University of Alberta, Edmonton, Canada
jabbari@ualberta.ca

Abstract. RNA secondary structure prediction by energy minimization is *the* central computational tool for the analysis of structural noncoding RNAs and their interactions. Sparsification has been successfully applied to improve the time efficiency of various structure prediction algorithms while guaranteeing the same result; however, for many such folding problems, space efficiency is of even greater concern, in particular for long RNAs and complex folding algorithms. So far, space-efficient sparsified RNA folding *with fold reconstruction* was solved only for simple pseudo-energy models. Here, we revisit the problem of space-efficient free energy minimization. Whereas the space-efficient minimization of the free energy has been sketched before, the reconstruction of the optimum structure has not even been discussed. We show that this reconstruction is not possible in trivial extension of the method for simple energy models. Then, we present the time- and space-efficient sparsified free energy minimization algorithm SPARSEMFEFOLD, which guarantees optimal structure prediction. In particular, this novel algorithm provides efficient fold reconstruction based on dynamically garbage collected trace arrows. We provide theoretical and empirical results on the efficiency of the method. SPARSEMFEFOLD is free software, available at http://www.bioinf.uni-leipzig.de/~will/Software/SparseMFEFold.

Keywords: Space efficient sparsification · Pseudoknot-free rna folding · RNA secondary structure prediction

1 Introduction

The manifold catalytic and regulatory functions of non-coding RNAs are mediated by the formation of inter-molecular structures with other RNAs or proteins, as well as their intra-molecular structures [3,5,10]. Currently computational RNA structure prediction methods mainly focus on predicting RNA secondary structure - the set of base pairs that form when RNA molecules fold. There is evidence that RNA molecules in their natural environments usually fold to their minimum free energy (MFE) secondary structure [16]. This motivates various

© Springer-Verlag Berlin Heidelberg 2015
M. Pop and H. Touzet (Eds.): WABI 2015, LNBI 9289, pp. 257–270, 2015.
DOI: 10.1007/978-3-662-48221-6_19

algorithms that predict MFE secondary structures of RNAs. Commonly, the free energy of a secondary structure is calculated by summing up the energies of its single features, where these energies are empirically determined [9]. MFE-based methods are applicable in cases of novel RNAs with unknown function, design applications in biotechnology and interacting RNAs.

Recently, sparsification techniques were applied to improve time and space efficiency of various RNA folding algorithms, while guaranteeing the same result. Wexler et al. [17] reduced the time complexity of standard MFE RNA folding by saving redundant recursion cases in the complexity-limiting step of the dynamic programming algorithm. For this purpose, they introduced candidates, which – by and large – are understood as sub-instances that cannot be optimally partitioned into two smaller sub-instances (cf. folding recursions of Fig. 1).

The approach of Wexler et al., which solely improves time efficiency, was implemented for the full free energy model by Dimitrieva and Bucher [4]. Beyond standard folding, these ideas have been studied for more complex folding algorithms, namely pseudoknot folding [11] and RNA-RNA-interaction [14].

Backofen et al. [2] showed that the concept of candidates can be extended to improve time *and* space of RNA folding in base pair-based (bp-based) pseudo-energy models (i.e. a generalized form of base pair maximization [12]). The two subproblems, energy minimization and fold reconstruction, are commonly solved by dynamic programming (DP) and trace-back through the DP matrix, respectively. Instead of storing the entire DP matrix, Backofen et al. [2] saved space by storing only a single matrix row (in the case of MFE prediction, several rows) as well as a list of candidates. This suffices to solve energy minimization subproblem, and at the same time allows efficient reconstruction of the optimal structure by recomputing matrix rows during trace-back. Note that [14] transferred Backofen et al.'s space savings to MFE RNA-RNA-interaction prediction, however only without space-efficient fold reconstruction.

Contributions: We show that the fold reconstruction method suggested by Backofen et al. cannot be trivially transferred beyond bp-based models. Consequently, we present a space-saving sparse MFE prediction algorithm with fold reconstruction. In preparation, we present space-efficient MFE folding without fold reconstruction for the MFE folding algorithm of [19], which extends [18] by multiloop penalties and is implemented by modern RNA folding software [6]; to the best of our knowledge, even this algorithm is presented here for the first time. Our efficient fold reconstruction algorithm keeps the additional information to a minimum using garbage collection. Whereas we present our techniques for the most common case of RNA MFE folding, they are intentionally more general; in particular, they can be transferred to complex sparsified folding algorithms (e.g., [11,14]), as well as simultaneous alignment and folding [15], which profit from sparsification even stronger than standard folding.

2 Time and Space Efficient Computation of the MFE

An *RNA sequence* $S = S_1, \ldots, S_n$ is represented as a sequence over the alphabet $\{A, C, G, U\}$. $S_{i,j}$ denotes the subsequence $S_i, \ldots, S_j$. Fix an RNA sequence S of length n. A base pair of S is an ordered pair $i.j$ with $1 \leq i < j \leq n$, such that ith and jth bases of S are complementary (i.e. $\{S_i, S_j\}$ is one of $\{A, U\}, \{C, G\}$, or $\{G, U\}$). A secondary structure R for S is a set of base pairs such that for all $i.j$, $i'.j' \in R$: $\{i, j\} \cap \{i', j'\} = \emptyset$. Base pairs of secondary structure R partition the unpaired bases of sequence S into loops [13] (i.e., hairpin loop, internal loop and multiloop). Hairpin loops have a minimum length of m; consequently, $j - i > m$ for all base pairs $i.j$ of R. A secondary structure R is pseudoknot-free if it does not contain $i.j$ and $i'.j'$ such that $i < i' < j < j'$.

2.1 Time and Space Efficient Bp-Based Folding

The simplest form of RNA folding minimizes a bp-based (pseudo-)energy; the most prominent special case is base pair maximization, in which (equivalently put), each base pair is assigned -1 if it is complementary and $+\infty$ otherwise. The bp-based energy E of a structure R is then: $E^{\mathrm{bp}}(R) = \sum_{i.j \in R} E^{\mathrm{bp}}(i.j)$. Since bp-based models by nature cannot capture thermodynamics of even stacked loops, here we reserve the term *free energy minimization* to refer to optimization in loop-based energy models. In a loop-based energy model, free energy of hairpin loop closed by $i.j$ and internal loop (including stacked and bulge loops) closed by $i.j$ and $k.l$ as external and internal base pairs are referred to by $\mathcal{H}(i, j)$, and $\mathcal{I}(i, j, k, l)$, respectively. Following common practice we limit the size of internal loops to M to cap the time complexity to $O(n^3)$. Free energy of multiloops is calculated from their numbers of inner base pairs, p, and unpaired bases, q, as follows: $\mathcal{ML}(i, j, p, q) = a + bp + cq$ [9]. Energy of a structure R is then calculated as: $E(R) = \sum_{\ell \in \mathrm{loops}(R)} E^{\mathrm{loop}}(\ell)$.

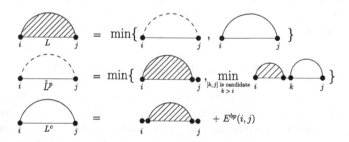

Fig. 1. Graphical representation of the sparse bp-based energy minimization recursions. A minimum energy general substructure (L, lined pattern) over region $[i, j]$ is a closed structure (L^c, solid arcs) or it is partitionable into two substructures ($\hat{L}^p$, dotted arcs). Sparsification restricts the minimization over the partitions in the second row to consider only candidates $[k, j]$ for the second fragment; candidates are defined as sub-instances that are not optimally partitionable, i.e. $L^c(i, j) < \hat{L}^p(i, j)$.

To find the minimum free energy of substructures for region $[1, n]$, $L(1, n)$, a dynamic programming algorithm, with similar recurrences as follows, is used; in which $L^c(i, j)$ corresponds to "closed" substructures (i.e., closed by base pair $i.j$) and $L^p(i, j)$ to "partitionable" substructures, which can be optimally decomposed into two independent subparts.

$$L(i, j) = \min\{L^p(i, j), L^c(i, j)\}$$
$$L^p(i, j) = \min\{ L(i, j - 1), \min_{i < k < j} L(i, k - 1) + L(k, j) \}$$
$$L^c(i, j) = L(i + 1, j - 1) + E^{\mathrm{bp}}(i.j),$$

where $1 \le i < j \le n$, $L^p(i, i) = L^c(i, i) = +\infty$ and $L(i, i) = 0$.

We obtain equivalent sparsified recursions after *replacing* $L^p(i, j)$ by $\hat{L}^p(i, j)$[1]:

$$\hat{L}^p(i, j) = \min\{ L(i, j - 1), \min_{[k, j] \text{ is candidate, } k > i} L(i, k - 1) + L^c(k, j)\}, \quad (\hat{L}^p)$$

where $[i, j]$ is an *L-candidate*, i.e., a candidate for recursion L, iff $L^c(i, j) < \hat{L}^p(i, j)$ (see Fig. 1). If, for $i < j$, $[i, j]$ is not an L-candidate, we call it *L-partitionable*. Note that here we consider $[i, i]$ as neither candidates nor partitionable, whereas in [2] they are considered as candidates. To prove the correctness one has to show $\hat{L}^p(i, j) = L^p(i, j)$; this follows the *triangle inequality* $L(i, j) \le L(i, k - 1) + L(k, j)$ (for all $1 \le i < k \le j \le n$) [2].

Backofen et al. [2] improved the time and space efficiency of $O(n^3)$ and $O(n^2)$ in the non-sparsified version to $O(n^2 + n \cdot Z_L)$ and $\Theta(n + Z_L)$ respectively, where Z_L is the total number of candidates; typically $Z_L \ll n^2$. The efficient implementation, which computes the matrix entries row by row starting with row n, is based on two further observations: (1) During the DP algorithm, one can maintain an appropriate data structure that allows traversing the candidates $[k, j]$ of Eq. ($\hat{L}^p$) in time linear to the number of candidates $[k, j]$. The data structure takes $\Theta(Z_L)$ space. (2) In addition to storing L^c for all candidates in $\Theta(Z_L)$ space, for computing row i, it suffices to store the rows i and $i + 1$, the latter for accessing $L(i + 1, j - 1)$, at any given time in the DP evaluation.

2.2   Time and Space Efficient Calculation of the MFE

Whereas Wexler et al. [17] sparsified the recursions of [18], we build on the recursions of [19], since they are used by modern folding software [6]. These recursions are explicitly given in [6] using matrix symbols C, F, and FM. For reference, we restate these original recursions in our notation and using respective symbols V, W, and WM; the symbols V and W are taken from [18], which makes the presentation closer to Wexler et al.

---

[1] i.e., we replace the recursion $L^p(i, j)$ by Eq. ($\hat{L}^p$) and replace the symbol $L^p(i, j)$ by $\hat{L}^p(i, j)$ in the recursion L.

Definition 1 (Free energy minimization recursions from [6]).

$$W(i,j) = \min\{\, V(i,j), \min_{i<k<j} W(i,k) + W(k+1,j) \,\}$$

$$V(i,j) = \min\{\, \mathcal{H}(i,j), \min_{\substack{i<p<q<j \\ p-i+j-q-2\leq M}} \mathcal{I}(i,j,p,q) + V(p,q),$$

$$\min_{i<k<j} WM(i+1,k) + WM(k+1,j-1) + a\}$$

$$WM(i,j) = \min\{\, V(i,j) + b, WM(i+1,j) + c, WM(i,j-1) + c,$$

$$\min_{i<k<j} WM(i,k) + WM(k+1,j) \,\}$$

$W(i,j)$ denotes the minimum free energy of any pseudoknot-free MFE secondary structure of the subsequence $S_i \ldots S_j$, such that the free energy minimization algorithm returns the energy $W(1,n)$. $V(i,j)$ is the free energy of the MFE structure for $S_i \ldots S_j$ that contains $i.j$; $\mathcal{H}(i,j)$ and $\mathcal{I}(i,j,p,q)$ are energy of a hairpin loop and internal loop closed by $i.j$, respectively. The parameter a is the penalty for multiloop initiation, b is the penalty for inner base pairs and c is the penalty for an unpaired base in a multiloop. Thus, $WM(i,j)$ is analogous to $W(i,j)$, but includes multiloop penalties and minimizes over structures that contain at least one base pair (initialization $WM(i,i) := \infty$).

For preparing the subsequent sparsification, but resulting in equivalent recursions, we introduce the matrices W^p, WM^p, and WM^2. $W^p(i,j)$ and $WM^p(i,j)$ correspond to the respective cases of $W(i,j)$ and $WM(i,j)$, where the substructure is partitionable. WM^2 represents multiloop fragments with at least two inner base pairs. The term $WM^2(i+1,j-1) + a$ corresponds to the energy of a MFE structure for $S_i \ldots S_j$ in which $i.j$ closes a multiloop.

Definition 2 (Sparsification-ready free energy minimization recursions).

$$W(i,j) = \min\{\, W^p(i,j), V(i,j) \,\}$$

$$W^p(i,j) = \min\{\, W(i,j-1), \min_{i<k<j} W(i,k-1) + W(k,j) \,\}$$

$$V(i,j) = \min\{\mathcal{H}(i,j), \min_{\substack{i<p<q<j \\ p-i+j-q-2\leq M}} \mathcal{I}(i,j,p,q) + V(p,q), WM^2(i+1,j-1) + a\}$$

$$WM(i,j) = \min\{\, WM^p(i,j), V(i,j) + b \,\}$$

$$WM^p(i,j) = \min\{\, WM(i+1,j) + c, WM(i,j-1) + c, WM^2(i,j) \,\}$$

$$WM^2(i,j) = \min_{i<k<j} WM(i,k-1) + WM(k,j)$$

where $i < j$, $W(i,i) = 0$; $V(i,j) = WM(i,j) = \infty$ for all $j - i \leq m$; and $WM^2 = \infty$ for all $j - i \leq 2m + 3$ (m is the minimum size of a hairpin loop).

We sparsify the recurrences by rewriting $W^p(i,j)$ to $\widehat{W}^p(i,j)$ and $WM^2(i,j)$ to $\widehat{WM}^2(i,j)$, where

$$\widehat{W}^p(i,j) = \min\{\, W(i,j-1),\!\!\!\!\!\!\min_{[k,j]\text{ is W-candidate},k>i}\!\!\!\!\!\! W(i,k-1)+V(k,j)\,\}$$

$$\widehat{WM}^2(i,j) = \min\{\, WM^2(i,j-1)+c,\!\!\!\!\!\!\min_{[k,j]\text{ is WM-candidate},k>i}\!\!\!\!\!\! WM(i,k-1)+V(k,j)+b\,\}$$

together with the candidate criteria

- $[k,j]$ is a *W-candidate* iff $V(k,j) < \widehat{W}^p(k,j)$ and
- $[k,j]$ is a *WM-candidate* iff $V(k,j)+b < WM^p(k,j)$.

We note that to have similar recurrences, we also rewrite $WM^p(i,j)$ recurrence as follows: $WM^p(i,j) = \min\{\, WM(i+1,j)+c, \widehat{WM}^2(i,j)\,\}$ which merges the second case of original WM^p into the $\widehat{WM}^2$ recurrence.

Lemma 1. *The sparsified version of W^p and WM^2 recurrences are equivalent to the non-sparsified recurrences.*

Proof.

1. Choose the largest k, $i < k < j$, s.t. $W(i,k-1)+W(k,j)$ is minimal. We show that $[k,j]$ is a W-candidate. Assuming the opposite, choose e ($e > k$), such that $W^p(k,j) = W(k,e-1)+W(e,j)$. Now $W^p(i,j) = W(i,k-1)+W(k,e-1)+W(e,j) \geq W(i,e-1)+W(e,j)$, which contradicts the choice of k such that $W(i,k-1)+W(k,j)$ is minimal. Therefore we must have $W(k,j) = V(k,j) < \widehat{W}^p(k,j)$, and $[i,j]$ is a *W-candidate*.

2. Choose the largest k, $i < k < j$, s.t. $WM(i,k-1)+WM(k,j)$ is minimal. We show that $[k,j]$ is a WM-candidate. Assuming the opposite, choose e ($e > k$), such that $WM^2(k,j) = WM(k,e-1)+WM(e,j)$. Now $WM^2(i,j) = WM(i,k-1)+WM(k,e-1)+WM(e,j) \geq WM(i,e-1)+WM(e,j)$, which contradicts the choice of k such that $WM(i,k-1)+WM(k,j)$ is minimal. Therefore we must have $WM(k,j) = V(k,j)+b < WM^p(k,j)$, and $[i,j]$ is a *WM-candidate*. □

Going beyond Wexler et al., these recursions handle multiloop energies correctly by introducing the matrices WM, WM^p and $\widehat{WM}^2$.

Analogous to [2], there is an algorithm that evaluates the above recursions efficiently, such that its time and space complexity depends on Z, where Z is the total number of *candidates* (which are W- or WM-candidates). We call this algorithm SPARSEENERGYMINIMIZATION.

Lemma 2. $W(1,n)$ *can be calculated in* $O(n^2 + nZ)$ *time and* $\Theta(n+Z)$ *space, where Z is the total number of candidates.*

Proof *Time.* SPARSEENERGYMINIMIZATION computes $O(n^2)$ entries and performs the minimizations over all candidates in the calculations of $\widehat{W}^p$ and $\widehat{WM}^2$.

These minimizations require $O(Z)$ steps per matrix row, resulting in $O(nZ)$ additional time.

Space. To calculate all $\widehat{W}^\mathrm{p}(i,j)$ and $\widehat{WM}^2(i,j)$ in row i it suffices to compute and store the entries in the same matrix row and store the matrix entries at the candidates of rows $i' > i$. For calculating the $V(i,j)$ in row i ($j : i < j \leq n$), it suffices to keep row $i + 1$ of WM^p and the rows $i + 1$ to $i + M + 1$ of V in memory, since the interior loop size is bounded by M. □

2.3 The Difficulty of MFE Fold Reconstruction Compared to Bp-Based Folding

The MFE structure in the bp-based model is efficiently reconstructed using the minimum energy, the energies of candidates, and $O(n)$ space by trace-back with recomputation of partitionable entries, which are not stored in the DP-matrix. We briefly recapitulate this result of [2].

Lemma 3. *The optimal structure in the bp-based model can be reconstructed from the candidates and the minimum free energy in $O(n + Z_L)$ space and $O(n^2 + nZ_L)$ time.*

Proofsketch. The algorithm starts similar to a regular trace-back from $W(1, n)$. Recursively, it derives the optimum recursion cases of the current matrix entry and continues to trace back from the identified successive trace entries. For finding the successive trace entries from a current entry (i, j), it suffices to know the entries (i, j') ($j' \leq j$) of the same row: if $[i, j]$ is a candidate, then the successive trace entry is $(i + 1, j - 1)$; otherwise, it can be split at some k, s.t. entry $(i, k - 1)$ is in the same row and $[k, j]$ is a candidate (unless $k = j$). On demand, the entries (i, j') can be recomputed from entries (i, j'') ($j'' < j'$) of this row and candidate entries. Note that access to non-candidates of rows $i' > i$ is never required. In particular, the algorithm utilizes that the candidates $[i, j]$ of row i do not have to be recomputed, because candidates necessarily trace back to $(i + 1, j - 1)$. Thus, the trace-back with recomputation takes $O(n \cdot Z_L)$ time and does not require additional space. □

After executing SPARSEENERGYMINIMIZATION, all candidates are calculated and stored in memory, analogously to the bp-based case. However, there is no trivial transfer of the bp-based trace-back algorithm of [2], FOLDING-TRACEBACK, to the loop-based case.

The main difference between the bp-based and the loop-based folding algorithm is the evaluation of interior loops. In both cases, bp-based folding and loop-based folding, the energy of a closed structure, respectively $L^c(i, j)$ and $V(i, j)$, depends only on a constant number of rows (resp., 1 row or M rows). However, FOLDING-TRACEBACK relies on the fact that the successive trace entry of candidates is known, whereas the MFE fold reconstruction has to infer the optimum recursion case of $V(i, j)$, even if $V(i, j) < \min\{W(i, j), WM(i, j)\}$ — corresponding to the optimum co-terminus criterion of [2].

Naively, this requires to recompute the (non-candidate) V entries of rows $i+1,\ldots,i+M+1$, which in turn rely on V and WM entries of larger rows. Consequently, the non-candidate entries of the whole V matrix have to be recomputed. This negates the sparsification benefits. Furthermore, there seems to be no simple way to overcome this problem. In particular, we cannot directly compute $V(i,j)$ by minimizing only over candidates, since there is no guarantee that the inner base pair of an interior loop corresponds to a candidate.

Lemma 4. *The minimization over inner base pairs in the recursion of V cannot be restricted to candidates.*

Proof. We show that there is a loop-based energy model (namely the Turner energy model [9]), a sequence S and $1 \leq i < j \leq n$, such that $V(i,j) < \min\{\mathcal{H}(i,j), WM^2(i+1,j-1)+a\}$, but there is no candidate $[p,q]$, $i < p < q < j$, such that $V(i,j) = \mathcal{I}(i,j,p,q) + V(p,q)$.

Consider the RNA sequence $S = \text{GCCAAAAGGGC}$ of length 11. In the Turner model, the optimal recursion case of $V(2,10)$ forms the interior loop closed by $(2,10)$ with inner base pair $(3,9)$, because $V(3,9) = \mathcal{H}(3,9) = 4.3$ kcal/mol and $\mathcal{I}(2,10,3,9) = -3.3$ kcal/mol. However, $[3,9]$ is not a candidate, since $W(3,9) = WM(3,9) = \mathcal{H}(3,8) = 4.1 < V(3,9)$, i.e. the MFE structure of $S_{3,9}$ forms the hairpin loop closed by $(3,8)$ – not by $(3,9)$.

The lemma holds for arbitrarily large instances. This can be seen by, for example, looking at the family of RNA sequences $S_k = GC_kA_4G_{k+1}C$, where X_k is the k-times repeat of X. Furthermore, this issue is not limited to stacked base pairs, since there are non-stacked interior loops with stabilizing energy contributions, in the Turner energy model.

3   MFE Folding with Fold Reconstruction

As discussed earlier direct transfer of FOLDING-TRACEBACK from [2] is not possible because the optimum case of the V-recursion cannot be determined efficiently by recomputation. Therefore we suggest to store trace arrows from all entries that cannot be recomputed efficiently. Subsequently, we discuss several space optimizations for this idea, such as avoiding trace arrows by rewriting the recursions and removing trace arrows as soon as they become inaccessible for the trace-back.

3.1   Adding Trace Arrows

As a first step towards efficient trace-back, we store trace arrows from each potential base pair $i.j$ to its optimum inner base pair during the DP evaluation. Here, a *trace arrow* is simply a directed edge connecting two matrix entries. By storing these arrows we avoid the recomputation of all V entries in the trace-back, by inferring their successive trace entries. If there is no trace arrow to an inner base pair and $V(i,j) \neq \mathcal{H}(i,j)$, we can simply continue to trace from $(i+1, j-1)$ in matrix WM^2.

Furthermore, the case $WM(i+1,j) + c$ of WM^p accesses entries beyond the current row i. As before, we cannot efficiently recompute row $i+1$, which could be resolved by recording trace arrows.

3.2 Avoiding Trace Arrows

One can avoid the trace arrows for the case $WM(i+1,j)+c$ of WM^p by rewriting the case equivalently as follows:

$$WM^p(i,j) = \min\{ WM(i+1,j) + c, WM^2(i,j) \}$$
$$= \min\{ \min_{i<k<j} (k-i) \times c + WM(k,j), WM^2(i,j) \}$$

Since $WM(i,i) = +\infty$, we can sparsify the recurrence as follows:

$$\widehat{WM}^p(i,j) = \min\{ \min_{\substack{i<k<j \\ [k,j] \text{ is } WM\text{-candidate}}} (k-i) \times c + V(k,j) + b, \widehat{WM}^2(i,j)\}.$$

We have previously stablished equivalence of $\widehat{WM}^2$ and WM^2 recurrences. We establish the correctness of this rewriting by the following lemma. This serves well as an example of a typical *small* change during sparsification of recursions, which is nevertheless non-trivial.

Lemma 5. *Replacing WM^p by $\widehat{WM}^p$ leaves the values of W, V, and WM entries unchanged.*

Proof. We have to show that restricting the minimization of $\min_{i<k<j}(k-i) \times c + WM(k,j)$ to only WM-candidates is admissible; this boils down to showing that non-candidates in the new minimization do not change the minimum values in the recursions. Assume that $[k,j]$ is WM-partitionable. By definition there exists a $k' > k$, where $[k',j]$ is a WM-candidate s.t. one of the following holds

1. $V(k,j) + b \geq (k'-k) \times c + V(k',j) + b$
2. $V(k,j) + b \geq WM(k, k'-1) + V(k',j) + b$

Case 1. $(k-i) \times c + V(k,j) + b \geq (k-i) \times c + (k'-k) \times c + V(k',j) + b \geq (k'-i) \times c + V(k',j) + b$, i.e. k' dominates k in this minimization. **Case 2.** $(k-i) \times c + WM(k, k'-1) + V(k',j) + b \geq WM(i, k'-1) + V(k',j) + b$; since the latter is a case of $\widehat{WM}^2$, k is again dominated. $\qquad\square$

Furthermore, we do not have to store trace arrows from V entries (i,j) to candidates. In such cases, the optimum interior loop case, $V^{\text{il-cand}}(i,j)$, can be reconstructed by minimizing over all candidates:

$$V^{\text{il-cand}}(i,j) = \min_{\substack{i<p<q<j \\ p-i+j-q-2 \leq M \\ [p,q] \text{ is candidate}}} \mathcal{I}(i,j,p,q) + V(p,q).$$

If $V^{\text{il-cand}}(i,j)$ is the MFE, we have reconstructed the trace arrow, however there is one catch: recall that we cannot use $WM^2(i+1,j-1)+a$ to decide whether $WM^2(i+1,j-1)+a < min\{ V^{\text{il-cand}}(i,j), \mathcal{H}(i,j) \}$, because it is neither stored nor can be recomputed efficiently. Thus, our strategy is to trace into the multiloop, iff no other case yields the MFE $V(i,j)$. However, as of this computation, we do not know this energy for non-candidate entries during trace-back. Therefore, we additionally keep track of this energy: each time we trace back from some (i,j) to a non-candidate (p,q), we recalculate the entry $V(p,q)$ due to $V(p,q) = V(i,j) - \mathcal{I}(i,j,p,q)$.

3.3   Garbage Collecting Trace Arrows

So far, our algorithm stores all trace arrows from V entries to non-candidates. However, most of those V entries are not on the MFE trace (rather far off from it). Identifying unnecessary arrows during the recursion evaluation, allows saving space for trace arrows, while still supporting MFE fold reconstruction.

Of course, during the evaluation we generally have only partial information about the MFE trace. Therefore, a safe strategy is to remove the trace arrows that are inaccessible from current and future accessible entries.

Definition 3. *An entry $V(p,q)$ is accessible, iff during recursion evaluation, after computing row i:*

- $p \leq i + M + 1$,
- $[p,q]$ is a candidate, or
- there is a trace arrow from some accessible entry to $V(p,q)$.

The trace arrows induce a directed graph on the set of matrix entries. Therefore, detecting inaccessible entries can be performed by garbage collection (GC) [7]. Since there is no cycle in our directed graph, we apply a simple reference counting GC technique. Each arrow ta receives a counter, which keeps track of the arrows that point to the source of ta. After computing row i, we scan through the arrows with source in row $i + M + 1$. Arrows from non-candidates in row $i + M + 1$ are removed, if their reference count is zero. In a recursive procedure, we detect all arrows pointing from inaccessible entries, remove them and update the appropriate counters.

3.4   Algorithm Summary

Employing the above two techniques, our algorithm SPARSEENERGYMINIMIZA-TION (Algorithm 1) keeps track of trace arrows and performs reference counting garbage collection (Procedure GARBAGECOLLECT). Note that for further space savings, the algorithm does not distinguish W- and WM-candidates; this does not affect our complexity bounds.

The final algorithm SPARSEMFEFOLD performs energy minimization and fold reconstruction. The fold reconstruction relies on the complete results of Algorithm 1, i.e. the minimum free energy, the candidates, and the trace arrows.

Algorithm 1. Space-efficient sparsified calculation of minimum free energy that keeps track of garbage-collected trace arrows to enable trace-back.

---

procedure SPARSEENERGYMINIMIZATION(S)

    allocate arrays $V[0..M][1..|S|]$, $W[1..|S|]$, $WM[1..|S|]$, $\widehat{WM}^2[1..|S|]$

    for $i \leftarrow |S|$ downto 1 **do**

        $i' \leftarrow i \bmod (M+1)$ $\triangleright$ row index for "rotating" matrix V

        $W[i] \leftarrow 0;\ WM[i] \leftarrow \infty\ ;\ \widehat{WM}^2[i] \leftarrow \infty\ ;\ \widehat{WM}^{2\prime} \leftarrow \infty$

        for $j \leftarrow i+1$ to $|S|$ **do**

            $(p^*, q^*) = \arg\min_{\substack{i<p<q<j \\ p-i+j-q-2\le M}} \mathcal{I}(i,j,p,q) + V[p \bmod (M+1)][q]$

            $VI = \mathcal{I}(i,j,p,q) + V[p^* \bmod (M+1)][q^*]$ $\triangleright$ interior loop cases

            $VO = \min\{\ \mathcal{H}(i,j), \widehat{WM}^{2\prime} + a\ \}$ $\triangleright$ other cases of V

            $V[i'][j] = \min\{\ VI, VO\ \}$

            $\widehat{WM}^{2\prime} \leftarrow \widehat{WM}^2[j]$ $\triangleright$ store $\widehat{WM}^2(i+1,j)$ for use in iteration $j+1$

            $\widehat{WM}^2[j] = \min\{\ \widehat{WM}^2[j-1] + c, \min_{(k,j,e)\in C} WM[k-1] + e\ \}$

            $\widehat{WM}^{\mathrm{P}} = \min\{\min_{\substack{i<k<j \\ (k,j,e)\in C}} (k-i)\cdot c + e + b, WM[j-1] + c, \widehat{WM}^2[j-1]\}$

            $WM[j] = \min\{\ \widehat{WM}^{\mathrm{P}}, V[i'][j] + b\ \}$

            $\widehat{W}^{\mathrm{P}} = \min\{\ W[j-1], \min_{(k,j,e)\in C} W[k-1] + e\ \}$

            $W[j] = \min\{\ \widehat{W}^{\mathrm{P}}, V[i'][j]\ \}$

            if $V[i'][j] < \widehat{W}^{\mathrm{P}}$ or $V[i'][j] + b < \widehat{WM}^{\mathrm{P}}$ **then** $\triangleright$ register candidate

                add the candidate $(i,j,V[i'][j])$ to C

            if $(p^*, q^*, \cdot) \notin C$ and $VI < VO$ **then** $\triangleright$ register trace arrow

                add the trace arrow $(i,j,0) \mapsto (p^*, q^*, V[p^*][q^*])$ to T

                MODIFYREFCOUNT$(p, q, +1)$

            if $i + M + 1 \le |S|$ **then**

                for $(i+M+1, j, 0) \mapsto (p, q, e) \in T$ **do**

                    GARBAGECOLLLECT(i+M+1,j)

    return (E, C, T)

procedure GARBAGECOLLECT(i, j)

    if $(i, j, c) \mapsto (p, q, e) \in T$ and $c = 0$ **then**

        release the trace arrow $(i, j, 0) \mapsto (p, q, e)$ from T

        MODIFYREFCOUNT$(p, q, -1)$

        GARBAGECOLLECT(p, q)

procedure MODIFYREFCOUNT(i, j, δ)

    if $(i, j, c) \mapsto (p, q, e) \in T$ **then**

        change counter c of $(i, j, c) \mapsto (p, q, e)$ in T to $c + \delta$

3.5   Complexity

Implementing the trace arrow data structure as a hash allows access to a trace arrow by its origin in amortized constant time. In the following, we assume constant time access.

Lemma 6. SPARSEENERGYMINIMIZATION *(including storing and garbage collection of trace arrows) calculates* $W(1, n)$ *in* $O(n^2 + nZ)$ *time and* $\Theta(n + Z + T)$ *space, where* Z *is the total number of* candidates *and* T *is the maximum number of accessible trace arrows to non-candidates.*

Proof. The number of trace arrows is quadratically limited and the criterion for storing a trace arrow is checked in constant time, such that the time complexity is not changed. The time for garbage collection of trace arrows is at most quadratic, because GARBAGECOLLECT is called at most once per matrix entry from SPARSEENERGYMINIMIZATION. Furthermore, each time it calls itself it removes one trace arrow. Each trace arrow can be inserted and removed only once by SPARSEENERGYMINIMIZATION. The space complexity depends on the maximum number of trace arrows that have to be stored simultaneously. Without garbage collection, this is the number of trace arrows to non-candidates. Due to the garbage collection, we reduce this to the maximum number of simultaneously accessible trace arrows to non-candidates.   □

4   Empirical Results

We implemented the algorithm SPARSEMFEFOLD in C++ utilizing the Vienna RNA library [8] for calculating the single loop energies. Consequently, we compute exactly the same energies and structures as RNAfold of the Vienna RNA package 2.x [8] (without dangling ends, i.e., option -d0). This implementation allows us to study the suggested strategies empirically. Tables 1 and 2 summarize results from folding 80 long RNA sequences from the RNA STRAND v2.0 database [1]. This set of 80 sequences comprises of all RNAs of length greater or equal 2500, for which a single molecule fold is available. (We note that the sparsification gain is pronounced for large values of n.) These sequences have a median length of 2904 and a maximum of 4381. Our comparison to RNAfold, currently the fastest RNA folding implementation, shows that our sparsified algorithm is significantly faster and uses significantly less space (Table 1). Experiments were performed on a Lenovo Thinkpad T431s with 12GB memory and Intel i5-3437U CPU. We measured run-time as user time and space consumption as maximum resident set size. Note that even the median resident set size of our method is about six-times lower than that of RNAfold. To empirically study the effects of our optimizations in SparseMFEFold, we further provide number of candidates and trace arrows (Table 2). For the trace arrows, we report minimum, median, and maximum of the final number of trace arrows (Final), passed to the fold reconstruction algorithm; the maximum number of trace arrows (Maximum), determining the memory foot print; the savings due to avoiding arrows to candidates (Avoided); and garbage collection of inaccessible arrows (GC-Removed).

Table 1. Time and space performance of SparseMFEFold compared to RNAfold

| | Run-time (s) | | Memory: resident set size (kB) | |
	RNAfold	SparseMFEFold	RNAfold	SparseMFEFold
Minimum	16.9	15.37	31800	5932
Median	29.7	22.89	42828	7262
Maximum	89.9	57.36	88548	9048

The latter two numbers show the importance of these two optimizations for the entire approach; together these strategies reduce the (median) number of stored trace arrows to only about 9 % (94443/1038525).

Table 2. Counts of candidates and trace arrows in SparseMFEFold.

| | Number of | Number of trace arrows | | | |
	candidates	Final	Maximum	Avoided	GC-Removed
Minimum	17032	49860	52293	137892	467230
Median	41215	92967	94443	237717	706365
Maximum	71508	147150	148947	419825	1748491

5   Conclusion and Future Work

We identified and solved the fundamental problem of efficient fold reconstruction in time- and space-efficient sparsified MFE folding of RNAs, while guaranteeing prediction of the MFE structure. This problem is not present in simple variants of RNA folding such as base pair maximization, but emerges only in realistic free energy minimization problems. Remarkably, Backofen et al. did not notice this problem when discussing the extension of their time and space-efficient base pair maximization algorithms to MFE prediction. Here, we provide an elegant and practical solution, which introduces garbage collection as a novel technique to RNA folding. The method is presented and studied for the most-common case of pseudoknot-free RNA secondary structure prediction using the Turner model.

Importantly, the introduced techniques are not specific to the presented folding scenario, but are applicable to many – even fundamentally more complex – variants of RNA folding, such as the MFE prediction of RNA-RNA-interactions and efficient pseudoknot folding algorithms. Similar to the case of time-efficient sparsification, the presented techniques will have even stronger impact on complex folding algorithms. Thus, we see the strongest potential of our method in reducing the often prohibitive space requirements of such algorithms.

References

1. Andronescu, M., Bereg, V., Hoos, H.H., Condon, A.: RNA STRAND: the RNA secondary structure and statistical analysis database. BMC Bioinform. **9**(1), 340 (2008)
2. Backofen, R., Tsur, D., Zakov, S., Ziv-Ukelson, M.: Sparse RNA folding: time and space efficient algorithms. J. Discrete Algorithms **9**(1), 12–31 (2011)
3. Dennis, C.: The brave new world of RNA. Nat. **418**(6894), 122–124 (2002)
4. Dimitrieva, S., Bucher, P.: Practicality and time complexity of a sparsified RNA folding algorithm. J. Bioinform. Comput. Biol. **10**(2), 1241007 (2012)
5. Hale, B.J., Yang, C.X., Ross, J.W.: Small RNA regulation of reproductive function. Mol. Reprod. Dev. **81**(2), 148–159 (2014)
6. Hofacker, I.L., Fontana, W., Stadler, P.F., Bonhoeffer, S., Tacker, M., Schuster, P.: Fast folding and comparison of RNA secondary structures. Monatshefte Chemie **125**, 167–188 (1994)
7. Jones, R., Lins, R.D.: Garbage Collection: Algorithms for Automatic Dynamic Memory Management. Wiley (1996)
8. Lorenz, R., Bernhart, S.H., Zu Siederdissen, C.H., Tafer, H., Flamm, C., Stadler, P.F., Hofacker, I.L.: ViennaRNA Package 2.0. Algorithms Mol. Biol. **6**(1), 26 (2011)
9. Mathews, D.H., Sabina, J., Zuker, M., Turner, D.H.: Expanded sequence dependence of thermodynamic parameters improves prediction of RNA secondary structure. J. Mol. Biol. **288**(5), 911–940 (1999)
10. Mattick, J.S., Makunin, I.V.: Non-coding RNA. Hum. Mol. Genet. **15**(suppl 1), R17–R29 (2006)
11. Möhl, M., Salari, R., Will, S., Backofen, R., Sahinalp, S.C.: Sparsification of RNA structure prediction including pseudoknots. Algorithms Mol. Biol. **5**(1), 39 (2010)
12. Nussinov, R., Jacobson, A.B.: Fast algorithm for predicting the secondary structure of single-stranded RNA. In: Proceedings of the National Academy of Sciences of the United States of America, vol. 77, issue no. 11, pp. 6309–6313 (1980)
13. Rastegari, B., Condon, A.: Parsing nucleic acid pseudoknotted secondary structure: algorithm and applications. J. Comput. Biol. **14**(1), 16–32 (2007)
14. Salari, R., Möhl, M., Will, S., Sahinalp, S.C., Backofen, R.: Time and space efficient RNA-RNA interaction prediction via sparse folding. In: Berger, B. (ed.) RECOMB 2010. LNCS, vol. 6044, pp. 473–490. Springer, Heidelberg (2010)
15. Sankoff, D.: Simultaneous solution of the RNA folding, alignment and protosequence problems. SIAM J. Appl. Math. **45**(5), 810–825 (1985). http://dx.doi.org/10.1137/0145048
16. Tinoco, I., Bustamante, C.: How RNA folds. J. Mol. Biol. **293**(2), 271–281 (1999)
17. Wexler, Y., Zilberstein, C., Ziv-Ukelson, M.: A study of accessible motifs and RNA folding complexity. J. Comput. Biol. J. Comput. Mol. Cell Biol. **14**(6), 856–872 (2007)
18. Zuker, M., Stiegler, P.: Optimal computer folding of large RNA sequences using thermodynamics and auxiliary information. Nucleic Acids Res. **9**(1), 133–148 (1981)
19. Zuker, M., Sankoff, D.: RNA secondary structures and their prediction. Bull. Math. Biol. **46**(4), 591–621 (1984). http://dx.doi.org/10.1007/bf02459506

A Sparsified Four-Russian Algorithm
for RNA Folding

Yelena Frid[(✉)] and Dan Gusfield

Department of Computer Science, U.C. Davis, Davis, USA
yafrid@ucdavis.edu

Abstract. The basic RNA Secondary Structure Prediction problem or Single Sequence Folding Problem (SSF) was solved thirty-five years ago by a now well-known $O(n^3)$-time dynamic programming method.

Recently three methodologies - Valiant, Four-Russians, and Sparsification - have been applied to speedup RNA Secondary Structure prediction.

In this paper we combine the previously independent speedups of Sparsification and Four-Russians.

The Sparsification method exploits two properties of the input: the number of subsequence Z with the endpoints belonging to the optimal folding set and the maximum number base-pairs L. These sparsity properties satisfy $0 \leq L \leq n/2$ and $n \leq Z \leq n^2/2$, and the method reduces the algorithmic running time to $O(LZ)$. In this paper, we first reformulate the SSF Four-Russians $\Theta(\frac{n^3}{\log^2 n})$-time algorithm, implied by Pinhas et al. [24], to utilize an *on-demand* lookup table. This formulation not only removes all extraneous computation and allows us to incorporate more realistic scoring schemes, but leads us to take advantage of the sparsity properties.

Our main result is a framework that combines the fastest Sparsification and fastest Four-Russians Methods. For SSF, this combined method has worst-case running time of $O(\tilde{L}\tilde{Z})$, where $\frac{L}{\log n} \leq \tilde{L} \leq min(L, \frac{n}{\log n})$ and $\frac{Z}{\log n} \leq \tilde{Z} \leq min(Z, \frac{n^2}{\log n})$.

Through asymptotic analysis and empirical testing on the base-pair maximization variant, we show that this framework is able to achieve a speedup on every problem instance, that is asymptotically never worse, and empirically better than achieved by the minimum of the two methods alone.

1 Introduction

Non-coding RNA (ncRNA) affects many aspects of gene expression, regulation of epigenetic processes, transcription, splicing, and translation [14]. It has been observed that in eukaryotic genomes the ncRNA function is more clearly understood from the structure of the molecule, than from sequence alone. While there have been advances in methods that provide structure experimentally, the need for computational prediction has grown as the gap between sequence availability and structure has widened. In general, RNA folding is a hierarchical process in

© Springer-Verlag Berlin Heidelberg 2015
M. Pop and H. Touzet (Eds.): WABI 2015, LNBI 9289, pp. 271–285, 2015.
DOI: 10.1007/978-3-662-48221-6_20

which tertiary structure folds on top of thermodynamically optimal[1] secondary structure, secondary structure is a key component of structure prediction [14].

Efficient $O(n^3)$-time dynamic programming algorithms were developed more than thirty years ago to find non-crossing secondary structure of a single RNA molecule with n bases [22,23,27,29,38,39]. We call this basic folding or single sequence folding (SSF) problem. In addition, McCaskill [19] created an $O(n^3)$-time algorithm for the *partition function* for RNA secondary structure. Based on these algorithms, software has been developed and widely used [15,18,25,36,37]. Probabilistic methods, employing Stochastic Context-Free Grammar (SFCG), were also developed to solve the basic folding problem [7,8].

The accuracy of all these methods is based on the parameters given by the scoring function. Thermodynamic parameters [16,17,28,33] and statistical parameters [6,7], or a combination of the two [2,13] are currently employed.

The Valiant [1,34], Sparsification [4,30], and the Four-Russians (FR) [9,24] methods where previously applied to improve on the computation time for secondary structure prediction. For SSF, the Valiant method achieves the asymptotic time bound of $O(\frac{n^3}{2^{\Omega \log(n)}})$ by incorporating the current fastest min/max-plus matrix multiplication algorithm [32,34]. The Four-Russians method was applied to single sequence [10,24], cofolding [11] and pseudoknotted [12] folding problems. The Sparsification method, was developed to improve computation time in practice for a family of RNA folding problems, while retaining the optimal solution matrix [4,20,21,26,30,35].

In this paper, we combine the Four-Russians method [24] and the Sparsification method [4]. While the former method reduces the algorithm's asymptotic running time to $\Theta(\frac{n^3}{\log^2 n})$, the latter eliminates many redundant computations. To combine these methods, we use an *on-demand* tabulation (instead of a preprocessing approach which is typically applied in FR algorithms), removing any redundant computation and guaranteing the combined method is at least as fast as each individual method, and in certain cases even faster. First, we reformulate SSF Four-Russians $\Theta(\frac{n^3}{\log^2 n})$-time algorithm [24] to utilizes *on-demand* lookup table creation. Second, we combine the fastest Sparsification and Four-Russians SSF speedup methods. The Sparse Four Russians speedup presented here leads to a practical and asymptotically fastest combinatorial algorithm(even in the worst-case). The new algorithm has an $O(\tilde{L}\tilde{Z})$ run time where $\frac{LZ}{\log^2 n} \leq \tilde{L}\tilde{Z} \leq \min(\frac{n^3}{\log^2 n}, LZ)$. In practice, when accounting for every comparison operation the Sparse Four Russians outperforms both the Four-Russians and Sparsification methods.

2   Problem Definition and Basic Algorithm

Let $s = s_0 s_1 \ldots s_{n-1}$ be an RNA string of length n over the four-letter alphabet $\Sigma = \{A, U, C, G\}$, such that $s_i \in \Sigma$ for $0 \leq i < n$. Let $s_{i,j}$ denote the substring $s_i s_{i+1} \ldots s_{j-1}$. We note that for simplicity of exposition substring $s_{i,j}$ does not

---

[1] Or close to optimal.

contain the nucleotide j. A *folding* (or a *secondary structure*) of s is a set M of position pairs (k, l), such that: (1) $0 \leq k < l < n$; (2) and there are no two different pairs $(k, l), (k', l') \in M$ such that $k \leq k' \leq l \leq l'$ (i.e. each position participates in at most one pair, and the pairs are non-crossing).

Let $\beta(i, j)$ return a score associated with position pair (i, j). Let $L(s, M)$ be the score associated with a folding M of RNA string s, and let $L(s)$ be the maximum score $L(s, M)$ over all foldings M of s. The **RNA Folding** or SSF problem is: given an RNA string s, compute $L(s)$, and find an optimal folding M such that $L(s, M) = L(s)$. In this work, we assume the following simple scoring scheme:

$$L(s, M) = \sum_{(i,j) \in M} \beta(i, j),$$

where $\beta(i, j) = 1$ if $(s_i, s_j) \in \{(A, U), (U, A), (C, G), (G, C)\}$, and $\beta(i, j) = 0$ otherwise. Richer scoring schemes allow more biologically significant information to be captured by the algorithm. However, the algorithms for solving the problem similar recurrences and other discrete scoring schemes may be accelerated in a similar way to what we present here.

For the folding M of $s_{i,j}$, an index $k \in (i, j)$ is called a *split point* in M if for every $(x, y) \in M$, either $y < k$ or $k \leq x$. A folding M is called a *partitioned folding* (with respect to $s_{i,j}$) if there exists at least one split point; otherwise M is called a *co-terminus folding*. Let the matrix L be a matrix such that $L[i, j] = L(s_{i,j})$. In addition, let $L^p[i, j]$ be the maximum value of $L(s_{i,j}, M)$ taken over all partitioned foldings M of $s_{i,j}$. Similarly, let $L^c[i, j]$ be the maximum value of $L(s_{i,j}, M)$ taken over all co-terminus foldings M of $s_{i,j}$. Let $L[i, i] = L[i, i + 1] = 0$. For all $j > i + 1$, $L[i, j]$ can be recursively computed as follows [23]:

$$L[i, j] = \max(L^p[i, j], L^c[i, j]), \tag{1}$$

$$L^p[i, j] = \max_{k \in (i,j)} (L[i, k] + L[k, j]), \tag{2}$$

$$L^c[i, j] = L[i + 1, j - 1] + \beta(i, j - 1). \tag{3}$$

For completeness, when $j < i$, define $L[i, j] = L^p[i, j] = L^c[i, j] = -\infty$.

The above recurrence may be efficiently implemented using a dynamic programming (DP) algorithm. Essentially, the DP algorithm computes and maintains values of the form $L[i, j], L^p[i, j]$ and $L^c[i, j]$ for every $0 \leq i \leq j \leq n$ in three $n + 1 \times n + 1$ matrices. The algorithm traverses the matrices in increasing column order index j from 1 to n. Within each column, the cell $L[k, j]$ is computed in decreasing index order k from $j - 1$ to 0. Once $L[k, j]$ is computed, $L^p[i, j]$ is updated for all $i < k$ such that $L^p[i, j] = max(L^p[i, j], L[i, k] + L[k, j])$. The solution $L(s, M)$ is stored in cell $L[0, n]$. Clearly, computing L^p is the bottleneck of the computation, since for a given i, j, there may be $\Theta(n)$ split points to examine.

2.1   Extending the Notation and Moving Towards a Vector by Vector Computation of L

For a matrix A and some integer intervals I, J, denote by $A[I, J]$ the sub-matrix of A obtained by projecting it onto the row interval I and column interval J. When $I = [i]$ or $J = [j]$, we simplify the notation by writing $A[i, J]$ or $A[I, j]$.

Definition 1. *For a set of integers K, define the notation $L^p_K[i, j]$, and the max-plus operation $\otimes$ as*

$$L^p_K[i, j] = L[i, K] \otimes L[K, j] = \max_{k \in K} (L[i, k] + L[k, j]).$$

For an interval $I = [i, i+1, \ldots i']$, define $L^p_K[I, j]$ to be the vector such that

$$L^p_K[I, j] = L[I, K] \otimes L[K, j] = \left[L^P_K[i, j] \text{ for all } i \in I \right]$$

We divide the solution matrix L in two ways: $q \times q$ submatrices (Fig. 1) and size q sub column vectors (the value of q will be determined later). Let $\boldsymbol{K_g}$ be the gth interval such that $K_g = \{q \cdot g, q \cdot g + 1, \ldots, q \cdot g + q - 1\}$. We call these sets *Kgroups*, and use K_g as the interval starting at index $g \cdot q$. For an index i, define $\boldsymbol{g_i} = \left\lfloor \frac{i}{q} \right\rfloor$. It is clear that $i \in K_{g_i}$.

Similarly, we break up the row indices into groups of size q, denoted by $\boldsymbol{I_g}$ where $I_g = \{k = q \cdot g, k + 1, \ldots k + q - 1\}$. (Clearly, row index set I_g is equivalent to the Kgroup K_g. We only introduce this extra notation for simplicity of the exposition).

Given this notation $L^P[i, j]$ can be rewritten as maximization $L^p_{K_g}[i, j]$ values for all K_g index Kgroups between i and j. However, in some cases, the indices $\{i+1, \ldots q \cdot g_{i+1} - 1\}$ do not form a full Kgroup K_{g_i}. Similarly indices $\{qg_j, qg_j + 1, \ldots j - 1\}$ do not form a full Kgroup K_{g_j}. Therefore, $L^P[i, j]$ can be computed by maximizing the full and non full Kgroups K_g. In Eq. 4 and the following sections we do not explicitly differentiate between full and non full groups.

$$L^P[i, j] = \max_{g_i \leq g \leq g_j} L^P_{K_g}[i, j] \tag{4}$$

We extend the notation further, to compute the matrix L^P not cell by cell but instead by vectors of size q corresponding to the $I_{g'}$ row sets, as follows.

$$L^P[I_{g'}, j] = \max_{g' \leq g \leq g_j} L^P_{K_g}[I_{g'}, j]. \tag{5}$$

The DP algorithm can be updated to incorporate the extended notation. Within each column, compute the matrices in vectors of size q. Once $L[K_g, j]$ is computed it is used in computation of $L^p_{K_g}[I_{g'}, j]$ for $g' < g$. When computing $L^p_{K_{g'}}[I_{g'}, j]$ we follow Eqs. 1–3 to complete the computation of cells $L[I_{g'}, j]$.

3   Sparsification of the SSF Algorithm

The Sparsification method achieves a speedup by reducing the number of split points examined during the computation of $L^P[i, j]$. In this section we give a brief overview of the Sparsification method applied to SSF [4,30].

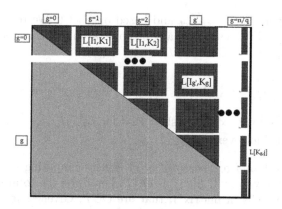

Fig. 1. An example of how a solution matrix L is broken down into submatrices. Using the extended vector notation we can say that cell $L[i, j]$ belongs to the vector $L[K_{g_i}, j]$ as well as submatrix $L[I_{g_i}, K_{g_j}]$. We partition the solution matrix L into $O(n^2/q)$ vectors of size $O(q)$ and $O(n^2/q^2)$ submatrices, of size $O(q^2)$.

3.1 OCT and STEP Sub-instances of Sequence s

Sub-instance $s_{i,j}$ is optimally co-terminus (OCT) if every optimal folding of $s_{i,j}$ is co-terminus. We introduce the extra notation below

if $L[i, j] = L^c[i, j] > L^p[i, j]$ then we say $L[i, j]$ is OCT.

Sub-instance $s_{i,j}$ is $STEP$, if $L[i, j] > L[i + 1, j]$ where $L[i, j] = L(s_{i,j})$ and $L[i + 1, j] = L(s_{i+1,j})$. For ease of exposition we also say $L[i, j]$ is $STEP$ when $s_{i,j}$ is $STEP$. A $STEP$ sub-instance $s_{i,j}$ implies that nucleotide i is paired in every optimal folding of $s_{i,j}$.

Fact 1. *For every sub-instance $s_{i,j}$ with $j > i$ there is an optimal split point $k \in (i, j)$ such that either $k = i + 1$ or $L[i, k]$ is $STEP$ and $L[k, j]$ is OCT [4].*

Notation: For the index set $K = \{k, k + 1, ...k'\}$ and column j, let K^{oct_j} be the set of indices such that $K^{oct_j} \subset K$ and $\forall_{k \in K^{oct_j}} L[k, j]$ is OCT. Given the row interval $I = \{i, i + 1, ...i'\}$, let I^{step_k} be the set of rows such that $I^{step_k} \subset I$, and for all $i \in I^{step_k} L[i, k]$ is $STEP$.

We further define operation $\otimes_{step-oct}$ such that given $I = \{i, i + 1, ..., i'\}$ and $K = \{k, k + 1, ..., k'\}$, $L[I, K] \otimes_{step-oct} L[K, j]$ results in $A[I, j]$ where $\forall_{i \in (I^{step_k} \cup I^{step_{k+1}} \cup ... I^{step_{k'}})} A[i, j]$ is computed by the following procedure:

for $k \in K^{oct_j}$ **do**
$\quad \forall_{i \in I^{step_k}} A[i, j] = \max(L[i, k] + L[k, j], A[i, j])$

Using the operation $\otimes_{step-oct}$ and based on Fact 1. We reduce the time to compute $L^p[I_{g'}, j]$ by considering a split-point k only if $k = i + 1$ or $L[i, k]$ is $STEP$ and $L[k, j]$ is OCT for $i \in I_{g'}$ and $k \in (i, j)$.

$$L^p[I_{g'}, j] = \max_{g' \leq g \leq g_j} L^p_{K_g}[I_{g'}, j] = \max_{g' \leq g \leq g_j} L[I_{g'}, K_g] \otimes_{step-oct} L[K_g, j]. \qquad (6)$$

Note Eq. 6 does not explicitly show that for $L^p_{K_{g'}}[I_{g'}, j]$ the split-point $i + 1$ must be examined for every $i \in I_{g'}$.

Asymptotic time bound of Sparsified SSF. When computing matrix $L^p[i, j]$ we examine value $L[i, k]$ only if $L[k, j]$ is OCT. Let Z, be the total number of sub-instances in s or cells in matrix L that are OCT. Given that $L[k, j]$ is OCT, $L^p[i, j]$ must examine the split point k, for all $i \in \{0, 1, ...k\}$ such that $L[i, k]$ is $STEP$. Let $\boldsymbol{L}$ be the total number of $STEP$ sub-instances in column k. More precisely $L = |\{0, 1, ...k\}^{step_k}|$ (Creating the list of split-points that correspond to $STEP$ incidence requires no additional computation time [4]). The total time to compute SSF when examining only $STEP$, OCT combinations (Sparsification method), is $O(LZ)$. As shown in Backofen et al. [4] Z is bounded by $Z \leq n^2$ and L is bounded by $L \leq \frac{n}{2}$. The overall asymptotic time bound of the Sparsification method is $O(LZ)$ remains $O(n^3)$.

4 On-demand Four Russians Speedup

Presented here is an *on-demand* version of the $\Omega(\log^2 n)$-time Four-Russians algorithm implied by Pinhas et al. [24].

Observation 4.1. *The scores stored in $L[k, j]$ and $L[k+1, j]$ differ by the effect of adding only one more nucleotide (i.e., s_k). Therefore, $L[k, j] - L[k + 1, j]$ belongs to a finite set of differences $\mathbb{D}$, where $\mathbb{D}$ is the set of scores created as the result of the scoring scheme β. The cardinality of the set of differences, $D = |\mathbb{D}|$, is $O(1)$ when β is discrete. For the simple β scoring function (+1 for every permitted pair, and 0 otherwise), the set $\mathbb{D}$ is equal to $\{0, 1\}$ and therefore $|\mathbb{D}| = 2$ [23].*

Let $\boldsymbol{x} = [x_0, x_1, \ldots, x_{q-1}]$ be an integer vector of length q. We say that $\boldsymbol{x}$ is D-discrete if $\forall_{l \in (0,q)} |x_{l-1} - x_l| \in \mathbb{D}$. We define the Δ-*encoding* of 2-discrete vector $\boldsymbol{x}$ to be a pair of integers (x_0, Δ_x) such that x_0 is the first element in $\boldsymbol{x}$ and Δ_x is the integer representation of the binary vector $[x_0 - x_1, x_1 - x_2, \ldots, x_{q-2} - x_{q-1}]$. Note that $0 \leq \Delta_x < 2^{q-1}$. For simplicity, we will interchangeably use $\boldsymbol{x}$ to imply either (x_0, Δ_x) or $[x_0, x_1, \ldots, x_{q-1}]$. Clearly, Δ-*encoding* takes $O(q)$ time to compute.

Δ-*encoding vector operations:*

- Let $(x_0, \Delta_x) + c = (x_0 + c, \Delta_x)$ be equivalent to $\boldsymbol{x} + c = [x_0 + c, x_1 + c, \ldots, x_{q-1} + c]$.
- Let $B \otimes (x_0, \Delta_x)$ be equivalent to $B \otimes \boldsymbol{x}$.
- Let $\max((x_0, \Delta_x), (y_0, \Delta_y))$ be equivalent to $\max(\boldsymbol{x}, \boldsymbol{y})$.

MUL Lookup Table. Based on Observation 4.1, any column vector in matrix L is 2-discrete. Given vector $L[K_g, j]$ and its Δ-encoding ($x_0 = L[gq, j]$, $\Delta_x = \Delta_{L[K_g, j]}$), it is clear that $\Delta_x \in [0, 2^q - 1]$.

Fact 2. $L[I_{g'}, K_g] \otimes L[K_g, j]$ *is equivalent to* $L[I_{g'}, K_g] \otimes (0, \Delta_{L[K_g, j]}) + L[gq, j]$ *[24].*

Let $MUL_B[i]$ be a lookup table, where given a $q{\times}q$ submatrix $B = L[I_{g'}, K_g]$ and $i = \Delta_{L[K_g, j]}$, the entry $MUL_{L[I_{g'}, K_g]}[\Delta_{L[K_g, j]}] = (y_0, \Delta_y)$ where $\mathbf{y} = L[I_{g'}, K_g] \otimes (0, \Delta_{L[K_g, j]})$. We could reformulate the computation of $L^p_{K_g}[I_{g'}, j]$ to utilize the MUL lookup table.

$$L^p_{K_g}[I_{g'}, j] = L[I_{g'}, K_g] \otimes L[K_g, j] = MUL_{L[I_{g'}, K_g]}[\Delta_{L[K_g, j]}] + L[gq, j]. \quad (7)$$

Equation 7, abstracts the detail that we still have to compute each referenced entry in the MUL lookup table. Each entry in the MUL lookup table is computed *on-demand* i.e. only when it corresponds to a required calculation. (This removes any extraneous calculation incurred when preprocessing all possible entries as in the typical Four-Russians implementation.) If entry $MUL_{L[I_{g'}, K_g]}[\Delta_{L[K_g, j]}]$ does not exist we compute $L[I_{g'}, K_g] \otimes (0, \Delta_{L[K_g, j]})$ directly in $O(q^2)$ time. If entry $MUL_{L[I_{g'}, K_g]}[\Delta_{L[K_g, j]}]$ exists then the operation is $O(1)$-time lookup.

There are $O(\frac{n^2}{q^2})$ submatrices within L. For each submatrix the maximum number of entries we compute for lookup table MUL is 2^{q-1}. In total, the asymptotic time bound to populate lookup table MUL is $O(\frac{n^2}{q^2} \cdot 2^{q-1} \cdot q^2) = O(n^2 \cdot 2^q)$.

MAX Lookup Table. Let the *max* of two 2-discrete q-size vectors $\mathbf{v}$ and $\mathbf{w}$, denoted $max(\mathbf{v}, \mathbf{w})$, result in a q-size vector $\mathbf{z}$, where $\forall_{0 \leq k < q}\, z_k = \max(v_k, w_k)$. Without loss of generality, let $w_0 \geq v_0$. Comparing the first element in each vector there are two possibilities either (1) $w_0 - v_0 > q - 1$ or (2) $w_0 - v_0 \leq q - 1$. In the first case, ($w_0 - v_0 > q - 1$), it is clear that $max(\mathbf{v}, \mathbf{w})$ is equal to $\mathbf{w}$. In the second case, we make use of the following fact [24].

Fact 3. *Given two vectors* (w_0, Δ_w) *and* (v_0, Δ_v), *if* $w_0 - v_0 \leq q - 1$ *then* $max(\mathbf{v}, \mathbf{w}) = \max((0, \Delta_v), (w_0 - v_0, \Delta_w)) + v_0$.

Lets define lookup table MAX such that entry $MAX[i, i', h] = \max((0, i), (h, i'))$. Hence, we reformulate Fact 3. To incorporate the MAX lookup table:

$$max(\mathbf{v}, \mathbf{w}) = MAX[\Delta v_0, \Delta w_0, (w_0 - v_0)] + v_0$$

We summarize these results in the function $\Delta \max$:
Function $\Delta \max ::$

    input: $\mathbf{v}, \mathbf{w}$ such that $w_0 \geq v_0$ and $\mathbf{v} = (v_0, \Delta_v)$ and $\mathbf{w} = (w_0, \Delta_w)$
    output: $\mathbf{z} = (z_0, \Delta_z)$ where $\forall_{i \in [0,q)} z_i = \max(v_i, w_i)$
    if($w_0 - v_0 \geq q - 1$) : $\mathbf{z} = \mathbf{w}$
    else :
                    $\mathbf{z} = MAX[\Delta v_0, \Delta w_0, (w_0 - v_0)] + v_0$

In Eq. 8, below, we integrate the vector comparison function $\Delta \max$. Each vector $L^p[I_{g'}, j]$ is computed by maximizing over $O(n/q)$ vectors. We will compute the lookup table MAX on-demand for every entry that does not exist an $O(q)$. Clearly the lookup table MAX will contain at most $2^{(q-1)} \cdot 2^{(q-1)} \cdot q$ entries. In worst case, the lookup table MAX computes in $O(2^{q^2} q)$ time.

$$L^p[I_{g'}, j] = \Delta \max_{g' \leq g \leq g_j} \left(MUL_{L[I_{g'}, K_g]}[\Delta_{L[K_g, j]}] + L[gq, j] \right) \tag{8}$$

The matrix L^p and hence L is solved by a total of $O(\frac{n^2}{q})$ computations of Eq. 8. In total, given lookup table MUL and MAX, the time to compute the Four-Russians SSF is $O(\underbrace{\frac{n^3}{q^2}}_{computation} + \underbrace{2^{q^2} q + n^2 2^q}_{\text{on-demand lookup table}})$.

Setting $q = \epsilon \log n$, where $\epsilon \in (0, .5)$ [31], the total computation time is equal to $\Theta(\frac{n^3}{\log^2 n})$, which achieves a speedup by a factor of $\Omega(\log^2 n)$, compared to the original $O(n^3)$-time solution method.

5   Sparse Four-Russian Method

With the Four-Russians method, a speedup is gained by reducing q split point index comparisons for q subsequences to a single $O(1)$ time lookup. The Sparsification method reduces the comparison to only those indices which correspond to $STEP\text{-}OCT$ folds.

5.1 $STEP - OCT$ Condition for Sets of Split Points

In this section, we achieve a Sparsified Four-Russian speedup for the computation of the L^p matrix. As in the Four Russians Method, we will conceptually break up the solution matrix L in two ways: in $q \times q$ size submatrices, and q size subcolumn vectors. The submatrices are indexed by g' and g such that the corresponding submatrix is $L[I_{g'}, K_g]$. The subcolumn vectors are indexed by g and j, such that the corresponding subcolumn vector is $L[K_g, j]$.

We augment the Four-Russians SSF to reduce the number of entries, and lookups into the MUL table. If and only if, the matrix $L[I_{g'}, K_g]$ contains at least one cell $L[i, k]$ that is $STEP$ and within vector $L[K_g, j]$ the cell $L[k, j]$ is OCT we will lookup $MUL_{L[I_{g'}, K_g]}[\Delta_{L[K_g, j]}]$. If such an entry does not exist we will compute $L[I_{g'}, K_g] \otimes (0, \Delta_{L[K_g, j]})$ and store the result into lookup table MUL.

The following notation will be used to help determine if a split point Kgroup should be examined in the computation.

OCT subcolumn vector. Given the vector $L[K_g, j]$ let m be a q size binary vector such that $\forall_{0 \leq x \leq q-1} m[x] = 1$ if $L[gq + x, j]$ is OCT. Let the $sigOct$ of the vector $L[K_g, j]$, written $sigOct(L[K_g, j])$, be equal to m the integer representation of the binary vector m. Clearly $0 \leq m < 2^q$, and if $m > 0$ then $L[K_g, j]$ contains at least one OCT instance. Let $O(\tilde{Z})$ be the total number of subcolumn vectors which contain an instance that is OCT. Clearly, $\frac{Z}{q} \leq \tilde{Z} \leq \min(\frac{n^2}{q}, Z)$.

$STEP$ submatrix. Given the submatrix $L[I_{g'}, K_g]$, let $\boldsymbol{m'}$ be a q size binary vector such that $\forall_{x \in [0,q)} m'[x] = 1$ if $\exists_{0 \leq i \leq q-1} L[qg' + i, qg + x]$ is $STEP$. Let $sigStep$ of a submatrix, written $sigStep(L[I_{g'}, K_g])$, be equal to $\boldsymbol{m'}$ the integer representation of the binary vector $\boldsymbol{m'}$. Clearly $0 \leq m' < 2^q$. Let $\tilde{L}$ be the total number of submatrices which contain an instance that is $STEP$ within $L[[0, n], K_g]$. Clearly, $\frac{L}{q} \leq \tilde{L} \leq \min(\frac{n}{q}, L)$.

Observation 5.1. *Suppose that, $s_{i,k}$ is $STEP$, and integer*
$m' = sigStep(L[I_{g'}, K_g])$ *such that $i \in I_{g'}$ (or $I_{g'} = I_{g_i}$) and $k \in K_g$ (or $K_g = K_{g_k}$). Then, the corresponding binary vector $\boldsymbol{m'}$ must be set to 1 in position x where x is an index such that $k = qg + x$. More precisely, if $L[i, k]$ is $STEP$ then $m'[x] = 1$ by the definition of sigStep.*

Observation 5.2. *Suppose $s_{k,j}$ is OCT, and suppose integer*
$m = sigOct(L[K_g, j])$ *such that $k \in K_g$. Then, the corresponding binary vector $\boldsymbol{m}$ must be set to 1 in position x, where x is an index such that $k = qg + x$. More precisely, if $s_{k,j}$ is OCT then $m[x]=1$ by the definition of sigOct.*

Given two binary vectors v and w the *dot product* of their integer representation is equal to a binary number x such that $x = v \odot w = v_0 \wedge w_0 \vee v_1 \wedge w_1 \vee \ldots \vee v_{q-1} \wedge w_q$ where $|v| = |w| = q - 1$.

Theorem 1. *For any subinstance $s_{i,j}$ either $i + 1$ is the optimal split point, or there is an optimal split point $k \in (i, j)$, such that $sigStep(L[I_{g_i}, K_{g_k}]) \odot sigOct(L[K_{g_k}, j])$ equals 1.*

Proof. Based on Fact 1 for any sub-instance $s_{i,j}$ there is an optimal split point k such that either $k = i + 1$ or $s_{i,k}$ is $STEP$ and $s_{k,j}$ is OCT. If $s_{i,k}$ is $STEP$ and $s_{k,j}$ is OCT then $L[i, k]$ is $STEP$ and $L[k, j]$ is OCT. The cell $L[i, k]$ belongs to submatrix $L[I_{g_i}, K_{g_k}]$ and the cell $L[k, j]$ belongs to the vector $L[K_{g_k}, j]$. Let x be an index such that $k = qg_k + x$. Let $\boldsymbol{m'}$ be a binary vector that corresponds to $sigStep(L[I_{g_i}, K_{g_k}])$. Based on Observation 5.1, $m'[x]$ must equal 1. Let $\boldsymbol{m}$ be the binary vector that corresponds to $sigOct(L[K_{g_k}, j])$. Based on Observation 5.2, $m[x]$ equals 1. Therefore, $m[x] \wedge m'[x] = 1$ and $sigStep(L[I_{g_i}, K_g]) \odot sigOct(L[K_g, j]) = 1$.

Notation: The index g is $STEP$-OCT if given the set of rows $I_{g'}$ and the column j if $sigStep(L[I_{g'}, K_g]) \odot sigOct(L[K_g, j]) = 1$.

We can reformulate the computation of $L^p[I_{g'}, j]$ by referencing the lookup table MUL only if g is $STEP$-OCT. This reduces the number of operations used in computing the bottleneck L^P matrix.

$$L^p[I_{g'}, j] = \underset{\substack{g \text{ is } STEP\text{-}OCT \\ \text{where } g \in [g', g_j]}}{\Delta \max} \left(MUL_{L[I_{g'}, K_g]}\left[\Delta_{L[K_g, j]}\right] + L[gq, j] \right) \qquad (9)$$

We update the DP algorithm to only access the MUL lookup table for matrix and vector combinations that satisfy the property
$sigStep(\, L[I_{g'}, K_g]\,) \odot sigOct(\, L[K_g, j]\,) = 1.$

Let G be a lookup table, where give an index $g \in [0, n/q]$ and integer $m \in [0, 2^q]$ the $G[g][m] \subset \{I_0, I_1, \ldots, I_g\}$ is a set of row index intervals. Each index $I_{g'}$ within G[g][m] satisfies the following condition:

$$\text{if } I_{g'} \in G[g][m] \text{ then } sigStep(\, L[I_{g'}, K_g]\,) \odot m\,) = 1.$$

Lookup table G (updated *on-demand*) allows us to implement Eq. 9. As $L[K_g, j]$ is computed, the corresponding $SigOct$ is also computed. Let $m = sigOct(L[K_g, j])$. By iterating through $I_{g'} \in G[g][m]$ set of row indices we access table MUL only when both of the following conditions hold at the same time: the submatrix $L[I_{g'}, K_g]$ contains at least one cell $L[i, k]$ where $s_{i,k}$ is $STEP$ and within vector $L[K_g, j]$ the cell $L[k, j]$ contains $s_{k,j}$ that is OCT (where $i \in I_{g'}$ and $k \in K_g$).

The Sparsified Four-Russian algorithm implements Eq. 9. The *complete* function will tabulate $STEP$, and OCT instances as well as $sigStep$ and $sigOct$ values. The G, MUL and MAX lookup tables will be computed *on-demand* (Fig. 2).

Sparsified Four-Russian Folding s
input: An RNA string s of length n.
outout: $L(s)$.

---

for $j \leftarrow 2$ to n do
    $L[K, j]$=complete $(L^p[K, j])$ where $K = \{qg_j, qg_j + 1, \ldots, j - 1\}$
    for $I \in \{I_{g_j-1}, I_{g_j-2}\ldots I_0\}$ and if $sigOct(L[K, j]) > 0$ do
        let $L^p[I, j]$= the Δ-encoding of $L[I, K] \otimes_{step-oct} L[K, j]$
    for $g \leftarrow g_j - 1$ to 0 do
        $L[K_g, j]$=complete$(L^p[K_g, j])$
        if $(sigOct(L[K_g, j]) > 0)$ then
            let $m = sigOct(L[K_g, j])$
            if $G[g][m]$ does not exist then
                compute entry $G[g][m]$ // *on-demand update*
            let $x = (x_0, \Delta_x)$ where $x_0 = L[gq, j]$ and $\Delta_x = \Delta_{L[K_g, j]}$
            for $I_{g'} \in G[g][m]$ do
                if $MUL_{L[I_{g'}, K_g]}[\Delta_x]$ does not exist then
                    compute $MUL_{L[I_{g'}, K_g]}[\Delta_x]$ // *on-demand update*
                $L^p_{K_g}[I_{g'}, j] = MUL_{L[I_{g'}, K_g]}[\Delta_x] + x_0;$
                $L^p[I_{g'}, j]= \Delta \max(L^p_{K_g}[I_{g'}, j], L^p[I_{g'}, j])$

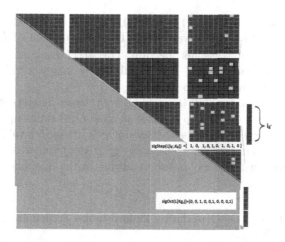

Fig. 2. whetherare $STEP-OCT$. The yellow cells indicate $STEP$ instances. The red cells indicate OCT instances. The $L[I_{g'}, K_g] \otimes L[K_g, j]$ operation is only performed on submatrices with $sigSTEP \odot sigOCT > 0$.

Complete

input: $L^P[K, j]$ where j is the column index, and $K = [k, k+1, ...k+q']$
   with $q' < q$

outout: $L(s_{k,j}), L(s_{k+1,j}), ..., L(s_{k+q',j})$ or $L[K, j]$

1: **for** $k' = k + q'$ to $k' = k$ **do**
2:      $L^c[k', j] = \quad L[k'+1, j-1] + \beta(k', j-1)$
3:      $L[k', j] = \quad \max(L^c[k', j], L^P[k', j])$
4:      **if** $(L[k', j] > L^P[k', j])$ **then**
5:          $s_{k',j}$ is OCT ; add k' to list OCT in column j
6:          update $sigOct(L[K_{g_k}, j])$ such that $m[k' - k] = 1$ *where* $k' \in K_{g_k}$
7:      **if** $(L[k', j] > L[k', k'] + L[k'+1, j])$ **then**
8:          $s_{k',j}$ is $STEP$; add k' to list $STEP$ in column j
9:          update $sigStep(L[I_{g'}, K_g])$ such that $k' \in I_{g'}$ and $j \in K_g$
10:     **if** $s_{k',j}$ is OCT **then**
11:         **for** $i \in \{k'-1, k'-2, ..., k\} \cap STEP$ in column k' **do**
12:             $L^P[i, j] = \Delta \max(L^P[i, j], L[i, k'] + L[k', j])$
13: **return** $L[K, j]$

Asymptotic Analysis of Sparsified Four-Russians. We assume $O(1)$-time RAM access for $\log(n)$ bits. The calculation for column j can be broken down into $L^P_{K=[qg_j,j)}[i, j]$ and $L^P_{K=[0,qg_j)}[i, j]$ for all $i < j$. The computation of $L^P_{[qg_j,j)}[[0, n], j]$ occurs when Kgroup K_{g_j} is not full, and follows the Sparsification algorithm maximizing over $STEP - OCT$ split points only. This reduces the comparisons made from $O(n \cdot q)$ to $O(L\tilde{q})$ where $\tilde{q} < q$ is the total number OCT instances within the interval $[qg,j)$. The computation of $L^P_{[0,qg_j)}[[0,n],j]$ employs Sparsified Four Russians speedup. The MUL table entries are created and references only for $STEP - OCT$ submatrix vector combinations. This reduces the comparisons made to $O(\tilde{L}\tilde{Z})$.

The helper function *complete* is called $O(n^2/q)$ times for the entire algorithm. The *complete* function outer-loop iterates at most $O(q)$ times updating the lists of OCT and $STEP$ split points, as well as $sigOct$ and $sigStep$ values. Overall the *complete* function takes $O(q + \tilde{x})$ where $\tilde{x} \leq q^2$ is the number of $STEP - OCT$ instance combinations. The asymptotic runtime of the Sparsified Four-Russian algorithm is

$$O(\tilde{L}\tilde{Z}) + O(\frac{n^2}{q} \cdot \tilde{x}) + O(\text{updating lookup tables on-demand}) = O(\tilde{L}\tilde{Z})$$

5.2   Asymptotic Analysis of On-demand Lookup Tables Calculation

We compute the lookup tables G, MUL, and MAX *on-demand*. For each vector $L[K_g, j]$ containing an OCT instance (where $m = sigOct(L[K_g, j])$), if $G[g][m]$ does not exist then we directly compute it. For the computation of a single entry into lookup table G, we iterate through $O(\tilde{L})$ submatrices and compute the dot product in $O(q)$ time[2]. In total, an update is called to lookup table G at most $O(\tilde{C} = min(2^q, \tilde{Z}))$ times. The entire G lookup table on-demand computation takes $O(\text{on-demand}G) = O(\tilde{L}\tilde{C} \cdot q)$ or $\boldsymbol{O(G)} \leq O(\min(\tilde{L}2^q, \tilde{L}\tilde{Z}) \cdot q) \leq O(min(\frac{n2^q}{q}, \frac{LZ}{q}))$.

For each vector containing an OCT instance if an entry doesn't exist in the lookup table MUL it is computed on-demand. Each entry takes $O(\tilde{L} \cdot q^2)$ time to compute. There are $min(2^q, \tilde{Z})$ such computation. In total, lookup table MUL takes $O(\tilde{L}q^2 \cdot min(2^q, \tilde{Z}))$-time. Setting $q = \epsilon \log n$ where $\epsilon \in (0, .5)$ the asymptotic run-time for *on-demand* computation is $O(\tilde{L}\tilde{Z})$.

The entire algorithm takes $O(\tilde{L}\tilde{Z})$ where $\frac{LZ}{\log^2 n} \leq \tilde{L}\tilde{Z} \leq min(\frac{n^3}{\log^2 n}, LZ)$.

5.3   Empirical Results

We tested 20 randomly generated sequences for each size $N = 64, 128, 256, 512$.

The empirical testing results are given not in seconds but in the number of operations including both lookup table creation and split-point comparisons. We do so to abstract from the effect compiler optimizations. Note that the testing does not account for memory access time, or extend the algorithm to $D > 2$ scoring schemes (Table 1).

For $N = 128$ the Sparse Four-Russians(SFR) algorithm performs 25% less comparisons than the Sparsified(SP) SSF algorithm and 80% less comparison than the Four-Russians (FR) algorithm. In all test cases, the Sparse Four-Russians performed better than the minimum of either method alone.

---

[2] Using some word tricks the dot product could be computed in $O(1)$-time.

Table 1. Number of all comparisons computed

Size	$O(n^3)$	FR	SP	SFR
64	43,680	12,014	2,733	1,837
128	349,504	49,456	13,196	9,982
256	2,796,160	346,692	79,544	41,393
512	22,500,863	5,746,853	650,691	503,425

5.4   Future Work

The *on-demand* tabling enables the Four-Russians method to be efficiently (in terms of time) applied to $D > 2$. It would be interesting to test this method for a more biologically informative scoring scheme. It would also be interesting to examine the ability to sparsify memory [3], as Four-Russians at worst case requires an additional factor of $2^{log(n)}$ in memory. Another open question is wether it is possible to apply the $\Omega(\log^3 n)$ [5] speedup of boolean matrix multiplication to RNA folding. Lastly, the Four-Russians speedup presented here easily lends itself to an $O(n^2/log^2n)$-time parallel formulation, by augmenting the algorithm to compute in parallel for each column j. It would be interesting to extend the Sparse Four-Russians to a parallel architecture.

Acknowledgement. We would like to sincerely thank Shay Zakov and Michal Ziv-Ukelson for their many helpful comments and suggestions. This research was partially supported by the IIS-1219278 grant.

References

1. Akutsu, T.: Approximation and exact algorithms for RNA secondary structure prediction and recognition of stochastic context-free languages. J. Comb. Optim. **3**(2–3), 321–336 (1999)
2. Andronescu, M., Condon, A., Hoos, H.H., Mathews, D.H., Murphy, K.P.: Efficient parameter estimation for RNA secondary structure prediction. Bioinformatics **23**(13), i19–i28 (2007)
3. Backofen, R., Tsur, D., Zakov, S., Ziv-Ukelson, M.: Sparse RNA folding: time and space efficient algorithms. In: Kucherov, G., Ukkonen, E. (eds.) CPM 2009 Lille. LNCS, vol. 5577, pp. 249–262. Springer, Heidelberg (2009)
4. Backofen, R., Tsur, D., Zakov, S., Ziv-Ukelson, M.: Sparse RNA folding: time and space efficient algorithms. J. Discrete Algorithms **9**(1), 12–31 (2011)
5. Chan, T.: Speeding up the Four Russians algorithm by about one more logarithmic factor. In: SODA, pp. 212–217 (2015)
6. Do, C.B., Woods, D.A., Batzoglou, S.: Contrafold: RNA secondary structure prediction without physics-based models. Bioinformatics **22**(14), e90–e98 (2006)
7. Dowell, R., Eddy, S.: Evaluation of several lightweight stochastic context-free grammars for RNA secondary structure prediction. BMC Bioinform. **5**(1), 71 (2004)

8. Durbin, R., Eddy, S., Krogh, A., Mitchison, G.: Biological Sequence Analysis: Probabilistic Models of Proteins and Nucleic Acids. Cambridge University Press, Cambridge (1998)

9. Frid, Y., Gusfield, D.: A simple, practical and complete $O(\frac{n^3}{\log n})$-time algorithm for RNA folding using the *Four-Russians* speedup. In: Salzberg, S.L., Warnow, T. (eds.) WABI 2009. LNCS, vol. 5724, pp. 97–107. Springer, Heidelberg (2009)

10. Frid, Y., Gusfield, D.: A simple, practical and complete O(n³/log(n))-time algorithm for RNA folding using the [Four-Russians] speedup. Algorithms Mol. Biol. **5**(1), 13 (2010)

11. Frid, Y., Gusfield, D.: A worst-case and practical speedup for the RNA co-folding problem using the *Four-Russians* idea. In: Moulton, V., Singh, M. (eds.) WABI 2010. LNCS, vol. 6293, pp. 1–12. Springer, Heidelberg (2010)

12. Frid, Y., Gusfield, D.: Speedup of RNA pseudoknotted secondary structure recurrence computation with the Four-Russians method. In: Lin, G. (ed.) COCOA 2012. LNCS, vol. 7402, pp. 176–187. Springer, Heidelberg (2012)

13. Juan, V., Wilson, C.: RNA secondary structure prediction based on free energy and phylogenetic analysis. J. Mol. Biol. **289**(4), 935–947 (1999)

14. Leontis, N.B., Westhof, E.: RNA 3D Structure Analysis and Prediction, vol. 27. Springer, Heidelberg (2012)

15. Markham, N.R., Zuker, M.: UNAFold. In: Keith, J.M. (ed.) Bioinformatics. Methods in Molecular Biology, vol. 453, pp. 3–31. Humana Press, New York (2008)

16. Mathews, D.H., Disney, M.D., Childs, J.L., Schroeder, S.J., Zuker, M., Turner, D.H.: Incorporating chemical modification constraints into a dynamic programming algorithm for prediction of RNA secondary structure. Proc. Natl. Acad. Sci. U.S.A. **101**(19), 7287–7292 (2004)

17. Mathews, D.H., Sabina, J., Zuker, M., Turner, D.H.: Expanded sequence dependence of thermodynamic parameters improves prediction of RNA secondary structure. J. Mol. Biol. **288**(5), 911–940 (1999)

18. Mathews, D.H., Andre, T.C., Kim, J., Turner, D.H., Zuker, M.: An updated recursive algorithm for RNA secondary structure prediction with improved thermodynamic parameters. Mol. Model. Nucleic Acids **682**, 246–257 (1998)

19. McCaskill, J.S.: The equilibrium partition function and base pair binding probabilities for RNA secondary structure. Biopolymers **29**(6–7), 1105–1119 (1990)

20. Møhl, M., Salari, R., Will, S., Backofen, R., Sahinalp, S.C.: Sparsification of RNA structure prediction including pseudoknots. Algorithms Mol. Biol. **5**, 39 (2010)

21. Möhl, M., Salari, R., Will, S., Backofen, R., Sahinalp, S.C.: Sparsification of RNA structure prediction including pseudoknots. In: Moulton, V., Singh, M. (eds.) WABI 2010. LNCS, vol. 6293, pp. 40–51. Springer, Heidelberg (2010)

22. Nussinov, R., Jacobson, A.B.: Fast algorithm for predicting the secondary structure of single-stranded RNA. PNAS **77**(11), 6309–6313 (1980)

23. Nussinov, R., Pieczenik, G., Griggs, J.R., Kleitman, D.J.: Algorithms for loop matchings. SIAM J. Appl. Math. **35**(1), 68–82 (1978)

24. Pinhas, T., Zakov, S., Tsur, D., Ziv-Ukelson, M.: Efficient edit distance with duplications and contractions. Algorithms Mol. Biol. **8**, 27 (2013)

25. Reuter, J., Mathews, D.H.: RNAstructure: software for RNA secondary structure prediction and analysis. BMC Bioinform. **11**(1), 129 (2010)

26. Salari, R., Möhl, M., Will, S., Sahinalp, S.C., Backofen, R.: Time and space efficient RNA-RNA interaction prediction via sparse folding. In: Berger, B. (ed.) RECOMB 2010. LNCS, vol. 6044, pp. 473–490. Springer, Heidelberg (2010)

27. Sankoff, D., Kruskal, J.B., Mainville, S., Cedergreen, R.J.: Fast algorithms to determine RNA secondary structures containing multiple loops. In: Sankoff, D., Kruskal, J.B. (eds.) Time Warps, String Edits and Macromolecules: The Theory and Practice of Sequence Comparison, pp. 93–120. Addison-Wesley, Reading (1983)
28. Tinoco, I., Borer, P.N., Dengler, B., Levine, M.D., Uhlenbec, O.C., Crothers, D.M., Gralla, J.: Improved estimation of secondary structure in ribonucleic-acid. Nat. New Biol. **246**(150), 40–41 (1973)
29. Waterman, M.S., Smith, T.F.: RNA secondary structure: a complete mathematical analysis. Math. Biosci. **42**, 257–266 (1978)
30. Wexler, Y., Zilberstein, C.: A study of accessible motifs and RNA folding complexity. J. Comput. Biol. **14**(6), 856–872 (2007)
31. Williams, R.: Matrix-vector multiplication in sub-quadratic time: (some preprocessing required). In: Bansal, N., Pruhs, K., Stein, C. (eds.) SODA, pp. 995–1001. SIAM (2007)
32. Williams, R.: Faster all-pairs shortest paths via circuit complexity. In: Symposium on Theory of Computing, STOC 2014, New York, NY, USA, May 31–June 03 2014, pp. 664–673 (2014)
33. Xia, T., SantaLucia, J., Burkard, M.E., Kierzek, R., Schroeder, S.J., Jiao, X., Cox, C., Turner, D.H.: Thermodynamic parameters for an expanded nearest-neighbor model for formation of RNA duplexes with watson-crick base pairs. Biochemistry **37**(42), 14719–14735 (1998)
34. Zakov, S., Tsur, D., Ziv-Ukelson, M.: Reducing the worst case running times of a family of RNA and CFG problems, using valiant's approach. In: Moulton, V., Singh, M. (eds.) WABI 2010. LNCS, vol. 6293, pp. 65–77. Springer, Heidelberg (2010)
35. Ziv-Ukelson, M., Gat-Viks, I., Wexler, Y., Shamir, R.: A faster algorithm for RNA co-folding. In: Crandall, K.A., Lagergren, J. (eds.) WABI 2008. LNCS (LNBI), vol. 5251, pp. 174–185. Springer, Heidelberg (2008)
36. Zuker, M.: The use of dynamic programming algorithms in RNA secondary structure prediction. In: Waterman, M.S. (ed.) Mathematical Methods for DNA Sequences, pp. 159–184. CRC Press Inc., Boca Raton (1989). Chapter 7
37. Zuker, M.: Mfold web server for nucleic acid folding and hybridization prediction. Nucleic Acids Res. **31**(13), 3406–3415 (2003)
38. Zuker, M., Sankoff, D.: RNA secondary structures and their prediction. Bull. Math. Biol. **46**(4), 591–621 (1984)
39. Zuker, M., Stiegler, P.: Optimal computer folding of large RNA sequences using thermodynamics and auxiliary information. Nucleic Acids Res. **9**(1), 133–148 (1981)

Higher Classification Accuracy of Short Metagenomic Reads by Discriminative Spaced k-mers

Rachid Ounit and Stefano Lonardi[✉]

Department of Computer Science and Engineering, University of California, Riverside, CA 92521, USA
{rouni001,stelo}@cs.ucr.edu

Abstract. The growing number of metagenomic studies in medicine and environmental sciences is creating new computational demands in the analysis of these very large datasets. We have recently proposed a time-efficient algorithm called CLARK that can accurately classify metagenomic sequences against a set of reference genomes. The competitive advantage of CLARK depends on the use of discriminative *contiguous* k-mers. In default mode, CLARK's speed is currently unmatched and its precision is comparable to the state-of-the-art, however, its sensitivity still does not match the level of the most sensitive (but slowest) metagenomic classifier. In this paper, we introduce an algorithmic improvement that allows CLARK's classification sensitivity to match the best metagenomic classifier, without a significant loss of speed or precision compared to the original version. Finally, on real metagenomes, CLARK can assign with high accuracy a much higher proportion of short reads than its closest competitor. The improved version of CLARK, based on discriminative *spaced* k-mers, is freely available at http://clark.cs.ucr.edu/Spaced/.

Keywords: Metagenomics · Microbiome · Classification · Discriminative spaced k-mers · Short metagenomic reads

1 Introduction

One of the primary goals of metagenomic studies is to determine the composition of a microbial community, which typically involves the analysis of short reads obtained from sequencing a heterogenous microbial sample. The analysis can reveal the presence of unknown bacteria and viruses in a newly explored microbial habitat (e.g., in marine environment [24]), or in the case of the human body, elucidate relationships between diseases and imbalances in the microbiome (see, e.g., [7,10]).

Classification tools such as NBC [21], KRAKEN [25], CLARK [19], among others, can be used to determine the composition of the microbial diversity from the sequenced reads for a microbial sample. We have recently proposed CLARK in [19] and demonstrated that its classification speed is currently unmatched.

© Springer-Verlag Berlin Heidelberg 2015
M. Pop and H. Touzet (Eds.): WABI 2015, LNBI 9289, pp. 286–295, 2015.
DOI: 10.1007/978-3-662-48221-6_21

Independently from us, it has been shown that CLARK's classification precision is comparable or better than best state-of-the-art classifiers [15]. However, CLARK's classification sensitivity is inferior compared to NBC [19].

The work presented in this manuscript describes a new approach to improve CLARK's classification sensitivity. The approach exploits the concept of (discriminative) spaced k-mers. We first describe the notion of spaced k-mers as implemented in a new mode called CLARK-S (S for "spaced"), then compare the performance of CLARK-S against two of the most sensitive classifiers in the literature (i.e., NBC and KRAKEN), on several simulated/real metagenomic datasets. We show that at the phylum/genus level CLARK-S outperforms both NBC and KRAKEN on all metrics.

2   Classification by Discriminative Spaced k-mers

2.1   Preliminaries

The concept and the utility of spaced seeds were initially described in context of a sequence-alignment tool called PATTERNHUNTER [17]. A *spaced seed s* is a string over the alphabet $\{1,*\}$, where '1' indicates that one should sample that position while '*' indicates that position should be ignored. The number of symbols in s is the *length* $|s|$ of s, while the number of 1s in s is the *weight* of s. A *spaced k-mer* is a spaced seed of length k. Let s be a spaced k-mer and weight w, and let m be a text of length k. We define $s(m)$ be the w-mer obtained from m using only the positions in s denoted by a 1. For example, if the text $m =$ AAGTCT and $s = $ 11*1*1 ($k = 6, w = 4$) then $s(m) = $ AATT. The same text processed using the spaced 6-mer $s = $ 1*11*1 would give the 4-mer $s(m) = $ AGTT.

The work of Ma *et al.* in [17] demonstrated that the use of single (and multiple) spaced seeds/k-mers significantly increased the chance of detecting a valid sequence alignment between the query and the target compared to contiguous seeds/k-mers, while incurring no additional computational cost. As a direct consequence of this work, spaced seeds are now used in the state-of-the-art homology search methods, such as BLAST [1] or MEGABLAST [26]. For more information about spaced seeds, we also refer the reader to [5,6,11–14] and references therein.

Consider now the following problem: we are given a read r and two target sequences g_1 and g_2, and we want to classify r to g_1 or g_2, *i.e.*, we want to know whether r is more likely to originate from g_1 or from g_2. As it is done in homology search methods, we can use seeds/k-mers as "witnesses" of possible valid alignments. A time-efficient solution is to count the number shared k-mers between r and targets g_1 and g_2, and assign r to the target that has the highest count. As said, spaced seeds/k-mers increases the probability of detecting a valid alignment compared to contiguous seeds/k-mers. It is always possible, however, that a shared seed/k-mer (whether it is spaced or not) may be a false positive. In order to compensate for false positives, we use discriminative spaced k-mers, as described next.

2.2   Discriminative Spaced k-mers

Given a set of reference sequences (or *targets*) $\{g_1, g_2, \ldots, g_p\}$, $i \in \{1, 2, \ldots, p\}$, the set D_i of discriminative k-mers for target g_i is the set of all k-mers in g_i that do not appear in any other reference sequences [19]. Given a spaced seed s of length k and weight w, we define $D_{i,s}$ to be the set of all w-mers obtained via s from k-mers in D_i. We then define the set $E_{i,s}$ of discriminative spaced k-mers as the set of all w-mers of $D_{i,s}$ that do not appear in any set $D_{j,s}$ where $j \neq i$. Thus, any w-mer in $E_{i,s}$ is a spaced k-mer of weight w that can be found in one and only one target.

As stated earlier, the concept of spaced k-mers is not new. Several popular metagenome classifiers, such as METAPHYLER [16], PHYMMBL [4] or MEGAN [9], as BLAST-based methods, have been implicitly using spaced seeds. In addition, other similarity-based methods that analyze genomic and metagenomic sequences use spaced k-mers, such as SEED [2]. However, to the best of our knowledge, the concept of discriminative spaced k-mers is novel and introduced for the first time in this manuscript.

2.3   Selection of Optimal Spaced Seeds and Index Creation

The selection of specific spaced seed is critical to achieve high precision and sensitivity (see, e.g., [5,6,11–14,17]). For contiguous k-mers, the classification precision increases as we increase k. However, the highest sensitivity occurs with somewhat shorter k-mers. CLARK is more precise for long contiguous k-mers (e.g., $k = 31$), but the highest sensitivity occurs for k-mers of length 19–22 [19]. As a consequence, we considered here spaced seeds of length $k = 31$ and weight $w = 22$. The choice of selecting a length of 31 is also motivated by a fair comparison against CLARK and KRAKEN, which achieve high accuracy thanks to long 31-mers in their default mode. However, we realize that a more exhaustive analysis of k and w would be necessary, but (i) the intent of this work is to show the advantage of replacing discriminative contiguous seed with discriminative spaced seed, (ii) an analysis of other choices of w will be reported in the journal version of this paper.

Given k and w, the second step is to determine the structure of the spaced seed. In order to determine the optimal structure we proceeded to model sequence similarly as it is done in alignments-based method (see, e.g., [17]). We considered that the succession of matches/mismatches follows a Bernoulli distribution with parameter p, where p represents the similarity level between the read and the reference sequence. If a short read belongs to a known reference sequence, then the similarity level should be high since the amount of mismatches dues to genomic variations or sequencing errors are low. This is why we assumed a high similarity level, and chose $p = 95\%$.

We searched exhaustively through all the spaced seeds of length $k = 31$ and weight $w = 22$ (starting/ending with '1') using a similarity level of 95%, and a random region of length 100 bp, by using the dynamic programming approach from [17] and implemented in [12]. The spaced seed with the highest hit probability [17], 0.998113, is 1111*111*111**1*111**1*11*11111. In addition, we have also selected two additional spaced seeds with the highest hit probability namely 11111*1**111*1*11*11**111*11111 (0.998099) and finally 11111*1*111**1*11*111**11*11111 (0.998093).

Before a read can be classified, CLARK-*S* builds a database of discriminative spaced *k*-mers for each target. CLARK-*S* can take advantage of multiple spaced seeds, thus multiple databases can be created. For each spaced seed, discriminative spaced *k*-mers were built from contiguous discriminative 31-mers. Once the three databases of discriminative spaced *k*-mers were computed, they are stored in disk so they can be loaded for classification.

The classification algorithm of the "Spaced" mode is identical to that of the "full" mode (extensively described in [19]), except for two differences, namely (i) CLARK-*S* queries against discriminative spaced *k*-mers instead of discriminative *k*-mers and (ii) CLARK-*S* does three queries for each *k*-mer in a read, because there are three different databases. Finally, as done in the full and other modes, the read is assigned to the target that has the highest amount of successful queries, and several statistics (such as the confidence score and gamma score, see [19]) are computed as well.

3 Results

3.1 Datasets

To evaluate numerically the performance of the classifiers we used simulated datasets. From the available literature, we have selected the following three simulated metagenomes, which we made available at http://clark.cs.ucr.edu/Spaced/. The first dataset is "A1.10.1000" which was derived from "A1", the first group of paired-end reads in the dataset "A" from [15]. According to authors, this dataset closely mimics the complexities, size and characterization of real metagenomes. The A1 dataset contains about 28.9M reads, 80 % of which correspond to known sequenced genomes (from bacterial, archaeal and eukaryotes genomes), and 20 % of which are randomized reads (from real genomes) that should not be assigned to any taxa. We have extracted 10,000 reads from A1 as follows. We have arbitrarily taken nine different genomes from the list of genomes used to build "A1" (see Supplementary Table 1 in [15]). Then, we took the first 1,000 reads for each selected genome, and also 1,000 "random" reads. The resulting dataset, called "A1.10.1000", contains 10,000 reads (each 100 bp long) and can be considered as medium/high complexity.

The second dataset is "B1.20.500" which was derived from "B1", the first group of reads in the dataset "B", from [15]. Similarly as done for A1.10.1000, we have extracted 10,000 reads from B1 as follows. We have arbitrarily taken 19 different genomes from the list of genomes used to build "B1" (see Supplementary Table 2 in [15]). Note that these 19 selected genomes are different from those selected in A1. Then we took the first 500 reads for each selected genome, and also 500 "random" reads. The resulting dataset, called "B1.20.500", contains 10,000 reads (each 100 bp long) and can be considered as medium/high complexity.

The third dataset "simBA-5" comes from the KRAKEN paper and is described in it. According to the authors, it was created using bacterial and archaeal genomes, and with an error rate five times higher than the default. It contains 10,000 reads, each read is 100 bp long, and can be considered as high complexity.

To classify these metagenomic datasets, we use the entire set of bacterial/archaeal genomes from NCBI/RefSeq as reference genomes. At the time of writing, they represent 2,644 genomes and distributed in 36 phyla. The cumulative length of these genomes is 9.1 billion base pairs, where the average genome length is 3.4 million base pairs.

3.2 Comparison with Other Tools

A large set of metagenomic classifiers exists in the literature. However, a comparison between CLARK and all existing classifiers is not necessary. An independent comprehensive evaluation of a wide range of metagenomics classifiers has been carried out recently using six large datasets of short paired-end reads [15]. On the data tested, KRAKEN is among the most accurate methods at the phylum level compared to other popular and used methods, such as MOTU [23], METAPHLAN [22], METAPHYLER or MEGAN. However, the experimental results in [25] shows that NBC is more sensitive than KRAKEN, MEGABLAST and PHYMMBL at the genus level. In our study [19], we have also shown that NBC is more sensitive than KRAKEN at the genus level. In addition, NBC is more sensitive than CLARK, at the genus level, even when the latter is run in its most sensitive settings (*i.e.*, "full" mode and $k = 20$) [19]. Note that the study [3] also shows the high sensitivity of NBC. As a consequence of this analysis, it appears sufficient to compare CLARK against NBC and KRAKEN, as they are the two most accurate classifiers among current published methods, at the phylum and genus level.

3.3 Classification Accuracy

In this section, we present the performance of CLARK (v1.2.1-beta), NBC (v1.1) and KRAKEN (v0.10.5-beta) on the three simulated datasets described above. Consistently with other published studies (e.g., [19,25] or [3]), the sensitivity is defined as the ratio between the number of correct assignments at a given taxonomy rank (e.g., phylum or genus) and the number of reads defined for that rank. The precision is defined as the ratio between the number of correct assignments at a given taxonomy rank (e.g., phylum or genus) and the number of assigned reads.

We present below results for the phylum and genus level. In Tables 1 and 2, the first three rows report results from KRAKEN CLARK, and NBC, all run in their default/recommended parameters. We ran KRAKEN and CLARK in the default mode, with $k = 31$, and NBC, with $k = 15$. The last two rows in these tables report the performance of CLARK-S. In the last row we report the precision and sensitivity when filtering only high confidence (HC) assignments (*i.e.*, assignment with confidence score ≥ 0.75 and gamma score ≥ 0.03).

Table 1. Phylum-level accuracy (%) of KRAKEN, NBC, CLARK, CLARK-S and CLARK-S (HC) on A1.10.1000, B1.20.500 and simBA-5

	A1.10.1000		B1.20.500		simBA-5	
	Precision	Sensitivity	Precision	Sensitivity	Precision	Sensitivity
KRAKEN	99.91	77.59	99.98	90.91	99.98	94.49
CLARK	**99.93**	76.87	**100.00**	90.12	99.99	93.46
NBC	79.86	79.86	94.91	94.91	99.89	**99.89**
CLARK-S	94.50	**79.99**	98.95	**94.98**	99.87	99.70
CLARK-S (HC)	99.63	79.97	99.99	94.93	**100.00**	99.29

Table 2. Genus-level accuracy (%) of KRAKEN, NBC, CLARK, CLARK-S and CLARK-S (HC) on A1.10.1000, B1.20.500 and simBA-5

	A1.10.1000		B1.20.500		simBA-5	
	Precision	Sensitivity	Precision	Sensitivity	Precision	Sensitivity
KRAKEN	**99.80**	70.61	99.94	90.55	**99.85**	91.97
CLARK	**99.80**	69.98	**99.95**	89.69	99.82	90.77
NBC	77.94	77.94	94.76	**94.76**	98.97	**98.97**
CLARK-S	92.71	**78.38**	98.76	94.74	98.58	98.22
CLARK-S (HC)	99.35	76.41	**99.95**	94.52	99.61	97.24

Observe in Table 1 that (i) CLARK-S (HC) and NBC achieve very high sensitivity, (ii) KRAKEN's sensitivity is lower than NBC or CLARK-S for all datasets, (iii) CLARK-S outperforms NBC's sensitivity in A1.10.1000 and B1.20.500, (iv) both CLARK and KRAKEN have high precision and achieve more than 99.9 % in all datasets (even though A1.10.1000 and B1.20.500 contain reads that do not belong to any bacterial/archaeal genomes), but (v) CLARK-S (HC) is as precise as them and outperforms NBC in all datasets.

Table 2 shows that (i) CLARK's sensitivity is lower than NBC, (ii) CLARK-S (HC) and NBC achieve the highest sensitivity and outperforms KRAKEN, (iii) CLARK-S is more NBC in A1.10.1000, (iv) KRAKEN and CLARK show high precision and achieve both more than 99.8 % in our datasets, (v) CLARK-S (HC) is as precise as KRAKEN and CLARK, it outperforms NBC in all datasets, especially for A1.10.100 or B1.20.500. For simBA-5, NBC achieves the best sensitivity with 98.97, less than 2 % more than the level performed by CLARK-S (HC).

3.4 Real Metagenomic Samples

In this section, we evaluate the performance of CLARK-S (HC) on a large real metagenomic dataset. We have selected the dataset from [18], which is a recently published study on the population dynamics in microbial communities present in surface seawater in Monterey Bay, CA.

This dataset contains 42M reads, and the average read length is 510 bp. We pre-processed the dataset of raw reads using the following trimming steps: (i) we removed the first five bases and kept the following 100 bases using FASTQ Trimmer[1], (ii) we removed reads containing sequencing adapters using Scythe[2], (iii) we trimmed the read ends if contained bases with a quality score below 30 and discarded reads containing any Ns using Sickle[3]. The resulting dataset contained 37M short reads.

We classified these 37M short reads using KRAKEN (default) and CLARK-S, using the bacterial/archaeal genomes from NCBI/RefSeq. KRAKEN was able to classify only 1,1M reads (or 3 % of the total). CLARK in its default mode also classifies about 1,1M reads. However, CLARK-S classifies 20M reads (or 54 % of the total), about 20 times more than KRAKEN. Among these 20M classified reads, there are 7M high confidence assignments (or 19 % of the total), which is about 6 times more than KRAKEN.

The fact that KRAKEN assigns only 3 % of the reads can be explained by the fact that (i) KRAKEN relies on matching exact k-mer, and (ii) the current database of bacterial/archaeal likely contains only a limited fraction of the bacterial/archaeal diversity in seawater. Seawater metagenomes are likely to contain a high proportion of organisms that are missing in NBCI/RefSeq database because while the marine environment is one of the most biologically diverse on the planet [8], the culture in laboratory of bacteria from seawater is difficult [20]. Since CLARK-S allows mismatches on the k-mers, it can identify at least the phylum/genus of unknown organisms.

KRAKEN identified, as dominant phyla, *Proteobacteria* (57 %) and *Bacteroides* (27 %). This is consistent with results reported in [18], as well as phyla in low-abundance such as *Actinobacteria* (1 %) or *Thaumarchaeota* (2 %). Within high confidence assignments of CLARK-S, the two dominant phyla are, as expected by estimations from [18], *Proteobacteria* (56 %) and *Bacteroides* (32 %). Consistently with [18], phyla in low-abundance were correctly identified, for example, *Actinobacteria* (1 %) and *Thaumarchaeota* (2 %).

Experimental results from KRAKEN and CLARK-S (HC) indicate the expected dominant phyla in the dataset (with the expected abundance for each). While KRAKEN and CLARK-S (HC) are consistent for this dataset, we do notice one significant disagreement. The expected abundance of *Cyanobacteria* is 0–2 %, according to [18], but KRAKEN reports 9 % and CLARK-S (HC) reports 3 %. Such discrepancies can be explained by our pre-processing to create this dataset, however, the estimation by CLARK-S (HC) is more accurate than KRAKEN. As a consequence, CLARK-S was able to assign about 20 times more short reads than KRAKEN, and its high confidence assignments show stronger consistency with expected results than KRAKEN's results.

---

[1] http://hannonlab.cshl.edu/fastx_toolkit/index.html.

[2] https://github.com/ucdavis-bioinformatics/scythe.

[3] https://github.com/ucdavis-bioinformatics/sickle.

3.5   Time and Space Complexity

All experiments presented in this study were run on a Dell PowerEdge T710 server (dual Intel Xeon X5660 2.8 Ghz, 12 cores, 192 GB of RAM). NBC's speed is the slowest at 8–9 reads per minute, KRAKEN's speed is 1.8–2M reads per minute, while CLARK (default mode) runs the fastest, at 2.8–3M reads per minute. However, CLARK-S runs slower than CLARK, and classifies about 150–200 thousand reads per minute. While CLARK is the fastest in the default mode, it does not provide the same classification accuracy of NBC or CLARK-S. The fact that CLARK-S computes spaced k-mers and uses several spaced seeds explains this difference of speed. However, CLARK-S is still several thousand of times faster than NBC.

NBC consumed less than 500 MB of RAM, while CLARK and KRAKEN used 70 and 77 GB respectively. Finally, CLARK-S used 110 GB. This larger RAM usage is due to the multiple databases corresponding to the three spaced seeds. However, this amount remains significantly lower than 160 GB, which is the amount needed to build/construct the database of discriminative k-mers.

4   Discussion

We have introduced for the first time the use of discriminative spaced k-mers for the classification problem of short metagenomic reads. To the best of our knowledge, CLARK is the first metagenome classifier using (multiple) discriminative spaced k-mers. We have tested CLARK-S against CLARK, KRAKEN and NBC.

Our results on several realistic metagenomic samples show that (i) CLARK/ KRAKEN achieves high precision while being less sensitive than NBC at the phylum/genus level, (ii) NBC achieves high sensitivity while being less precise than the other tools, however, (iii) CLARK-S (HC) can be *both* as precise as (or more precise than) KRAKEN and as sensitive as NBC. While CLARK-S is slower than CLARK because its uses mutiple spaced seeds, it is still faster than NBC by several order of magnitude. Finally, in the context of real metagenomic data, we proved that CLARK-S (HC) can classify with high accuracy a much higher proportion of short reads than CLARK/KRAKEN.

We are currently improving the speed and the RAM usage of CLARK-S. A public release of CLARK-S is available at http://clark.cs.ucr.edu/Spaced/.

Acknowledgments. This work was supported in part by the U.S. National Science Foundation [IIS-1302134]. We are thankful to the anonymous reviewers for their constructive feedback.

References

1. Altschul, S.F., Gish, W., Miller, W., Myers, E.W., Lipman, D.J.: Basic local alignment search tool. J. Mol. Biol. **215**(3), 403–410 (1990)
2. Bao, E., Jiang, T., Kaloshian, I., Girke, T.: Seed: efficient clustering of next-generation sequences. Bioinformatics **27**(18), 2502–2509 (2011)

3. Bazinet, A.L., Cummings, M.P.: A comparative evaluation of sequence classification programs. BMC Bioinformatics **13**(1), 92 (2012)
4. Brady, A., Salzberg, S.: PhymmBL expanded: confidence scores, custom databases, parallelization and more. Nat. Methods **8**(5), 367–367 (2011)
5. Brown, D.G., Li, M., Ma, B.: A tutorial of recent developments in the seeding of local alignment. J. Bioinform. Comput. Biol. **2**(04), 819–842 (2004)
6. Choi, K.P., Zeng, F., Zhang, L.: Good spaced seeds for homology search. In: Proceedings of Fourth IEEE Symposium on Bioinformatics and Bioengineering, BIBE 2004, pp. 379–386. IEEE (2004)
7. Human Microbiome Project Consortium: A framework for human microbiome research. Nature **486**(7402), 215–221 (2012)
8. Felczykowska, A., Bloch, S.K., Nejman-Falenczyk, B., Baranska, S.: Metagenomic approach in the investigation of new bioactive compounds in the marine environment. Acta Biochim. Pol. **59**, 501–505 (2012)
9. Huson, D.H., Auch, A.F., Qi, J., Schuster, S.C.: MEGAN analysis of metagenomic data. Genome Res. **17**(3), 377–386 (2007)
10. Huttenhower, C., Gevers, D., Knight, R., Abubucker, S., Badger, J., Chinwalla, A., et al.: Structure, function and diversity of the healthy human microbiome. Nature **486**(7402), 207–214 (2012)
11. Ilie, L., Ilie, S.: Multiple spaced seeds for homology search. Bioinformatics **23**(22), 2969–2977 (2007)
12. Ilie, L., Ilie, S., Bigvand, A.M.: Speed: fast computation of sensitive spaced seeds. Bioinformatics **27**(17), 2433–2434 (2011)
13. Li, M., Ma, B., Kisman, D., Tromp, J.: Patternhunter ii: highly sensitive and fast homology search. J. Bioinform. Comput. Biol. **2**(03), 417–439 (2004)
14. Li, M., Ma, B., Zhang, L.: Superiority and complexity of the spaced seeds. In: Proceedings of the Seventeenth Annual ACM-SIAM Symposium on Discrete Algorithm. Society for Industrial and Applied Mathematics, pp. 444–453 (2006)
15. Lindgreen, S., Adair, K.L., Gardner, P.: An Evaluation of the Accuracy and Speed of Metagenome Analysis Tools. Cold Spring Harbor Laboratory Press (2015). doi:10.1101/017830
16. Liu, B., Gibbons, T., Ghodsi, M., Treangen, T., Pop, M.: Accurate and fast estimation of taxonomic profiles from metagenomic shotgun sequences. BMC Genomics **12**(Suppl 2), S4 (2011)
17. Ma, B., Tromp, J., Li, M.: Patternhunter: faster and more sensitive homology search. Bioinformatics **18**(3), 440–445 (2002)
18. Mueller, R.S., Bryson, S., Kieft, B., Li, Z., Pett-Ridge, J., Chavez, F., Hettich, R.L., Pan, C., Mayali, X.: Metagenome sequencing of a coastal marine microbial community from Monterey Bay, California. Genome Announc. **3**(2), e00341-15 (2015)
19. Ounit, R., Wanamaker, S., Close, T.J., Lonardi, S.: Clark: fast and accurate classification of metagenomic and genomic sequences using discriminative k-mers. BMC Genomics **16**(1), 236 (2015)
20. Pace, N.R.: Mapping the tree of life: progress and prospects. Microbiol. Mol. Biol. Rev. **73**(4), 565–576 (2009)
21. Rosen, G.L., Reichenberger, E.R., Rosenfeld, A.M.: NBC: the naive bayes classification tool webserver for taxonomic classification of metagenomic reads. Bioinformatics **27**(1), 127–129 (2011)
22. Segata, N., Waldron, L., Ballarini, A., Narasimhan, V., Jousson, O., Huttenhower, C.: Metagenomic microbial community profiling using unique clade-specific marker genes. Nat. Methods **9**(8), 811–814 (2012)

23. Sunagawa, S., Mende, D.R., Zeller, G., Izquierdo-Carrasco, F., Berger, S.A., Kultima, J.R., Coelho, L.P., Arumugam, M., Tap, J., Nielsen, H.B., et al.: Metagenomic species profiling using universal phylogenetic marker genes. Nat. Methods **10**(12), 1196–1199 (2013)
24. Venter, J.C., Remington, K., Heidelberg, J.F., Halpern, A.L., Rusch, D., Eisen, J.A., Wu, D., Paulsen, I., Nelson, K.E., Nelson, W., et al.: Environmental genome shotgun sequencing of the Sargasso Sea. Science **304**(5667), 66–74 (2004)
25. Wood, D., Salzberg, S.: Kraken: ultrafast metagenomic sequence classification using exact alignments. Genome Biol. **15**(3), R46 (2014)
26. Zhang, Z., Schwartz, S., Wagner, L., Miller, W.: A greedy algorithm for aligning DNA sequences. J. Comput. Biol. **7**(1–2), 203–214 (2000)

Graph-Theoretic Modelling of the Domain Chaining Problem

Poly H. da Silva[1], Simone Dantas[1], Chunfang Zheng[2],
and David Sankoff[2 (✉)]

[1] Institute of Mathematics and Statistics, Fluminense Federal University,
Niterói, Brazil
[2] Department of Mathematics and Statistics, University of Ottawa,
Ottawa, Canada
sankoff@uottawa.ca

Abstract. Methods for the clustering of genes into homologous families
(sets of genes descending from a single gene in an ancestral organism)
are susceptible to the inappropriate merging of unrelated families, called
domain chaining. We give formal criteria for the chaining effect by defining multiple alternative clique relaxation and path relaxation models and
the relationships among them, involving different graph characteristics.
We implement these definitions and apply them to 45 flowering plant
genomes in order to compare the Markov Cluster Algorithm (MCL) and
Soft Cliques with Backbones (SCWiB) clustering method. In the process
we note the extreme behavior of the *Amborella trichopoda* genome.

1 Introduction

A gene family is a set of genes, in one genome or several, that includes all
descendants of a single gene in an ancestral organism. The genes in a family are
called homologous. The goal of gene family classification is to partition a set of
sequences into homologous families. In practice, gene families are constructed
on the basis of DNA or protein sequence similarities under the assumption that
genes in the same family will retain more sequence similarity than unrelated
genes. Many methods are currently available for the clustering of genes into
families. However these methods are susceptible to the inappropriate merger
of unrelated families, due to the multiple domain structure of many proteins.
Some domains, more or less lengthy sequence fragments, recur in many different
families with largely distinct histories and functions, blurring the boundaries
between these families. This problem is called the *domain chaining effect* [4,5];
it stems from the evolutionary acquisition of widespread protein modules that
may help in the binding or movement of the protein but generally not its specific
primary enzymatic, synthetic, signaling or regulatory function.

Gene family classification has often been studied using graph concepts
[3,13,14]. A theoretical model of this problem can be obtained in the following

Partially supported by FAPERJ, CNPq, CAPES and NSERC. DS holds the Canada
Research Chair in Mathematical Genomics.

M. Pop and H. Touzet (Eds.): WABI 2015, LNBI 9289, pp. 296–307, 2015.
DOI: 10.1007/978-3-662-48221-6_22

way: each gene is identified with a vertex v of undirected graph $G_S = (V, E)$, where there exists an edge $\{u, v\} \in E$ between two vertices u and v if the pair of genes exceeds a threshold similarity score. Graph G_S is called a *similarity graph*; ideally each maximal clique in G_S corresponds to a single gene family, and vice versa, and a long path may represent a chaining effect (Fig. 1).

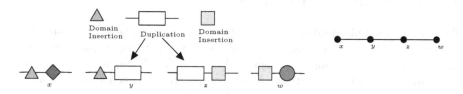

Fig. 1. The evolutionary history of a hypothetical multidomain family showing both gene duplication and domain insertions. Genes y and z share a common ancestor but do not have identical domain composition. Genes x and w share homologous domains with genes y and z, respectively, but there is no gene that is ancestral in all of x, y, z and w.

There is a large literature on clique relaxation models, where not all the elements are "directly connected" to each other [9]. These models are useful in applied contexts where connections between members of a group need not be direct and could be meaningfully accomplished through intermediaries. Clique relaxation models are obtained by allowing the clique property to be relaxed in various ways. Some examples are: *s-clique* — where the distance between vertices within the group must be at most s [6]; *s-club* — where the diameter of the graph induced by the group must be at most s [2]; *s-plex* — where the number of non-neighbors among elements of the group is bounded [11]; *k-core* where a certain minimum number k of neighbors within the group is guaranteed [10]; *s-defective clique* — which differs from a clique by at most s missing edges [12]; γ-*quasi-clique* — ensures a certain minimum ratio γ of the number of existing links to the maximum possible number of links within the group [1]. Although clique relaxation models have proved useful in many applications, there has been no formal study of domain chaining associated to these methods.

In this present paper, we also give formal criteria for the chaining effect in terms of a number of alternative *path* relaxation models and the relationships among them, involving different graph characteristics. We define the α-*quasi-path* that ensures a certain minimum ratio α of the diameter to number of existing links; *k-chain* that is relative to average degree; and (x, y)-*damaged-path* that takes into account the number of missing and extra edges to be dealt with in order to turn the graph into a path.

We use these cluster and path relaxation definitions in comparing two methods for generating gene families: *Markov Cluster Algorithm (MCL)* [3], one of the most widely used procedures for inferring gene families; and *Soft Cliques With Backbones (SCWiB)* [14], a new method that ensures that clusters satisfy a tolerant edge-density criterion that takes into account cluster size. We perform the comparisons on 45 published angiosperm genome sequences.

In Sect. 2 we give some basic graph theory definitions and formalize our proposed new path relaxation definitions. In Sect. 3 we implement the definitions and apply them to 45 genomes in order to compare the MCL and SCWiB methods. Finally, in Sect. 4, we summarize our results.

2   Definitions and Notations

We denote a simple undirected graph by $G = (V, E)$, where V is the set of vertices and E is the set of edges. Given a vertex $v \in V$, we denote the degree of v by $d(v)$ and the minimum degree of G by $\delta(G)$. A clique of a graph is a set of mutually adjacent vertices. A path of length r between vertices u and v in G is a subgraph of G defined by an alternating sequence of distinct vertices $u \equiv v_0, v_1, ..., v_{r-1}, v_r \equiv v$ such that $\{v_i, v_{i+1}\} \in E$ for all $1 \leq i \leq r - 1$. We denote by P_n the graph that is a path with n vertices. The distance $dist(u, v)$ between two vertices u and v of a connected graph is the length of the shortest path connecting them. The eccentricity of a vertex v, denoted by $\varepsilon(v)$, in a connected graph G, is defined to be the maximum distance between v and any other vertex u of G. Then we say the diameter of G, $diam(G)$, is the maximum value of $\varepsilon(v)$ over all vertices $v \in V$. The density $\rho(G)$ of G is the ratio of the number of edges to the total number of possible edges, i.e., $\rho = \frac{2|E|}{|V|(|V|-1)}$.

Next, some of clique relaxation models, which were already mentioned in the previous section, are formally defined. We assume that $G = (V, E)$ is connected, $\gamma \in (0, 1]$ is real and the constant s is positive integer.

Definition 1 (*s*-plex). *G is an s-plex if* $\delta(G) \geq |V| - s$.

Definition 2 (γ-quasi-clique). *G is a γ-quasi-clique if* $\rho(G) \geq \gamma$.

Definition 3 (*s*-defective-clique). *G is a s-defective-clique if G contains at least* $\frac{|V|(|V|-1)}{2} - s$ *edges.*

In order to study the domain chaining effect, we introduce some path relaxation definitions, each of which measures, in some sense, how close a graph is to being a path. We assume that $G = (V, E)$ is connected, constants k, x and y are positive integers and $\alpha \in (0, 1]$ is real.

Definition 4 (α-quasi-path). *G is an α-quasi-path if* $\frac{diam(G)}{|E|} \geq \alpha$.

Definition 4 ensures a certain minimum ratio α of the diameter to the number of edges. Note that this definition is more pertinent to chaining than the definition of an *s*-club, since in addition to considering the diameter, it also considers the total number of edges and the ratio between them.

Observe that for $\alpha = 1$, G is an α-quasi-path if and only if it is a path, and that the minimum value for α is $\frac{2}{|V|(|V|-1)}$, which occurs when G is a complete graph. If G contains an α-quasi-path, where α is close to 1, the graph is highly chained, i.e., similar in structure to a path.

Definition 5 (*k*-chain). *G is a k-chain if k is the smallest integer such that*

$$k \geq \frac{\sum_{v \in V \backslash \{u,w\}} d(v)}{|V|-2} = \frac{2|E|-d(u)-d(w)}{|V|-2}, \text{ where } u \text{ and } w \text{ are the vertices of smallest}$$

degree and $d(u) \leq d(w) < k$.

Definition 5 ensures that the average degree of a graph G, without the two vertices of smallest degree, is less than or equal to k. Furthermore, as $\delta(G) < k$ if G is a k-chain, all k-chain graphs are at most $(k-1)$-connected. Note that all p-regular graphs are $p+1$-chains. In particular a complete graph is a $|V|$-chain. Thus $2 \leq k \leq |V|$, and when k approaches 2, the graph is highly chained (it is structurally similar to a tree).

Definition 6 ((x,y)-damaged-path). *G is an* (x,y)-*damage-path if* $x = |V| - 1 - diam(G)$ *and* $y = |E| - diam(G)$.

Definition 6 involves two parameters (x, y). The former is the difference between the length of $P_{|V|}$ and the diameter of G; the latter is the difference between the number of edges and the diameter of G. Here the idea is, given a graph G and a path P in G with length equal to the diameter of G, y represents the number of edges that are not in P (extra edges), and x indicates the number of edges needed to complete P (missing edges) in order to obtain a path of length $|V| - 1$. Note that x and y could be defined in different ways, as long as the difference $y - x = |E| - |V| + 1$ always remained the same; it is convenient here to define x and y in terms of the diameter of G so that $x = y = 0$ if and only if G is a path. In this case, we have $0 \leq x \leq |V| - 2$ and $0 \leq y \leq \frac{|V|(|V|-1)}{2} - 1$. Note that $x = |V| - 2$ and $y = \frac{|V|(|V|-1)}{2} - 1$ if G is a complete graph. Thus, given two graphs G_1 and G_2, such that G_1 is (x_1, y_1)-damaged-path and G_2 is (x_2, y_2)-damaged-path, we say that G_1 is more chained than G_2 if $y_1 - x_1 < y_2 - x_2$, or $y_1 - x_1 = y_2 - x_2$ and $x_1 < x_2$. Some examples are depicted in Fig. 2.

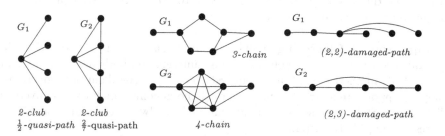

Fig. 2. Examples of path relaxation models, where G_1 is more chained than G_2 according to the respective definition.

We will compare two methods for generating gene families, SCWiB and MCL, in terms of the parameters we have defined.

The SCWiB method ensures that a gene family is determined by strong similarities connecting each of its members, by setting a high similarity threshold

U, and requiring that a cluster be connected, in the graph theoretical sense, solely in terms of similarities exceeding U. Also to control chaining, this method sets a less stringent threshold W, and requires that the elements in the cluster form an s-plex in terms of similarities exceeding W.

MCL is one of the most widely used methods for inferring gene families. Its basic principle is the iteration of a procedure that strengthens certain heavily weighted edges and weakens those with lesser weight. With appropriate parameter settings, MCL and SCWiB can produce very similar distributions of cluster sizes. The lack of any cluster quality criterion influencing the MCL process, however, results in many of its clusters, including some of the largest ones, having very few internal edges, while the SCWiB construction explicitly prohibits this.

3   Results

Data Source. In order to compare the SCWiB and MCL methods, we calculate the parameters introduced in Sect. 2 in 45 genomes. We extracted the data on these genomes from the CoGe database [7,8]. We require genomes to be published, publicly available, and have associated structural gene annotations. The genomes include *Amborella*, sacred lotus, rice, *Brachypodium*, maize, sorghum, millet, banana, duckweed, date palm, grape, eucalyptus, clementine, sweet orange, cacao, papaya, *Arabidopsis thaliana*, *Arabidopsis lyrata*, turnip, *Capsella rubella*, *Leavenworthia alabamica*, *Sisymbrium irio*, *Aethionema arabicum*, *Thellungiella parvula*, *Eutrema parvulum*, watermelon, cucumber, peach, strawberry, lotus, common bean, pigeonpea, soybean, coffee, poplar, flax, cassava, *Ricinus communis*, kiwifruit, tomato, potato, pepper, *Utricularia*, *Mimulus* and *Medicago* [15–56].

We analyze these two methods for each genome individually, calculating the average of each parameter (diameter, α, k, (x, y)) separately, for clusters with $|V|$ in each the following bins: 2, 3–4, 5–8, 9–16, 17–24, 25+. We consider each parameter as a function of bin size.

Comparison of Clustering Methods. In Figs. 3 and 4, we note that, in general, the diameter is a non-decreasing function for both methods and, in the SCWiB method it is always bounded above by 2 (by definition, given our choice of parameters). Furthermore, for all the genomes analyzed the average diameter of SCWiB clusters with the same $|V|$ is always less than that of the corresponding MCL clusters.

For α-quasi-paths, we observe that α is a decreasing function and, starting at bin 5-8, α decreases faster for SCWiB than for MCL.

For k-chains, k is a increasing function and, starting at bin 5-8, k increases in the SCWiB clusters faster than in those obtained by MCL. Therefore for the parameters "diameter", α and k, we find that MCL leads to more chaining than the SCWiB method.

Turning to the relaxation clique criteria, γ-quasi-clique and s-defective-clique, we also compare the similarity to cliques of the gene families generated by both

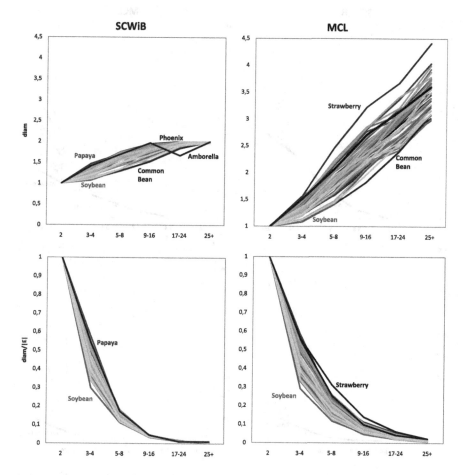

Fig. 3. Average of diameter (top) and α(bottom) separately for clusters with $|V|$ in each the following bins: 2, 3–4, 5–8, 9–16, 17–24, 25+. Singletons not included. Left: the SCWiB. Right: MCL. *Amborella* results highlighted in black.

the two clustering processes. We observe that SCWiB clusters are denser, and more uniform in density, than those of MCL, and the former have fewer missing edges than the latter in comparison with a complete graph. The range, across all genomes, of the average number of missing edges in SCWiB-generated families in every bin, starting at bin 9-16, is less than and actually disjoint from that for MCL method. SCWiB clusters are thus more clique-like than MCL clusters.

Comparison of Genomes. We previously observed that *Amborella trichopeda* has an anomalously small number of moderate and large-sized clusters [14]. Here, in comparing all 45 genomes by both clustering methods, we observe in Figs. 3, 4 and 5 that *Amborella* also demonstrates extreme behaviour with respect to high levels of chaining and non-clique-like families. Only papaya and strawberry have comparable behavior, but less consistently. The common bean, soybean and *Theobroma* are at the other extreme, with little chaining and more clique-like clusters.

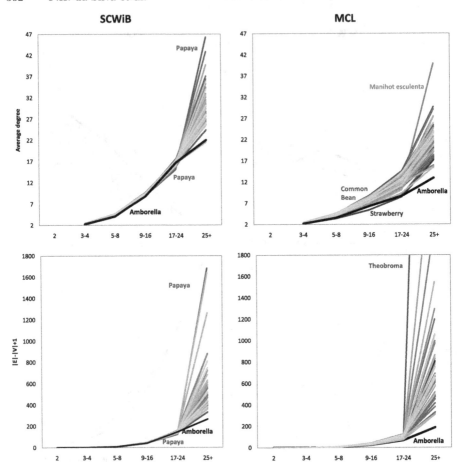

Fig. 4. Average of k (top) and $y - x$(bottom) separately for clusters with $|V|$ in each the following bins: 2, 3–4, 5–8, 9–16, 17–24, 25+. Singletons not included. On the left the SCWiB clusters and in right the clusters from MCL. *Amborella* always highlighted in black.

Comparison of Criteria. We computed the Pearson correlation coefficient among the parameters from an array of average values over all 45 genomes, for each bin (Table 1). We did this separately for the SCWiB clusters and the MCL clusters. All pairs of parameters are significantly correlated, either positively and negatively. To find the overall pattern, we submitted the correlations to a multidimensional scaling (MDS) procedure using the XLSTAT package from Addinsoft[TM]. For coherence, we used $-\alpha$ and $-\rho$ instead of α and ρ, so that all the correlations would have the same sign, and larger values of all parameters would indicate increased chaining.

Figure 6 shows that the parameters are largely disposed in a single dimension, although a two-dimensional space was specified in the scaling settings. $-\alpha$, $-\rho$ and "diameter" are at opposite sides from s and $|y - x|$, while k is intermediate.

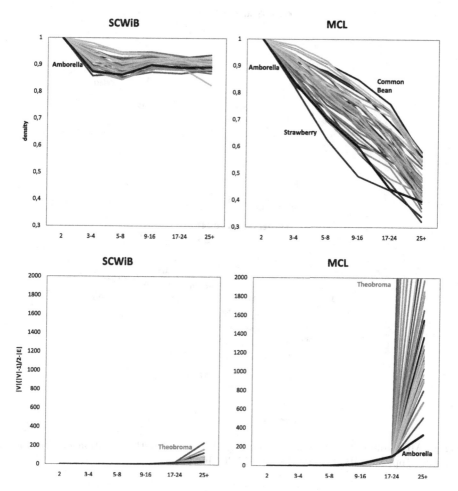

Fig. 5. Average of density (top) and average of missing edges to be complete graph (bottom) separately for clusters with $|V|$ in each the following bins: 2, 3–4, 5–8, 9–16, 17–24, 25+. Singletons not included. On the left the SCWiB clusters and in right the clusters from MCL. *Amborella* always highlighted in black.

Table 1. Pearson correlation coefficient among: diameter, α, k, $|y - x|$, density ρ and s missing edges.

SCWiB	diam	α	k	$|y - x|$	ρ	s	MCL	diam	α	k	$|y - x|$	ρ	s
diam	1						diam	1					
α	-0.94	1					α	-0.84	1				
k	0.84	-0.68	1				k	0.95	-0.75	1			
$|y - x|$	0.64	-0.46	0.96	1			$|y - x|$	0.75	-0.41	0.93	1		
ρ	-0.76	0.93	-0.28	-0.35	1		ρ	-0.99	0.86	-0.97	-0.76	1	
s	0.64	-0.45	0.96	0.99	-0.35	1	s	0.69	-0.37	0.89	0.99	-0.72	1

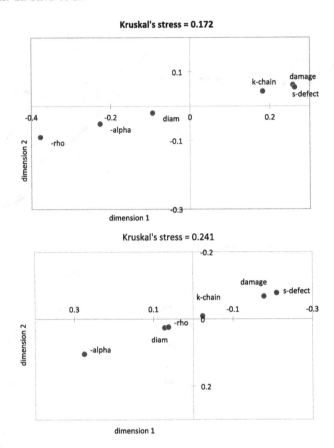

Fig. 6. MDS analysis of correlations among clique- and path-relaxation criteria for chaining. Top: SCWiB clusters. Bottom: MCL clusters.

There are clear differences between SCWiB and MCL, in the ordering of $-\alpha$, $-\rho$ and "diameter", for example, reflecting the constraints on these quantities in their SCWiB definitions.

4   Conclusion

Our application to plant genomes shows that *Amborella* clusters show less chaining than other flowering plants, extending the previous discovery, in the same set of angiosperm genomes surveyed here, of the special nature of this basal flowering plant [14] in having exceptionally few large and moderate-size gene families.

The different uses of clustering in genomics suggest that no one definition is universally useful. For partitioning the set of genes into disjoint gene families, as we have done here, allowing minor relaxation from a clique is probably more appropriate. On the other hand, for the investigation of the evolution of genes through the accumulation, loss and exchange of protein domains, it may be interesting to balance clique-like behaviour with a degree of chaining. For functional studies of gene networks, still other concepts and definitions of cluster shape may be preferable.

Although "chaining" is a generally understood concept in cluster analysis and domain chaining is familiar to all who work on automated gene family construction, there is no one single formal definition of chaining. We have suggested a range of formalizations that turn out to differ (empirically) along an axis measuring relaxation from a clique at one extreme, to relaxation from a path at the other. Our tests show that in general SCWiB yields clusters with less chaining than MCL, not only according to the clique relaxation criteria, but also by the path relaxation ones.

That the clustering criteria are disposed in an almost one-dimensional subspace when applied to our database is more than just an artifact of the clustering method is confirmed by similar results with the two methods. It is also unlikely to reflect only the properties of our database, but this should be confirmed by simulation studies. These observations reinforce our suggestion of more general research into how to operationalize various concepts of cluster shape. We could hope that eventually this would lead to an understanding of the statistical nature of the evolution of gene families.

References

1. Abello, J., Resende, M.G.C., Sudarsky, S.: Massive quasi-clique detection. In: Rajsbaum, S. (ed.) LATIN 2002. LNCS, vol. 2286, pp. 598–612. Springer, Heidelberg (2002)
2. Alba, R.D.: A graph theoretic definition of a sociometric clique. J. Math. Soc. 3(1), 113–126 (1973)
3. Enright, A.J., Van Dongen, S., Ouzounis, C.A.: An efficient algorithm for large-scale detection of protein families. Nucleic Acids Res. 30, 1575–1584 (2002)
4. Joseph, J.M.: On the Identification and investigation of homologous gene families, with particular emphasis on the accuracy of multidomain families. Lane Center for Computational Biology. Technical report CMU-CB-12-103.pdf (2012)
5. Joseph, J.M., Durand, D.: Family classification without domain chaining. Bioinformatics 25(12), i45–i53 (2009)
6. Luce, R.: Connectivity and generalized cliques in sociometric group structure. Psychometrika 15, 169–190 (1950)
7. Lyons, E., et al.: Finding and comparing syntenic regions among Arabidopsis and the outgroups papaya, poplar and grape: CoGe with rosids. Plant Physiol. 148, 1772–1781 (2008)
8. Lyons, E., Freeling, M.: How to usefully compare homologous plant genes and chromosomes as DNA sequences. Plant J. 53, 661–673 (2008)
9. Pattillo, J., Youssef, N., Butenko, S.: On clique relaxation models in network analysis. Eur. J. Oper. Res. 226(1), 9–18 (2013)
10. Seidman, S.B.: Network structure and minimum degree. Soc. Netw. 5, 269–287 (1983)
11. Seidman, S.B., Foster, B.L.: A graph theoretic generalization of the clique concept. J. Math. Soc. 6, 139–154 (1978)

12. Yu, H., Paccanaro, A., Trifonov, V., Gerstein, M.: Predicting interactions in protein networks by completing defective cliques. Bioinformatics **22**, 823–829 (2006)
13. Zahn, C.: Graph-theoretical methods for detecting and describing gestalt clusters. IEEE Trans. Comput. **C–20**, 68–86 (1971)
14. Zheng, C., Kononenko, A., Leebens-Mack, J., Lyons, E., Sankoff, D.: Gene families as soft cliques with backbones: Amborella contrasted with other flowering plants. BMC Genomics **15**(Suppl 6), S8 (2014)

Reference for Genome Data

15. Amborella-Genome-Project: The Amborella genome and the evolution of flowering plants. Science **342**(6165), 1241089 (2013)
16. Ming, R., et al.: Genome of the long-living sacred lotus (Nelumbo nucifera Gaertn.). Genome Biol. **14**(5), 41 (2013)
17. Yu, J., et al.: A draft sequence of the rice genome (Oryza sativa L. ssp. indica). Science **296**(5565), 79–92 (2002)
18. Vogel, J.P., et al.: Genome sequencing and analysis of the model grass Brachypodium distachyon. Nature **463**(7282), 763–768 (2010)
19. Schnable, P.S., et al.: The B73 maize genome: complexity, diversity, and dynamics. Science **326**(5956), 1112–1115 (2009)
20. Paterson, A.H., et al.: The Sorghum bicolor genome and the diversification of grasses. Nature **457**(7229), 551–556 (2009)
21. Bennetzen, J.L., et al.: Reference genome sequence of the model plant Setaria. Nat. Biotechnol. **30**(6), 555–561 (2012)
22. D'Hont, A., et al.: The banana (Musa acuminata) genome and the evolution of monocotyledonous plants. Nature **488**(7410), 213–217 (2012)
23. Wang, W., et al.: The Spirodela polyrhiza genome reveals insights into its neotenous reduction fast growth and aquatic lifestyle. Nat. Commun. **5**, 3311 (2014)
24. Al-Dous, E.K., et al.: De novo genome sequencing and comparative genomics of date palm (Phoenix dactylifera). Nat. Biotechnol. **29**(6), 521–527 (2011)
25. Jaillon, O., et al.: The grapevine genome sequence suggests ancestral hexaploidization in major angiosperm phyla. Nature **449**(7161), 463–467 (2007)
26. Myburg, A., et al.: The Eucalyptus grandis genome project: genome and transcriptome resources for comparative analysis of woody plant biology. BMC Proc. **5**, 20 (2011)
27. Wu, G.A., et al.: Sequencing of diverse mandarin, pummelo and orange genomes reveals complex history of admixture during citrus domestication. Nat. Biotechnol. **32**(7), 656–662 (2014)
28. Xu, Q., et al.: The draft genome of sweet orange (Citrus sinensis). Nat. Genet. **45**(1), 59–66 (2013)
29. Argout, X., et al.: The genome of Theobroma cacao. Nat. Genet. **43**(2), 101–108 (2011)
30. Ming, R., et al.: The draft genome of the transgenic tropical fruit tree papaya (Carica papaya Linnaeus). Nature **452**(7190), 991–996 (2008)
31. Arabidopsis-Genome-Initiative, et al.: Analysis of the genome sequence of the flowering plant Arabidopsis thaliana. Nature **408**(6814), 796–815 (2000)
32. Hu, T.T., et al.: The Arabidopsis lyrata genome sequence and the basis of rapid genome size change. Nat. Genet. **43**(5), 476–481 (2011)
33. Wang, X., et al.: The genome of the mesopolyploid crop species Brassica rapa. Nat. Genet. **43**(10), 1035–1039 (2011)

34. Slotte, T., et al.: The Capsella rubella genome and the genomic consequences of rapid mating system evolution. Nat. Genet. **45**(7), 831–835 (2013)
35. Haudry, A., et al.: An atlas of over 90,000 conserved noncoding sequences provides insight into crucifer regulatory regions. Nat. Genet. **45**(8), 891–898 (2013)
36. Dassanayake, M., et al.: The genome of the extremophile crucifer Thellungiella parvula. Nat. Genet. **43**(9), 913–918 (2011)
37. Yang, R., et al.: The reference genome of the halophytic plant Eutrema salsugineum. Front. Plant Sci. **4**, 46 (2013)
38. Guo, S., et al.: The draft genome of watermelon (Citrullus lanatus) and resequencing of 20 diverse accessions. Nat. Genet. **45**(1), 51–58 (2013)
39. Huang, S., et al.: The genome of the cucumber, Cucumis sativus L. Nat. Genet. **41**(12), 1275–1281 (2009)
40. Verde, I., et al.: The high-quality draft genome of peach (Prunus persica) identifies unique patterns of genetic diversity, domestication and genome evolution. Nat. Genet. **45**(5), 487–494 (2013)
41. Shulaev, V., et al.: The genome of woodland strawberry (Fragaria vesca). Nat. Genet. **43**(2), 109–116 (2011)
42. Sato, S., et al.: Genome structure of the legume, Lotus japonicus. DNA Res. **15**(4), 227–239 (2008)
43. Schmutz, J., et al.: A reference genome for common bean and genome-wide analysis of dual domestications. Nat. Genet. **46**, 707–713 (2014)
44. Varshney, R.K., et al.: Draft genome sequence of pigeonpea (Cajanus cajan), an orphan legume crop of resource-poor farmers. Nat. Biotechnol. **30**(1), 83–89 (2012)
45. Schmutz, J., et al.: Genome sequence of the palaeopolyploid soybean. Nature **463**(7278), 178–183 (2010)
46. Tuskan, G.A., et al.: The genome of black cottonwood, Populus trichocarpa (Torr. & Gray). Science **313**(5793), 1596–1604 (2006)
47. Wang, Z., et al.: The genome of flax (Linum usitatissimum) assembled de novo from short shotgun sequence reads. Plant J. **72**(3), 461–473 (2012)
48. Prochnik, S., et al.: The cassava genome: current progress, future directions. Trop. Plant Biol. **5**(1), 88–94 (2012)
49. Chan, A.P., et al.: Draft genome sequence of the oilseed species Ricinus communis. Nat. Biotechnol. **28**(9), 951–956 (2010)
50. Huang, S., et al.: Draft genome of the kiwifruit Actinidia chinensis. Nat. Commun. **4**, 2640 (2013)
51. Tomato-Genome-Consortium, et al.: The tomato genome sequence provides insights into fleshy fruit evolution. Nature **485**(7400), 635–641 (2012)
52. Potato-Genome-Sequencing-Consortium, et al.: Genome sequence and analysis of the tuber crop potato. Nature **475**(7355), 189–195 (2011)
53. Qin, C., et al.: Whole-genome sequencing of cultivated and wild peppers provides insights into Capsicum domestication and specialization. Proc. Natl. Acad. Sci. **111**(14), 5135–5140 (2014)
54. Ibarra-Laclette, E., et al.: Architecture and evolution of a minute plant genome. Nature **498**(7452), 94–98 (2013)
55. Hellsten, U., et al.: Fine-scale variation in meiotic recombination in Mimulus inferred from population shotgun sequencing. Proc. Natl. Acad. Sci. **110**(48), 19478–19482 (2013)
56. Young, N.D., et al.: The Medicago genome provides insight into the evolution of rhizobial symbioses. Nature **480**(7378), 520–524 (2011)

Efficient Design of Compact Unstructured RNA Libraries Covering All k-mers

Yaron Orenstein[1] and Bonnie Berger[1,2]([✉])

[1] Computer Science and Artificial Intelligence Laboratory,
Massachusetts Institute of Technology, Cambridge, MA, USA
{yaronore,bab}@mit.edu
[2] Department of Mathematics, Massachusetts Institute of Technology,
Cambridge, MA, USA

Abstract. Current microarray technologies to determine RNA structure or measure protein-RNA interactions rely on single-stranded, unstructured RNA probes on a chip covering together all k-mers. Since space on the array is limited, the problem is to efficiently design a compact library of unstructured ℓ-long RNA probes, where each k-mer is covered at least p times. Ray *et al.* designed such a library for specific values of k, ℓ and p using ad-hoc rules. To our knowledge, there is no general method to date to solve this problem. Here, we address the problem of finding a minimum-size covering of all k-mers by ℓ-long sequences with the desired properties for any value of k, ℓ and p. As we prove that the problem is NP-hard, we give two solutions: the first is a greedy algorithm with a logarithmic approximation ratio; the second, a heuristic greedy approach based on random walks in de Bruijn graphs. The heuristic algorithm works well in practice and produces a library of unstructured RNA probes that is only ~ 1.1-times greater in size compared to the theoretical lower bound. We present results for typical values of k and probe lengths ℓ and show that our algorithm generates a library that is significantly smaller than the library of Ray *et al.*; moreover, we show that our algorithm outperforms naive methods. Our approach can be generalized and extended to generate RNA or DNA oligo libraries with other desired properties. The software is freely available on curlcake.csail.mit.edu.

Keywords: de Bruijn graph · RNA secondary structure · Microarray library design

1 Introduction

RNAs play vital roles in many processes in the living cell. Through interaction of RNAs with other RNAs or proteins, they perform specific functions. RNA-RNA interactions play a role in many pathways of RNA metabolism, including pre-mRNA splicing, ribosome synthesis, and the regulation of mRNA stability by microRNAs [1]. RNA-binding proteins interact with RNAs to modulate and affect a wide variety of cellular processes, including RNA replication, repair and

© Springer-Verlag Berlin Heidelberg 2015
M. Pop and H. Touzet (Eds.): WABI 2015, LNBI 9289, pp. 308–325, 2015.
DOI: 10.1007/978-3-662-48221-6_23

recombination [2]. Both types of interactions are mediated through the structure and sequence of the RNA molecule. Typically, interactions occur with RNA accessible regions through either base-pairing to nucleotides of another RNA or hydrogen bonding to a protein's residues [3,4].

A given RNA may fold into different conformations, which vary in accessible regions [5]; therefore, relying on *in silico* prediction of its structure may lead to incorrect predictions for the accessible region of interest. Researchers would thus like to experimentally measure accessible regions in RNAs.

Numerous experimental methods have been developed to study the secondary structure of RNAs in a high-throughput (HTP) manner [6–8]. Microarray technologies measure RNA secondary structure through the hybridization of accessible regions to a set of oligos on a chip. An array covering all RNA k-mers (a contiguous RNA word of length k) can robustly and accurately measure the structure of many RNAs. Examples for such arrays covering all 6-mers and 7-mers include [7,8], respectively. In both experimental setups, each oligo contains a unique k-mer. Despite the fact that microarrays are limited in throughput compared to deep-sequencing based methods, they are still often being used to overcome limitations in sequencing methods [9].

RNA-binding proteins (RBPs) regulate gene translation post-transcriptionally via their binding to RNA molecules. More than 1,500 genes in the human genome are thought to code for RBPs, making this family one of the largest families in the human proteome [10]. Many of these proteins have sequence-specific RNA-binding properties and thus regulate genes by binding only to site-specific elements. Better characterization of RBPs sequence-specific binding preferences can improve our understanding of post-transcriptional gene regulation.

New experimental high-throughput (HTP) techniques have been developed to uncover protein-RNA interactions on a genome-wide scale at single-nucleotide resolution. For example, HITS-CLIP, CLIP-seq and RIP-seq [11] measure protein-RNA interactions *in vivo* in a HTP manner. However, much like protein DNA-binding, protein RNA-binding is influenced by a variety of factors, such as other RBPs (that either compete for the same binding site or co-bind as a complex) and RNA secondary structure, which determines if a binding site is accessible or not [12]. While the end goal is to understand and predict *in vivo* binding, *in vitro* experiments currently have higher resolution and lower noise and thus provide valuable complementary information to protein RNA-binding preferences.

Towards this aim, high-throughput *in vitro* methods have been developed to study the binding preferences of RBPs [13,14]. In RNAcompete [13], a specific protein binds to a set of pre-designed oligos, and binding is measured using a florescence tag. The binding of the protein to a set of more than 200,000 probe sequences is reported. A recent study by the authors presents the binding of more than 200 human RBPs and provides a compendium of RBPs [15]. RNA Bind-n-Seq is a new technology that measures protein RNA-binding based on HTP sequencing [14]. Since the initial library is composed of random oligos, these may be structured and as a result include k-mers that are likely to be base-paired in RNA secondary structure.

The oligo library used in RNAcompete experiments has unique properties that allow it to effectively measure protein RNA-binding in a universal and unbiased manner. The complete oligo set is designed such that each 9-mer is covered at least 16 times. This property guarantees the ability to infer accurate binding scores for 9-mers and shorter k-mers. Another key property is that the probe sequences are unstructured, which makes them accessible to the protein for binding [16].

In this paper, we address the problem of designing better microarray probe libraries for enhanced exploration of both RNA structure through base-pairing of the target RNA to the probes as well as protein RNA-binding through affinity between a protein and the probes. Note that array designs that consist of a single k-mer for each probe are disadvantageous: the space on the microarray is limited, while the number of probes grows exponentially with k (the number of possible RNA k-mers is 4^k). A small value of k is also undesirable, since the likelihood of having a k-mer appear more than once in a target RNA sequence, and thus preventing unique identification of accessible sites, increases as k gets smaller. Hence, we aim to increase the size of k, while maintaining a small number of oligos on the chip. This goal can be achieved by covering a number of k-mers on each oligo. In this scenario, the k-mers are no longer covered by a unique sequence. Alternatively, if a k-mer is covered multiple times, an aggregate score for its accessibility or affinity can be inferred.

There are numerous methods to design sequences with complete coverage of all k-mers. De Bruijn sequences are the most compact sequences to cover all k-mers [17]. They can be generated in linear time in various ways, including Euler tours in complete de Bruijn graphs [18], linear-feedback shift registers [19] and in a recursive manner [20]. De Bruijn sequences have been successfully used in HTP technologies that measure protein DNA-binding, such as protein-binding microarrays [21–23] and MITOMI [24].

However, the coverage of all k-mers is not enough, as RNAs may form structure. In RNAcompete, the authors used ad-hoc greedy rules to generate an oligo library with the desired properties [13]; however, their method cannot be generalized. To our knowledge, there is currently no method to generate an RNA oligo library such that each k-mer occurs at least p times in ℓ-long unstructured probe sequences. Such a method would be highly useful for current and future technologies that measure protein-RNA interactions or RNA secondary structure. In addition, the freed space on the device may be used to cover longer k-mers or sequences with other specific properties.

Here, we solve the problem of designing an RNA oligo library such that each k-mer occurs at least p-times in ℓ-long unstructured probe sequences. We prove that for a given set of ℓ-long probes, the problem of covering all k-mers by a minimum-size subset is NP-hard. Thus, we formulate the problem as a minimum m-set cover problem and give an approximation algorithm with guaranteed logarithmic ratio. We also present a heuristic greedy algorithm based on random walks in de Bruijn graphs which performs very well in practice; it produces an oligo library that is only $\sim$ 1.1-times greater in size than the theoretical lower bound. In our results, we analyze the fraction of unstructured

RNA oligos as a function of their length and show that traditional methods to cover all k-mers do not work. We conclude with an analysis of the computational performance of our heuristic algorithm over different values of k and ℓ and in comparison to the design by Ray *et al.* [15]. The software is freely available on curlcake.csail.mit.edu.

2   Preliminaries

2.1   de Bruijn Graphs

A *de Bruijn graph* of order k over alphabet Σ is a directed graph in which every vertex has an associated label (a string over Σ) of length k (k-mer) and every edge has an associated label of length $k + 1$. There are exactly $|\Sigma|^k$ vertices in the graph, each representing a unique k-mer. If an edge (u, v) has an associated label l, then the label associated with u must be a k-prefix of l and the label associated with v must be a k-suffix of l. A complete de Bruijn graph contains all possible edges, which represent together all $(k + 1)$-mers over Σ.

Every path in a de Bruijn graph represents a sequence. A path $v_1, e_1, v_2, \ldots, v_n$ of length n spells a sequence s of length $n + k - 1$ such that the label associated with v_i occurs in s at position i for all $1 \le i \le n$, and the label associated with e_i occurs in s at position i for all $1 \le i \le n - 1$.

2.2   Unstructured RNA Probes and Self-structured k-mers

We followed the definition of *structuredness* due to Ray *et al.* [13]. The authors use RNAshapes [5] to enumerate all secondary structures with free energies within 70 % of the minimum free energy. The exact command line is: `RNAshapes -s -c 70.0 -r -M 30 -t 1 -o 2`.

The sum of the probabilities of structures with free energies less than $-2.5\,\text{kcal/mol}$ quantifies structuredness. A value below 0.5 is considered *unstructured*. For any sequence, we prepend the linker used in the RNAcompete technology (AGG or AGA) [13]. From the two linkers, we selected the one that gave the smaller sum of probabilities.

A *self-structured* k-mer forms structure in itself. It follows that no probe can contain it without being structured. Thus, to cover all k-mers in a microarray, structured probes must be included. For the structure definition above, self-structured k-mers exist for $k \ge 9$. Smaller values of k do not require structured probes to cover all k-mers. We refer to k-mers which are not self-structured as *unstructured* k-mers.

2.3   Problems Definition

We first define the notion of *k-mer coverage* over alphabet Σ.

Definition 1. *A set L of sequences is a k-mer coverage over Σ if for every $w \in \Sigma^k$, there exists a sequence $L_i \in L$ s.t. $w \in L_i$.*

We generalize the definition of k-mer coverage with a *p-multi k-mer coverage*.

Definition 2. *A set L of sequences is a p-multi k-mer coverage over Σ if for every $w \in \Sigma^k$, $\sum_{L_i \in L} o(w, L_i) \geq p$, where $o(w, L_i)$ is the number of times w occurs in sequence L_i.*

We can now state our optimization problem:

THE MINIMUM K-MER COVERAGE BY ℓ-LONG SEQUENCES PROBLEM

INSTANCE: A set S of ℓ-long sequences that is a k-mer coverage over $\Sigma = \{A, C, G, U\}$.
VALID SOLUTION: A subset $S' \subseteq S$ that is a k-mer coverage over Σ.
GOAL: Minimize $|S'|$.

And a similar NP-hard problem that we reduce from and use for an approximation algorithm:

THE MINIMUM M-SET COVER PROBLEM

INSTANCE: A set S of subsets of $E = \{e_1 \ldots e_n\}$ s.t. for any $S_i \in S$, its size $|S_i| \leq m$.
VALID SOLUTION: A subset $S' \subseteq S$ s.t. for every $e_i \in E$ there exists $S_i \in S'$ s.t. $e_i \in S_i$.
GOAL: Minimize $|S'|$.

We generalize the k-mer coverage problem by requiring multiple k-mer occurrences. Note that multisets may contain an element multiple times. We use $distinct(S)$ to denote the set of unique elements in multiset S.

THE MINIMUM P-MULTI K-MER COVERAGE BY ℓ-LONG SEQUENCES PROBLEM

INSTANCE: A set S of ℓ-long sequences that is a k-mer coverage over $\Sigma = \{A, C, G, U\}$ and p.
VALID SOLUTION: A multiset S' s.t. $distinct(S') \subseteq S$ and S' is a p-multi k-mer coverage over Σ.
GOAL: Minimize $|S'|$.

Dealing with Self-structured k-mers. Note that since self-structured k-mers may exist, covering all k-mers by ℓ-long unstructured probes may be impossible. The coverage problem may be redefined as two sub-problems to handle self-structured k-mers:

1. Cover all **unstructured** k-mers by a minimum size set of ℓ-long unstructured RNA probes.
2. Cover all **self-structured** k-mers by a minimum size set of ℓ-long RNA probes.

The union of these sets covers all k-mers, since each k-mer is either unstructured or self-structured by definition (see Sect. 2.2).

3   Methods

Since the minimum k-mer coverage by ℓ-long sequences problem is NP-hard (as we prove in Sect. 3.3), we provide an approximation algorithm and heuristic, which performs very well in practice, to address this problem.

3.1   Approximation Algorithm Through the Minimum m-set Cover Problem

The problem of covering all k-mers in unstructured RNA probes can be formulated as a minimum m-set cover problem. The problem can be approximately solved by a greedy algorithm. The algorithm starts with an empty set and adds to the solution the set that has the most uncovered elements in it. The algorithm achieves an approximation ratio of $H(m) - \frac{196}{360}$, where m is the maximum cardinality of a set in S, and H is the harmonic number $H(n) = \sum_{i=1}^{n} 1/i \leq ln(n) + 1$ [25,26]. The algorithm can be highly accurate in some instances [27]. This leads us to the next corollary:

Corollary 1. *Algorithm 1 is an $(H_{\ell-k+1} - \frac{196}{390})$-approximation to the minimum k-mer coverage by ℓ-long sequences problem.*

If self-structured k-mers exist, Algorithm 1 can be modified to first handle the coverage of unstructured k-mers by unstructured RNA probes, and then re-run to cover uncovered self-structure k-mers by structured RNA probes. Thus, since the approximation ratio is valid for each sub-problem, Corollary 1 is valid for covering all k-mers (see Sect. 2.3 for definition of sub-problems).

Algorithm 1. Solve k-coverage by ℓ-long unstructured RNA probes problem as a set cover problem

---

1: For each ℓ-long RNA sequence:
2:     Test if the sequence is unstructured. If so, add it to the list of unstructured sequences.
3: Apply the greedy set cover algorithm:
4:     The elements are the k-mers.
5:     The sets are the unstructured sequences, and their elements are the k-mers they cover.

The running time of the algorithm is exponential in the oligo length ℓ. The first step iterates over all possible ℓ-long sequences, and for each one runs an RNA secondary structure prediction algorithm. Denote $f(\ell)$ to be the running time of the prediction algorithm on an ℓ-long sequence; then Step 1 takes $\Theta(4^\ell \cdot f(\ell))$. The second step can be implemented using a priority queue, whose keys are the number of uncovered elements of each unstructured sequence not in the solution. Since the keys are integers bounded by $\ell - k + 1$, queue operations can be implemented in $O(1)$ time. A dictionary is used to hold uncovered k-mers and pointers to the sequences that contain them. The dictionary can be implemented using an array of size 4^k, as our k-mers can be represented as integers from 0 to $4^k - 1$. Each cell contains a list of pointers to sequences containing the k-mer. The length of the list is bounded by $(\ell - k + 1) \cdot 4^{\ell-k}$, the number of possible ℓ-long sequences containing the k-mer. Therefore, we get that the running time for the second step is $\Theta(4^\ell + 4^k \cdot (\ell - k + 1) \cdot 4^{\ell-k})$. The first term consists of delete-minimum operations on the queue, and the second, the update operations. In total, Algorithm 1 takes time $\Theta(4^\ell \cdot f(\ell))$. (Since input size is ℓ, then $f(\ell) = \Omega(\ell)$. Predicting minimum free-energy structure can be done in $O(\ell^2)$ [28]. Predicting all possible structures takes $O(4^\ell \cdot \ell)$ as there is an exponential number of structures, and heuristics are used to estimate representative structures up to a given energy threshold [5]). Unfortunately, the running time is infeasible for most instances, e.g. $\ell = 35$ in RNAcompete's implementation [13]. Thus, we turn to a heuristic greedy algorithm.

3.2   A Heuristic Greedy Algorithm Based on Random Walks in de Bruijn Graphs

Our greedy algorithm, summarized as Algorithm 2, is based on the following two key ideas:

1. Using random walks in a de Bruijn graph to find unstructured oligos.
2. Backtracking strategy in cases where the random walk reaches a structured oligo.

The algorithm tries to find a set of disjoint ℓ-long paths in a de Bruijn graph, each representing an unstructured probe, and together covering all the edges. To cover each k-mer p times, $p-1$ copies are added to each edge. During the search for the desired paths, structured paths may be found. To address this problem, the algorithm backtracks and searches for a different path. An illustration of this process is depicted in Fig. 1. Through its random walk, the algorithm doubles the length of the explored path by possible extensions and selects the first unstructured path it encounters. The rationale behind this search process follows from two ideas related to RNA secondary structure prediction:

1. If a subsequence is structured, it is most likely that a sequence containing it is structured (for experimental support see Sect. 4.1).
2. A structure may form between one half of a sequence to the other half.

Thus, the algorithm does not waste time by trying to extend structured sub-paths. Indeed, it considers all possible path-extensions of length double the current path to test if it is unstructured. This fact is also beneficial in terms of running time: the number of extensions in a doubling scheme is $O(\log(\ell))$ instead of $O(\ell)$.

We bound the running time of the algorithm. The number of possible extensions at each vertex is at most 4^i, where i is the length of the current probe.

Algorithm 2. Generate a set of ℓ-long unstructured RNA sequences covering all k-mers p times. Input: k (coverage), ℓ (oligo length), p (multiplicity), c (a limit on the number of attempts)

```
 1: Generate a complete de Bruijn graph of order k − 1. For each edge add p − 1 copies.
 2: Initialize a list L of unfinished vertices with all vertices.
 3: Set current_vertex to the first element in the list.
 4: while there are edges in the graph do
 5:     probe = label of current_vertex.
 6:     extension_length = ℓ.
 7:     while |probe| < extension_length do
 8:         Try to extend probe to length minimum{2 · |probe|, extension_length}.
 9:         if unstructured extension was not found in c attempts then
10:             extension_length = extension_length − 1.
11:         end if
12:     end while
13:     if |probe| = k − 1 then
14:         Extend probe by a random extension of size 1.
15:     end if
16:     Output probe and delete the edges of its k-mers from the graph.
17:     if current_vertex has no outgoing edges AND |L| > 1 then
18:         remove it from L.
19:         Set current_vertex to a random vertex from L.
20:     end if
21: end while
```

Since the maximum number of extensions at any vertex is $4^{\lfloor \ell/2 \rfloor}$, the sum of possible extensions examined for each probe is $\Theta(4^{\ell/2})$. Denote by $f(\ell)$ the running time of the prediction algorithm and the number of probes by X, then the total time is $O(X 4^{\ell/2} f(\ell))$. This may be prohibitive in some instances, depending on the value of ℓ. Thus, for practical reasons, we replace the search of all possible extensions by a search of a limited number of random extensions. Denote this number c (given as input) and remember that the extensions are performed in a doubling scheme. Hence, the total running time is $O(X f(\ell) c \log(\ell))$. Results show that $X = \Theta(4^k/(\ell - k + 1))$ (see Table 1).

In some cases, no extension forms an unstructured oligo with the current sub-path. In these cases, we look for an extension shorter by one nucleotide, and continue shortening until an unstructured path is found or the searched extension is of size 1. This process incurs an additional factor of $O(f(\ell) c \ell)$ per probe in the running time, since in the worst case $\ell/2$ shortening may occur. Thus, the total running time is $O(X f(\ell) c \ell)$.

The final result of this process is a set of unstructured probe sequences of length at most ℓ. In this set each k-mer is represented exactly p times. The probes may be of length shorter than ℓ in two cases:

1. The path closed a cycle (i.e. reached a vertex with no outgoing edges.)
2. The path had to be shortened to become unstructured, since no unstructured extension was found in c attempts.

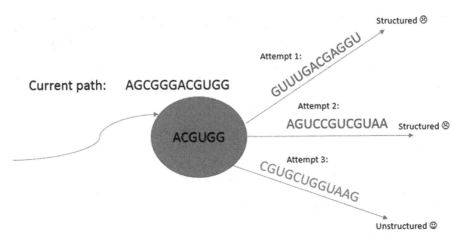

Fig. 1. An illustration of the search process for unstructured paths. In the example, the current path started from vertex $AGCGGG$. It was extended to the unstructured path $AGCGGGACGUGG$. Then, it attempted to extend the path, and succeeded in the third attempt to find an unstructured path. The de Bruijn graph is of order 6 to cover all 7-mers.

If the technology requires that all probes have the same length, then an additional process, Algorithm 3, is run to extend these probes into ℓ-long unstructured probes. Other methods may be used for this step, such as RNAinverse [28]. The total set in the end is the *complete set*.

If self-structured k-mers exist, the algorithm can be used to solve the two sub-problems (see Sect. 2.3). The algorithm as is solves the problem of covering all k-mers at the expense of having a few structured RNA probes. If structured probes are forbidden, the edges corresponding to self-structured k-mers can be removed, and the algorithm can be run on the remaining graph.

3.3 NP-Hardness of the Minimum k-mer Coverage by ℓ-long Sequences Problem

We prove that the following problem is NP-hard: covering all k-mers by a minimum-size subset of a restricted set of ℓ-long sequences. For sake of simplicity, we study the problem on the RNA alphabet, but it can be easily generalized to any finite alphabet Σ.

The problem is easy in two extreme instances. Clearly, when set S contains all possible ℓ-long sequences, the problem can be solved in linear time. A de Bruijn sequence can be generated in linear time. Cutting it into ℓ-long subsequences with $(k-1)$-overlaps covers all k-mers in the most compact manner. On another extreme, when $\ell = k$ the problem is trivial.

We reduce a known NP-hard problem, the minimum m-set cover [26], to our problem. While the problems look similar, one is not a private instance of the other and the reduction is not immediate. Here we describe the reduction.

Algorithm 3. Extend set S of RNA sequences covering all k-mers, each p times, to length ℓ. Input: k (coverage), ℓ (oligo length), S (incomplete set), c (a limit on the number of attempts)

```
 1: for Each Sᵢ ∈ S do
 2:     if |Sᵢ| = ℓ then
 3:         Output Sᵢ
 4:     else
 5:         attempts = 0
 6:         do
 7:             attempts = attempts + 1
 8:             Create ℓ-long sequence S′ᵢ:
 9:                 Pick a random index 1 ≤ j ≤ ℓ − |Sᵢ| + 1 for Sᵢ.
10:                 Assign random nucleotides in the positions outside Sᵢ.
11:         while (S′ᵢ is structured AND (attempts < c OR (attempts < 100 · c AND
            |Sᵢ| = k)))
12:         if S′ᵢ is structured AND |Sᵢ| > k then
13:             Continue recursively on the (|Sᵢ|/2+k/2)-prefix and (|Sᵢ|/2+k/2)-suffix
            of Sᵢ.
14:             Output a union of the returned sets.
15:         else
16:             Output S′ᵢ.
17:         end if
18:     end if
19: end for
```

Theorem 1. *The minimum k-mer coverage by ℓ-long sequences problem is NP-hard.*

Proof. Given an input to the minimum m-set cover problem, and a set S of subsets of $E = \{e_1 \ldots e_n\}$, we generate an input to the minimum k-mer coverage by ℓ-long sequences problem in polynomial time. We choose $k = \lceil \log_2(n) \rceil$ and $\ell = 3$ km. We map each element $e_i \in E$ to a k-long binary representation of i, where instead of bits we use A and U. We call this representation the element's $\{A, U\}$-representation and denote it by $f_{AU}(e_i)$.

We generate three sequence sets whose union is the input to the k-mer coverage problem.

1. L_1: For each set $S_i \in S$ we generate a sequence that contains all of its elements' $\{A, U\}$-representation, each buffered by G^k before and C^k after. Formally, for a set $S_i = \{e_{i_1}, \ldots, e_{i_m}\}$ we create the sequence: $\prod_{j=1}^{m} G^k \cdot f_{AU}(e_{i_j}) \cdot C^k$. If $|S_i| < m$, we append the sequence by C's, so that its total length is ℓ.
2. L_2: we add sequences that cover all the k-mers over $\{A, U\}$ that are not covered by L_1. For each k-mer w over $\{A, U\}$ that is not in L_1 we create a sequence $G^k \cdot f_{AU}(w) \cdot C^{\ell-2k}$.
3. L_3: we cover all non-$\{A, U\}$ k-mers. Formally, for each k-mer $w \in \Sigma^k \setminus \{G^i\{A,U\}^{k-i} \cup \{A,U\}^j C^{k-j} | 0 \le i, j \le k\}$ create the sequence $G^k \cdot w \cdot C^{\ell-2k}$.

The input to the minimum k-mer coverage problem is the set $L = L_1 \cup L_2 \cup L_3$.

Denote by L^{OPT} the optimal solution to the k-mer coverage problem and by $L_1^{OPT} = L^{OPT} \cap L_1$. The solution to the m-set cover problem are the sets corresponding to the sequences in L_1^{OPT}. The running time of the reduction is bounded by $O((4^k + |S|) \cdot \ell)$ to generate the input sequences, which is $O((n^2 + |S|) \cdot m \cdot log(n))$.

We now prove the correctness of the reduction. We start with proving a couple of properties of the solution.

Lemma 1. *Any k-mer coverage must include all L_2 sequences.*

Proof. Each sequence in L_2 contains a unique k-mer over $\{A, U\}^k$ that does not appear in L_1. In addition, by the design of L_3, there are no k-mers over $\{A, U\}^k$ in L_3. Thus, to cover all k-mers, all of L_2 sequences must be included. □

Lemma 2. *The selection of sets in L_3 is independent of L_1 and L_2.*

Proof. The set of k-mers covered by the selected sequences in L_1 and L_2 is $\{G^i\{A, U\}^{k-i} \cup \{A, U\}^j C^{k-j} \cup C^g G^{k-g} | 0 \leq i, j, g \leq k\}$. The selected sequences in L_3 are constructed to cover all other k-mers. It follows that their selection is independent of the input to the problem.

□

1. **k-mer coverage $\Rightarrow$ m-set cover:** all k-mers are covered by sequences in L^{OPT}. The selected sequences from L_2 and L_3 in L^{OPT} are independent of the input by Lemmas 1 and 2. Each sequence in L_1^{OPT} corresponds to a unique set in S. The set of corresponding sets is the optimal solution to the m-set cover problem. Assume the contrary, i.e. that there exists a smaller solution to the m-set cover problem. Then, the set of sequences corresponding to the sets in the solution together with $L^{OPT} \cap \{L_2 \cup L_3\}$ form a smaller solution to the k-mer coverage problem, in contradiction to the fact that L^{OPT} is a minimum k-mer coverage.

2. **m-set cover $\Rightarrow$ k-mer coverage:** denote S^{OPT} to be an optimal solution to the m-set cover problem. Denote L_1' as the set of sequences corresponding to the sets in S^{OPT}. Then, an optimal solution to the k-mer coverage problem is the set $L_3' \cup L_2 \cup L_1'$, where L_3' is the minimum-size set to cover $\Sigma^k \setminus \{G^i\{A, U\}^{k-i} \cup \{A, U\}^j C^{k-j} \cup C^g G^{k-g} | 0 \leq i, j, g \leq k\}$. All the elements in E were covered by S^{OPT}, and so their $\{A, U\}$-representations are covered by L_1'. By Lemmas 1 and 2, L_2 sequences are in any optimal solution, and the selection of L_3 sequences is independent of the input. Assume to the contrary that there exists a smaller solution to the k-mer coverage problem. $L_2 \cup L_3'$ are in any solution, so L_1' must be smaller. L_1' covers all the k-mers corresponding to the elements in E, so there is a smaller solution to the m-set cover problem, in contradiction to the fact that S^{OPT} is an optimal solution. □

Clearly, if we could solve the p-multi k-mer coverage problem in polynomial time, then we could solve the k-mer coverage problem. Thus, we get:

Corollary 2. *The minimum p-multi k-mer coverage by ℓ-long sequences is NP-hard.*

4   Results

4.1   Traditional Methods Won't Solve Our Problem

We sought to test whether traditional methods to cover all k-mers, such as random oligos or overlapping subsequences of a de Bruijn graph, could solve the minimum k-mer coverage problem. Towards this aim, we analyzed the properties of unstructured probe sequences. Here we followed the definition of Ray *et al.* [13]. Predicting a single minimum folding energy structure may be misleading, as many RNAs may fold into different structures. Thus, for each RNA sequence an ensemble of structures is predicted. The oligo is considered structured if its probability of forming a low energy structure is more than half. Figure 2A depicts the structuredness test. Unfortunately, this property cannot be elegantly formulated in combinatorial terms. For a formal definition and technical details see Sect. 2.2.

To better understand the problem at hand, we calculated the percentage of unstructured RNA probes. Ideally, we would iterate over all possible RNA ℓ-long sequences and test if each is structured. While for small values of ℓ this strategy is feasible, for greater values it is not, as it requires 4^ℓ iterations. To overcome this problem, we generated 10,000 random ℓ-long sequences for each value of ℓ, where each nucleotide is uniformly picked at each position in the sequence. The fraction of structured probes quickly converged (data not shown), and hence we are confident that these estimates are accurate.

The results are shown in Fig. 2B. As expected, the fraction of structured probes is higher for longer probes. More surprisingly, the decrease in the fraction of unstructured RNA probes as a function of length is fast, and for length 45, less than 10 % of the probes are unstructured. Thus, using random oligos is sub-optimal and requires many more probe sequences to cover all k-mers. de Bruijn sequences, which are the most compact sequences to cover all k-mers, are uniformly distributed over all k-mers [29], and are therefore prone to having many structured subsequences. Indeed, in the report by Ray *et al.* [13], in a de Bruijn sequence of order 11 over $\{A, C, G, U\}$, only 36,837 probes out of 167,773 were unstructured. Note also that for $k \geq 9$ the fraction is smaller than 1 due to self-structured k-mers (see Sect. 2.2). To conclude, neither random oligos nor probes generated by overlapping subsequences of a de Bruijn sequence are likely to provide an optimal or near-optimal solution to our problem.

To support our assumption that structured subsequences are likely to be extended to structured sequences (see Sect. 3.2), we calculated the fraction of structured probes given that their first half is structured. For each probe length $28 \leq \ell \leq 52$, we generated 100 random $\lfloor \ell/2 \rfloor$-long structured sequences and extended them by 100 random extensions to a probe of length ℓ. The fraction of structured probes out of the 10,000 probes is the reported value. We compared this value to the fraction of structured probes among random ℓ-long sequences. Results show that, given that the first half is structured, there is a chance of more than 95 % that a probe starting with it is structured (see Fig. 2C). The fraction of structured oligos among random oligos is much smaller, supporting

A

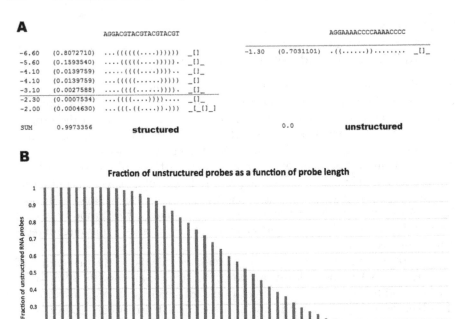

B

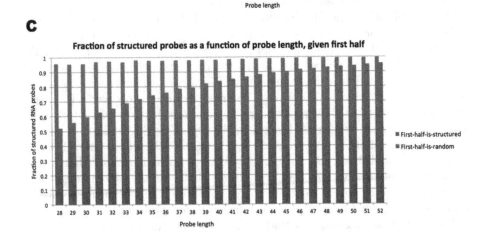

Fig. 2. Properties of unstructured oligos. (A) An output of RNAshapes. For each probe sequence, an ensemble of structures is predicted. If the sum of probabilities of structures with energy smaller than -2.5 is greater than 0.5, the oligo is considered *structured*. On the left is a highly structured oligo, while on the right an unstructured oligo. (B) Fraction of unstructured RNA probes as a function of their length. For each probe length, the fraction of unstructured RNA probes was empirically estimated using 10,000 randomly generated sequences of this length. (C) Fraction of structured RNA probes as a function of their length given their first half. For each probe length, the fraction of structured RNA probes was empirically estimated using 100 structured (blue)/random (red) first halves. For each first half, 100 random extensions were appended to generate a complete probe (Color figure online).

our assumption that structured subsequences are more likely to be extended to structured sequences.

4.2   A Theoretical Lower Bound for the Number of Oligos

We give a simple lower bound for the number of oligos needed to cover all k-mers based on k-mer counts. Since we do not know the optimal solution to the theoretical problem, we will use this lower bound as a baseline to compare to.

Denote the minimum number of oligos to be $n(k, \ell)$, where k is the desired k-mer coverage and ℓ the length of the probe. Then,

$$n(k, \ell) \geq \lceil \frac{4^k}{\ell - k + 1} \rceil \qquad (1)$$

It follows immediately from the fact that the number of k-mers to be covered is 4^k and each probe of length ℓ covers $\ell - k + 1$ k-mers.

For the p-multi k-mer coverage, the bound is:

$$n(k, \ell, p) \geq \lceil \frac{4^k \cdot p}{\ell - k + 1} \rceil \qquad (2)$$

4.3   Computational Results

We implemented and ran our heuristic algorithm on $5 \leq k \leq 10$ and $\ell = 30, 35, 40$, typical values used for library design [9,13]. Multiplicity was set to 1, number of random attempts to 100 and randomization seed to 0. The results are summarized in Table 1. On average, our method generates a library that is only $1.1 - 1.3$-times greater in size than the theoretical lower bound. Moreover, as expected, the ratio compared to the lower bound increases with oligo length. It is more difficult to find unstructured probes since the fraction of unstructured probes decreases with oligo length (see Sect. 4.1). Note that for $k \geq 9$, there are a few structured probes in the set. These cannot be avoided due to self-structured k-mers (see Sect. 4.1). Running times were benchmarked on a single CPU of a 20-CPU Intel Xeon E5-2650 (2.3 GHz) machine with 384 GB 2133 MHz RAM.

In addition, we implemented a naive algorithm to compare the performance with our algorithm. We generated random sequences of length ℓ and added them if they included uncovered k-mers until all k-mers were covered. We report the average set size over 100 runs. As can also be seen in Table 1, the naive algorithm produces much larger sets than our heuristic.

4.4   Comparison to the Library Design of Ray *et al.*

To compare our solution to the library design of Ray *et al.* [15], we ran the algorithm with $k = 9$, $\ell = 35$ and $p = 16$, as their library is required to cover each 9-mer at least 16 times. Notably, our solution is significantly more compact. Our library contains a total of 166,649 oligos of length 35. Compared to the

Table 1. Computational results for different oligos libraries. For a pair of oligo length ℓ and k to cover, we ran Algorithms 2 and 3 to generate an unstructured RNA library covering all k-mers in ℓ-long sequences. We report the number of oligos in the output of each run and the ratio compared to a theoretical lower bound. Algorithm 2 outputs the incomplete set (oligo length $\leq \ell$) and Algorithm 3 outputs the complete set (oligo length $= \ell$). Reported run times are elapsed times of running Algorithms 2 and 3 consecutively. The naive set is based on generating random sequences until all k-mers are covered.

ℓ	k	Lower bound	Incomplete set	Incomplete ratio	Complete set	Complete ratio	Structured	Naive set	Runtime (hh:mm:ss)
	5	40	50	1.25	51	1.27	0	149	00:02:11
	6	164	182	1.11	182	1.11	0	766	00:07:43
30	7	684	737	1.08	739	1.08	0	3 308	00:41:40
	8	2 850	3 081	1.08	3 106	1.09	0	13 801	02:58:52
	9	11 916	12 940	1.09	13 069	1.10	59	57 154	14:42:27
	10	49 934	55 882	1.12	56 526	1.13	670	236 477	82:18:01
	5	34	41	1.21	41	1.21	0	131	00:03:13
	6	138	158	1.14	162	1.17	0	670	00:21:20
35	7	566	635	1.12	648	1.15	0	2 884	01:17:43
	8	2 342	2 670	1.14	2 744	1.17	0	11 961	06:03:05
	9	9 710	11 022	1.14	11 439	1.18	60	49 289	26:47:31
	10	40 330	47 139	1.17	49 225	1.22	609	202 763	137:33:27
	5	30	37	1.23	38	1.27	0	117	00:02:44
	6	118	140	1.19	148	1.25	0	598	00:36:31
40	7	482	561	1.16	611	1.27	0	2 561	02:33:16
	8	1 986	2 362	1.19	2 627	1.32	0	10 597	11:24:15
	9	8 192	9 745	1.19	10 966	1.34	60	43 492	48:02:15
	10	33 826	41 798	1.24	47 457	1.40	557	178 187	246:05:17

theoretical lower bound of 155,346 oligos, our library is only 1.07-times greater in size. In comparison, the library of Ray *et al.* contains 214,948 probes, which is 1.38-times greater in size than the theoretical lower bound. Moreover, in our complete library, all oligos have the same length, as opposed to the library of Ray *et al.*, where oligo lengths vary. A more flexible length requirement may enable us to construct an even smaller library. More importantly, their library includes 2,858 structured probes due to self-structured 9-mers, while in our library there are only 841, a very small fraction of the total number of probes.

5   Conclusion

In this work, we have presented, for the first time, a general algorithm to generate a compact set of unstructured RNA probes that together cover all RNA k-mers. The algorithm's good performance can be attributed to the key ideas of generating probe sequences using de Bruijn graphs, but taking a random walk on those and backtracking when we encounter a structured sequence.

De Bruijn graphs and linear-feedback shift registers (LFSR) are commonly used to generate de Bruijn sequences. Euler tours over de Bruijn graphs have the advantage that all possible $(4!)^{4^{k-1}}/4^k$ de Bruijn sequences can be generated [30]. On the other hand, linear-shift feedback registers for generating de Bruijn sequences are limited by the number of primitive polynomials over $GF(4)$ with degree k [19]. There are only $\phi(4^{k-1})/k$ primitive polynomials, where ϕ is the totient function. For example, for $k = 11$ there are only 240,064 de Bruijn sequences that can be generated by an LFSR. In addition, LFSR-generated sequences have uniform properties [31], which are counter-productive to the problem at hand, since it requires local properties of unstructuredness. Thus, de Bruijn graphs provide a much more flexible mechanism than LFSRs to generate sets of sequences with specific properties covering all k-mers.

Our implementation deals cleverly with prohibitive running times. Our backtracking approach is particularly suited to the monotone property of RNA secondary structure. That is, having a structured subsequence highly influences the probability of the whole sequence being structured. In addition, the random walk works in a way that tries to double the length of the path in each attempt, and in so doing reduces the running time of the extension process by a factor of ℓ. We applied several practical heuristics, such as a limited number of attempts and shortening extensions, to avoid dead-end paths.

The potential downside of our approach is its heuristic nature, which intrinsically does not guarantee any ratio over the optimal solution. Unfortunately, the structuredness property of RNA sequences is not easily translated into combinatorial properties which can be targeted by short paths in a de Bruijn graph. Properties that proximate these features, such as not having a k-mer and its reverse complement in the same probe sequence, are not good enough to ensure that the probe is unstructured by the prediction algorithm.

While in this work we focused on one application, we see the substantial potential benefit of our algorithm in other applications. Our general scheme can be used to design sequence libraries with other desired properties or other definitions of structuredness. For example, RNA secondary structure can be defined by minimum free-energy instead of an ensemble of structures [32]. On the DNA front, DNA oligos with specific DNA shape features are desirable as shape plays a significant role in protein DNA-binding [33]. Moreover, our algorithm can be modified to cover only a subset of the k-mers, or have different multiplicities for each k-mer, by keeping the edges in the de Bruijn graph that represent those k-mers and add different numbers of edge copies for each k-mer. For example, in the RNAcompete technology two 7-mers are excluded as they are restriction sites of an enzyme used in the protocol [13].

To conclude, we have demonstrated the ability of our algorithm the meet the highly desired goal of generating compact sets of unstructured RNA probes that cover all k-mers. High-throughput technologies that measure RNA accessibility as part of the secondary structure or protein RNA-binding *in vitro* will greatly benefit from this design. The generated library set is only slightly larger than the theoretical lower bound, and thus achieves near-optimal results. The algorithms can be easily applied to other sequence design problems. Any design that requires complete coverage of all k-mers, with specific sequence properties, can utilize our general scheme of random path search in de Bruijn graphs.

Acknowledgments. This work was supported by NIH grant R01GM081871.

References

1. Kudla, G., Granneman, S., Hahn, D., Beggs, J.D., Tollervey, D.: Cross-linking, ligation, and sequencing of hybrids reveals RNA-RNA interactions in yeast. Proc. Natl. Acad. Sci. **108**, 10010–10015 (2011)
2. Rinn, J.L., Ule, J.: Oming in on RNA-protein interactions. Genome Biol. **15**, 401 (2014)
3. Wan, Y., Kertesz, M., Spitale, R.C., Segal, E., Chang, H.Y.: Understanding the transcriptome through RNA structure. Nat. Rev. Genet. **12**, 641–655 (2011)
4. Kertesz, M., Iovino, N., Unnerstall, U., Gaul, U., Segal, E.: The role of site accessibility in microRNA target recognition. Nat. Genet. **39**, 1278–1284 (2007)
5. Steffen, P., Voß, B., Rehmsmeier, M., Reeder, J., Giegerich, R.: RNAshapes: an integrated RNA analysis package based on abstract shapes. Bioinformatics **22**, 500–503 (2006)
6. Kertesz, M., Wan, Y., Mazor, E., Rinn, J.L., Nutter, R.C., Chang, H.Y., Segal, E.: Genome-wide measurement of RNA secondary structure in yeast. Nature **467**, 103–107 (2010)
7. Mandir, J.B., Lockett, M.R., Phillips, M.F., Allawi, H.T., Lyamichev, V.I., Smith, L.M.: Rapid determination of RNA accessible sites by surface plasmon resonance detection of hybridization to DNA arrays. Anal. Chem. **81**, 8949–8956 (2009)
8. Kierzek, E., Kierzek, R., Turner, D.H., Catrina, I.E.: Facilitating RNA structure prediction with microarrays. Biochemistry **45**, 581–593 (2006)
9. Kierzek, R., Turner, D.H., Kierzek, E.: Microarrays for identifying binding sites and probing structure of RNAs. Nucleic Acids Res. **43**, 1–12 (2015)
10. Gerstberger, S., Hafner, M., Tuschl, T.: A census of human RNA-binding proteins. Nat. Rev. Genet. **15**, 829–845 (2014)
11. König, J., Zarnack, K., Luscombe, N.M., Ule, J.: Protein-RNA interactions: new genomic technologies and perspectives. Nat. Rev. Genet. **13**, 77–83 (2012)
12. Fu, X.D., Ares Jr, M.: Context-dependent control of alternative splicing by RNA-binding proteins. Nat. Rev. Genet. **15**, 689–701 (2014)
13. Ray, D., Kazan, H., Chan, E.T., Castillo, L.P., Chaudhry, S., Talukder, S., Blencowe, B.J., Morris, Q., Hughes, T.R.: Rapid and systematic analysis of the RNA recognition specificities of RNA-binding proteins. Nat. Biotechnol. **27**, 667–670 (2009)
14. Lambert, N., Robertson, A., Jangi, M., McGeary, S., Sharp, P.A., Burge, C.B.: RNA Bind-n-Seq: quantitative assessment of the sequence and structural binding specificity of RNA binding proteins. Mol. Cell **54**, 887–900 (2014)
15. Ray, D., Kazan, H., Cook, K.B., Weirauch, M.T., Najafabadi, H.S., Li, X., Gueroussov, S., Albu, M., Zheng, H., Yang, A., et al.: A compendium of RNA-binding motifs for decoding gene regulation. Nature **499**, 172–177 (2013)
16. Stefl, R., Skrisovska, L., Allain, F.H.T.: RNA sequence-and shape-dependent recognition by proteins in the ribonucleoprotein particle. EMBO Rep. **6**, 33–38 (2005)

17. Berger, B., Peng, J., Singh, M.: Computational solutions for omics data. Nat. Rev. Genet. **14**, 333–346 (2013)
18. West, D.B., et al.: Introduction to Graph Theory, vol. 2. Prentice Hall, Upper Saddle River (2001)
19. Lempel, A.: On a homomorphism of the de Bruijn graph and its applications to the design of feedback shift registers. IEEE Trans. Comput. **100**, 1204–1209 (1970)
20. Alhakim, A., Akinwande, M.: A recursive construction of nonbinary de Bruijn sequences. Des. Codes Crypt. **60**, 155–169 (2011)
21. Berger, M.F., Philippakis, A.A., Qureshi, A.M., He, F.S., Estep, P.W., Bulyk, M.L.: Compact, universal DNA microarrays to comprehensively determine transcription-factor binding site specificities. Nat. Biotechnol. **24**, 1429–1435 (2006)
22. Philippakis, A.A., Qureshi, A.M., Berger, M.F., Bulyk, M.L.: Design of compact, universal DNA microarrays for protein binding microarray experiments. J. Comput. Biol. **15**, 655–665 (2008)
23. Orenstein, Y., Shamir, R.: Design of shortest double-stranded DNA sequences covering all k-mers with applications to protein-binding microarrays and synthetic enhancers. Bioinformatics **29**, i71–i79 (2013)
24. Fordyce, P.M., Gerber, D., Tran, D., Zheng, J., Li, H., DeRisi, J.L., Quake, S.R.: De novo identification and biophysical characterization of transcription-factor binding sites with microfluidic affinity analysis. Nat. Biotechnol. **28**, 970–975 (2010)
25. Berman, P., DasGupta, B., Sontag, E.D.: Randomized approximation algorithms for set multicover problems with applications to reverse engineering of protein and gene networks. In: Jansen, K., Khanna, S., Rolim, J.D.P., Ron, D. (eds.) RANDOM 2004 and APPROX 2004. LNCS, vol. 3122, pp. 39–50. Springer, Heidelberg (2004)
26. Levin, A.: Approximating the unweighted k-set cover problem: greedy meets local search. SIAM J. Discrete Math. **23**, 251–264 (2008)
27. Grossman, T., Wool, A.: Computational experience with approximation algorithms for the set covering problem. Eur. J. Oper. Res. **101**, 81–92 (1997)
28. Lorenz, R., Bernhart, S.H., Zu Siederdissen, C.H., Tafer, H., Flamm, C., Stadler, P.F., Hofacker, I.L., et al.: ViennaRNA package 2.0. Algorithms Mol. Biol. **6**, 26 (2011)
29. MacWilliams, F.J., Sloane, N.J.: Pseudo-random sequences and arrays. Proc. IEEE **64**, 1715–1729 (1976)
30. de Bruijn, N.: A combinatorial problem. Proceedings of the Koninklijke Nederlandse Akademie van Wetenschappen. Series A **49**, 758 (1946)
31. Hurd, W.J.: Efficient generation of statistically good pseudonoise by linearly interconnected shift registers. IEEE Trans. Comput. **100**, 146–152 (1974)
32. Churkin, A., Weinbrand, L., Barash, D.: Free energy minimization to predict RNA secondary structures and computational RNA design. In: Picardi, E. (ed.) RNA Bioinformatics, pp. 3–16. Springer, New York (2015)
33. Burgess, D.J.: DNA elements: shaping up transcription factor binding. Nat. Rev. Genet. **16**, 258–259 (2015)

Author Index

Printed in the United States
By Bookmasters